KU-288-907

POLYMERS AT SURFACES AND INTERFACES

This text deals with the behaviour of polymers at surfaces and interfaces from a fundamental point of view. It covers in an integrated way both current experimental results, and the most important theoretical approaches to understanding these findings.

Topics covered include the nature and properties of the surface of a polymer melt, the structure of interfaces between different polymers and between polymers and non-polymers, adsorption from polymer solutions, the molecular basis of adhesion and the properties of polymers at liquid surfaces. Emphasis is placed on the common physical principles underlying this wide range of situations. Statistical mechanics based models of the behaviour of polymers near interfaces are introduced, with the emphasis on theory that is tractable and applicable to experimental situations. Experimental techniques for studying polymer surfaces and interfaces are reviewed and compared.

Advanced undergraduates, graduate students and research workers in physics, chemistry and materials science with an interest in polymers will find this book of interest.

POLYMERS AT SURFACES
AND INTERFACES

RICHARD A. L. JONES

University of Sheffield

and RANDAL W. RICHARDS

University of Durham

CAMBRIDGE
UNIVERSITY PRESS

PUBLISHED BY THE PRESS SYNDICATE OF THE UNIVERSITY OF CAMBRIDGE
The Pitt Building, Trumpington Street, Cambridge, United Kingdom

CAMBRIDGE UNIVERSITY PRESS
The Edinburgh Building, Cambridge CB2 2RU, UK http://www.cup.cam.ac.uk
40 West 20th Street, New York, NY 10011-4211, USA http://www.cup.org
10 Stamford Road, Oakleigh, Melbourne 3166, Australia

© Cambridge University Press 1999

This book is in copyright. Subject to statutory exception
and to the provisions of relevant collective licensing agreements,
no reproduction of any part may take place without
the written permission of Cambridge University Press.

First published 1999

Printed in the United Kingdom at the University Press, Cambridge

Typeset in Times 11/14pt, in 3B2 [KT]

A catalogue record for this book is available from the British Library

Library of Congress Cataloguing in Publication data

Jones, Richard A. L. (Richard Anthony Lewis), 1961–
    Polymers at surfaces and interfaces/Richard A.L. Jones and
Randal W. Richards.
        p.   cm.
    Includes bibliographical references and index.
    ISBN 0 521 47440 X (hb).   ISBN 0 521 47965 7 (pb)
    1. Polymers   Surfaces. 2. Interfaces (Physical sciences)
I. Richards, R. W. (Randal William), 1948–.   II. Title.
QD381.9.S97J66   1999
620.1'92–dc21 98-6554 CIP

ISBN 0 521 47440 X hardback
ISBN 0 521 47965 7 paperback

UNIVERSITY
OF SHEFFIELD
LIBRARY D

Contents

Preface			*page* vii
1	Introduction and overview		1
2	The surface of a simple polymer melt		8
	2.0	Introduction	8
	2.1	Surface tension and surface energy	8
	2.2	Interfacial tensions, contact angle and wetting	12
	2.3	More precise theories of the surface tension of simple fluids	25
	2.4	Theories of the surface tension of polymers	33
	2.5	Surface tension of polymer melts: the experimental situation and comparison with theory	36
	2.6	The polymer surface at a microscopic level	44
	2.7	Dynamics of the polymer surface	50
	2.8	References	54
3	Experimental techniques		56
	3.0	Introduction	56
	3.1	Reflection of waves from interfaces	58
	3.2	Ion beam methods	87
	3.3	Surface spectroscopy	101
	3.4	Direct measurement of forces between surfaces	107
	3.5	References	123
4	Polymer/polymer interfaces		127
	4.0	Introduction	127
	4.1	Thermodynamics of polymer mixtures	128
	4.2	Interfaces between weakly immiscible polymers: square gradient theories	136
	4.3	Self-consistent field theory	144
	4.4	Kinetics of formation of polymer/polymer interfaces	152

4.5	The morphology of immiscible polymer blends	172
4.6	Appendix to chapter 4. The random phase approximation	181
4.7	References	184
5	Adsorption and surface segregation from polymer solutions and mixtures	187
5.0	Introduction	187
5.1	Surface segregation in polymer mixtures	188
5.2	Adsorption from polymer solutions	199
5.3	Wetting and surface-driven phase separation in polymer solutions and mixtures	227
5.4	References	241
6	Tethered polymer chains in solutions and melts	244
6.0	Introduction	244
6.1	Polymer brushes in solution	245
6.2	Polymer brushes in melts	261
6.3	Block copolymers at polymer/polymer interfaces	268
6.4	Other interfacially active species at polymer/polymer interfaces	277
6.5	Phase behaviour of neat block copolymers in bulk and in thin films	282
6.6	References	290
7	Adhesion and the mechanical properties of polymer interfaces at the molecular level	293
7.0	Introduction	293
7.1	The strength of interfaces involving glassy polymers	295
7.2	The strength of interfaces involving rubbery polymers	309
7.3	References	315
8	Polymers spread at air/liquid interfaces	317
8.0	Introduction	317
8.1	Surface pressure isotherms: classification, theory and scaling laws	317
8.2	Organisation of polymers at the air/liquid interface	328
8.3	Interfacial dynamics of polymers at fluid interfaces	347
8.4	A generalised approach to capillary wave phenomena	356
8.5	Other effects and future applications	364
8.6	References	369
Index		373

Preface

Reputedly an oft used curse among the literati was 'May you write a book!'. Equally, among scientists prolific authors who still complain bitterly to their editors about the trials and tribulations of completing the latest commission are legion. In the area of scientific texts it is generally true that as soon as you have committed something to paper it is out of date. Consequently such books are at the mercy of the reviewers, who can easily point to the built-in obsolescence of the product. In view of these factors one might ask what prompted us to write this book; were we courageous or just foolhardy? The answer is neither.

We formed the view that in recent years sufficient new experimental evidence concerning polymers at interfaces had been obtained by the application of new techniques (and there had been an equivalent increase in the range of theoretical descriptions available) that it was worth trying to write a monograph in the attempt to bring these two strands together in a unified way. The combination of the two authors (one a physicist, one a physical chemist) also seemed useful, in that the rigour of either discipline could be tempered by intelligent questions from the non-expert. Certainly we had some long discussions on the importance and relevance of some of the work reported in the literature and where it fitted in our overall strategy. A constant question was 'Does this work aid a clear view, or obscure it?'. We found surprisingly often that something which obscured insight in the original section in which it had been placed was a clarifying example when placed in another where originally it would have seemed out of place.

Our intention has not been to write an exhaustive treatment of polymer surfaces and interfaces but rather to produce a compact, reasonably complete treatment of the subject discussed at the level of the molecular length scales associated with polymers. This seemed appropriate because our approach was based on the statistical mechanics of coarse-grained models; a constant theme is the competition between enthalpy and entropy in polymers at or near interfaces of any kind. Thus we deal very lightly with the details of surface chemistry at the atomic level of resolution; for example our treatment of the high-resolution surface spectroscopic methods is far more brief than their overall importance to the field would indicate. Books more angled towards these aspects form a considerable proportion of the literature devoted to polymer surfaces. Similarly, a detailed account of the surface properties of specific polymers has not been attempted. We have adopted a more general approach, whereby the emphasis is on defining guiding principles and illustrating these with experimental data or applications where these principles apply (either by default or design).

More detailed texts on specific aspects of polymers at interfaces are available and some of these are referred to in the succeeding chapters. However, we are not aware of any other book whose authors have attempted the scope presented here. The choice of breadth at the expense of some depth was again deliberate; we wished to make clear the large number of areas in which the interfacial behaviour of polymers is relevant and important and to point out the close parallels between different aspects of the subject. Whether or not we have succeeded is a judgement for the reader to make. We have attempted to write a book that is relatively self-contained, though we do assume an acquaintance with elementary statistical mechanics at the level that a senior physics, chemistry or materials science undergraduate would have and a knowledge of polymer physics, again at the level of an undergraduate introductory course.

Thanks are due to Drs Mark Taylor and Stella Peace (both of Durham University), with whom R. W. R. had long discussions about capillary waves, thus contributing greatly to what understanding he has. In this respect Professor John Earnshaw of Queen's University, Belfast, should also be thanked since his was the original insight which has allowed surface quasi-elastic light scattering to become such a powerful method of observing capillary waves at fluid interfaces. He has always been a willing confidant when discussing some of the more obscure features of the results. Others to whom R. W. R. is grateful for educating him in aspects of polymer surfaces and interfaces (by experiment and/or discussion) include Dr Ian Hopkinson, Dr Brian Rochford and Dr Ian Reynolds. Dr R. K. Thomas of Oxford University and Dr Jeff Penfold of the Rutherford Appleton Laboratory should also be cited for their 'character building' encouragement of R. W. R. to persevere with neutron reflectometry. R. A. L. J. is indebted to Jacob Klein, Athene Donald and Ed Kramer, who were inspirational guides at the start of his scientific career in this field. What understanding R. A. L. J. has of polymers at interfaces has been developed in interactions with many other colleagues, friends and collaborators, including (but not limited to) Miriam Rafailovich, Jonathon Sokolov, Costantino Creton,

Ken Shull, Laura Norton, Taco Nicolai, Mark Geoghegan, Chris Clarke, Joe Keddie, Peter Mills and Easan Sivaniah. Professor David Tabor made perceptive comments on drafts of chapters 2 and 7.

R. A. L. J., Cambridge
R. W. R., Durham

1

Introduction and overview

The study of the properties of any material must begin with the development of an understanding of its properties in the bulk state. However, pure, bulk phases are idealisations of the physicist rather than widely encountered realities; all around us we find complicated assemblies of materials mixed on an intimate, but not molecular, scale. It is an old truism that a human being is 70% water, but it would be just as accurate and probably more helpful to characterise ourselves as being almost all interface. What makes us different from 50 l of water and a few kilograms of solid, much of it polymeric, is the fantastically intricate hierarchy of structures, from sub-cellular organelles of sub-micrometre size through cells up to organs of macroscopic size, all these structures being demarcated by interfaces of one kind or another. This is an extreme example of a common situation; most of the materials we encounter, from the food we eat to the structural materials of our technology, are composed of many different phases and the interfaces between these phases are often very important in determining the overall properties of these materials.

That there are differences between the properties of materials in bulk and close to interfaces is well known for many materials. The density and composition of liquids is different close to surfaces or interfaces (this can be thought of as the effect of 'lost molecules'), whereas in solids there may also be qualitative differences in structure; the surfaces of semi-conductors may be 'reconstructed' and have a different crystal structure from the bulk. For all these situations, the most important question to be clarified is that of what actually constitutes the interfacial region – how far does one need to be from a surface or interface before essentially bulk properties are recovered? In many cases – including the examples of liquid density and semiconductor surface reconstruction just mentioned – this distance is set by atomic or molecular length scales. Hence in a macroscopic sample, even of a polycrystalline solid, the fraction of molecules that have properties characteristic of the interface rather than the bulk is very small (although in many cases their influence on the behaviour of the material may be out of proportion to this small fraction). As we shall see, this statement is also broadly true for polymer molecules, but here the molecular length scale is many times longer than the few ångström units associated with molecules of low relative molecular mass. The additional factor for polymer systems arises from the *connectivity* between the units which make up a polymer molecule.

We understand the molecular level properties of gases (either pure or mixed) at low pressures sufficiently well enough to make confident predictions of bulk properties. The relative simplicity of gases is due to the low cohesive energies between the molecules and between the molecules and container walls. In contrast, relating bulk properties to molecular properties in condensed phases (liquids and solids) becomes more complex, because of the greatly increased cohesive energies between the molecules. Nonetheless, the macroscopic properties of condensed matter in the form of pure phases are now well understood from first principles.

Although such idealised pure phases are useful to develop theoretical or experimental methods, developments in society during the last 200 years have not depended significantly on the use of pure materials (with the possible exception of the steam engine and even here, despite the use of pure water and its gas, steam, the structure depends on the use of heterogeneous metals). This is even more true nowadays, as the costs of developing new single-component materials, or new methods to process such materials, become increasingly daunting. Improvements are sought by mixing or combining existing molecules to produce materials with improved properties. Such mixtures may be true molecular 'solutions', but are more likely to be dispersions of one component in another. In such circumstances, particularly when the average size of particles of the dispersed material is small, the role of the interface between the two phases becomes important; increasingly so as the characteristic dimension of the dispersed phase becomes smaller.

This book is solely concerned with polymers in the amorphous state, that is polymer molecules in solution, the melt or that are intrinsically amorphous in the solid state by virtue of their chemical structure. We discuss surfaces and interfaces involving pure polymeric phases and interfaces between simple liquids and solids or air that are modified by an accumulation of polymeric molecules. The situation is in one sense more complicated than that for materials composed of atoms or small molecules. For these systems, as hinted at above, there is a single length scale characterising the range of forces between molecules and this molecular length scale dictates the range over which the perturbation imposed by an interface persists. For polymers there are at least two controlling length scales, which may differ substantially in magnitude. There are relatively strong interactions between segments on different polymer chains that are spatially adjacent. These interactions decay in strength rapidly as the distance between molecules increases but they are the source of the cohesion in the bulk state and in their essentials are identical to those between small molecules; they are mainly enthalpic in character and determine such properties as compressibility and surface tension.

On the other hand, many of the unusual and useful properties of polymers arise from the contiguity of segments joined together by covalent bonds to form a long chain. Thus polymer molecules are spatially extended objects, with a new length scale – the polymer chain dimension (generally expressed as a radius of gyration or end-to-end distance). For flexible polymers the connectivity of covalently bonded chains means that configurational entropy plays a leading role in determining the equilibrium, minimum free energy, state. A flexible polymer can adopt any one of a large number of configurations of equal energy; at thermodynamic equilibrium it is the configuration of maximum entropy that is chosen. If the chain is somehow constrained to some other configuration the entropy is reduced from the maximum value associated with equilibrium and the free energy of the system is increased. A testament to the greatly increased significance of entropy in macromolecular systems is provided by comparing the elasticity of polymers (above their glass transition temperature) and metals. Using classical thermodynamics the elastic (stress–strain) behaviour of polymers can be shown to be almost totally due to the entropy changes in the system. (We note that this conclusion does not in itself require the predefinition of polymer molecules which are able to explore many configurations of equal energy; in common with all classical thermodynamics we do not have to presuppose the existence of molecules or atoms at all!) The stress–strain properties of metals, on the other hand, mainly result from the strong energetic interactions between metal atoms on crystal lattices. The influence of entropy on interfacial properties is most clearly seen in mixed polymer systems, for example polymer solutions or blends. Here the short-range cohesive forces impose an essentially constant density throughout the system. However, within that overall constant density, gradients in composition are controlled by a length scale derived from overall chain dimensions. Thus we should expect that polymer/polymer interfaces and layers of adsorbed polymers at solid/liquid and liquid/air interfaces will be much more spatially extended than their analogues in exclusively small-molecule systems; dimensions of such regions will be of the order of the polymer molecule dimensions rather than a few ångström units. It is this competition between enthalpy and entropy which underlies many of the subjects discussed in the ensuing pages.

In chapter 2 we consider perhaps the single most important quantity characterising a surface; the surface tension. Here the focus is on cohesive energies; after considering some of the classical thermodynamics of interfaces we consider how statistical mechanical models can relate surface tension to intermolecular forces. Our main tool is the theory of van der Waals; this yields the important insight that the statistical mechanics of an inhomogeneous system, such as an interface, is determined not only by the density and the composition, but also by the spatial gradients of these quantities. Near the surface of a simple liquid, the density changes smoothly and continuously from the bulk value to the vapour phase density and the distance characterising this density profile is related to the length characterising the range of intermolecular forces. Technically, the theory is constructed by writing down the free energy of the system as the integral of a functional of the density. In the simplest such theory which accounts for non-local effects, the free energy functional has one term depending on the local density and another term proportional to the square of the gradient of the density profile. This type of theory, known as a 'square gradient theory', reappears as the simplest realistic theory of inhomogeneous polymer systems and is used both in chapter 4 to discuss the polymer/polymer interface and in chapter 5 to discuss surface segregation and adsorption.

Since the energetic interactions between polymer segments are very similar to those between small molecules, the statistical mechanics of a polymer surface is rather similar to that of small-molecule liquids. The role of connectivity is minor and a theory of polymer surface tension can be constructed in the spirit of van der Waals, which is rather successful at predicting the observed surface tension of polymer melts. This is at first sight slightly surprising, for it must be the case that the presence of a surface imposes a serious perturbation on the conformation of a nearby chain. Both the theory of these conformational effects and experiments to measure them are in a rather rudimentary state, but we discuss what is known, together with the even more uncertain area of the effect of surfaces on polymer dynamics.

Theory and experiment always offer challenges to each other in any subject that is vibrant. This has been especially true for polymer surfaces and interfaces in the last few years. Experimentally what is needed are methods that provide data pertaining to microscopic levels. In chapter 3 we outline a range of techniques that can be used; throughout the emphasis is on obtaining molecular-level information. Advantages and drawbacks for each method are summarised. The techniques chosen are usually selected on the basis of their appropriateness for a particular task.

Interfaces between polymer melts are discussed in chapter 4. Here the importance of the connectivity of polymer chains becomes paramount. One effect of the large size of polymer molecules is that their entropy of mixing is small compared with that of small-molecule liquids. Immiscibility between chemically different polymers is thus very common and the nature of an interface between coexisting phases becomes more than usually important. The simplest theoretical approach is a square gradient theory of the type introduced in chapter 2. For the polymer/polymer interface, however, the important length scale is *not* the range of the force between segments but the overall *size of the polymer chain*. This means that polymer/polymer interfaces are typically broader than interfaces between simple liquids. However, for highly incompatible polymers interfacial widths are still smaller than the overall chain size and the simple assumptions underlying square gradient theory are not really valid. More powerful theoretical methods are required, which are provided by self-consistent field theory, which reduces the many-body problem of a chain interacting with all its neighbours to a simpler mean-field theory of one chain interacting with an effective potential that is averaged from all the segment–segment interactions in the system, the mean-field potential. This type of theory provides another level of accuracy in treating the statistical mechanics of inhomogeneous polymer systems and is referred to again in chapter 5, in the context of polymer adsorption, and in chapter 6, where it proves ideally suited to problems involving tethered polymer chains.

This kind of statistical mechanics deals with equilibrium, but many important situations deal with the kinetics of approach to equilibrium and it is this problem that is considered in the second half of chapter 4. Even though at equilibrium miscible polymers will form a single phase, bringing together two such polymers will result in an interface that will broaden in time by a process of interdiffusion. Diffusion of polymers is slow due to the entanglement of chains and will control, for example, the development of strength when polymers are welded or coextruded. The reverse situation is the demixing of two polymers that are initially in a single phase, but which are subsequently quenched into the two-phase part of the phase diagram. Domains of the coexisting phases form, interfaces between them approach their equilibrium width and the domains slowly coarsen, driven by the tendency of the system to lower its energy by reducing the total interfacial area.

Chapter 5 is concerned with interfaces between mixed polymer systems and non-polymer phases. This includes a large variety of different systems; the free surfaces of polymer blends and solutions and interfaces between blends and solutions and solids. The predominant feature here is segregation, whereby the component of a mixture or solution that has a lower surface or interfacial energy preferentially accumulates at that interface. This is a general phenomenon for all mixtures, but it is particularly important when polymers are involved. As mentioned above, the free energy of mixing of polymers is small because of their large molecular size, so even very small differences in surface energy between components can lead to very large amounts of segregation. Once again, because the natural length scale characterising compositional inhomogeneities in such mixed polymer systems is the chain dimension, this sets the distance from the interface over which the composition is perturbed from bulk values.

Segregation at the surface of blends of miscible polymers can be dealt with using the square gradient theory introduced in section 4.2. For concentrated polymer solutions such approaches are also possible, but it is more usual to use numerical versions of the self-consistent field theory described in section 4.3. However, we must be aware that both approaches are essentially mean-field ones in character and will fail in situations in which the polymer concentration in the solution falls into the semi-dilute regime, for which strong spatial concentration fluctuations are always present. Here progress has to be made using scaling theories. Finally, we consider the situation in which the two components of the mixture approach immiscibility. Here we consider the problem of whether one coexisting phase wets another one and what influence this has on the kinetics of phase separation in the vicinity of an interface.

In chapter 6 the choice of polymer architecture and the incorporation of interfacially active functional groups to make polymer molecules especially effective at modifying interfaces is outlined. In many cases this leads to a situation in which a strand of polymer is tethered by one end to the interface; an array of such polymers, when they are densely grafted enough to interact strongly, is known as a 'brush'. The properties of such brush layers (thickness, composition etc.) have been studied extensively using scaling and self-consistent field approaches. Examples of such brushes in use to modify interfaces are found when polymers are chemically end-grafted to polymer/solid interfaces, adsorbed by one end to an interface and in the selective adsorption of block copolymers. Block copolymers are used to modify polymer/polymer interfaces, in which role they can act as polymeric surfactants. Pure block copolymers *microphase* separate, resulting in a system in which the interfacial area between microphase-separated domains is very large and can be thought of as being made up entirely from stretched polymer brushes (at least for the case of lamellar domains).

The final two chapters develop special topics that draw on the ideas discussed in the earlier chapters. In chapter 7, the extent to which the macroscopic property of adhesion can be accounted for in terms of the microscopic structure of the interfaces involved is discussed. Although it is very far from completely understood, this area represents one place where fundamental science and technological practice are beginning to approach quite closely. Chapter 8 considers the structure and dynamics of liquid surfaces modified by adsorbed or spread polymer layers. This area provides a bridge between the statistical mechanical theories outlined in earlier chapters and the classical aspects discussed in chapter 2, and has some relation to certain biological systems. The description is by no means complete, especially insofar as capillary wave dynamics are concerned. The essentials of this complex subject are dealt with here for the case in which one of the fluids is a material of low relative molecular mass; this should be a starting point for the consideration of the more difficult case of polymer melt interfaces, for which the importance of capillary waves is only now being considered (see chapter 4).

Several important areas relating to polymers at surfaces and interfaces are not discussed in this book. Almost nothing is said about crystalline polymers. Many engineering polymers are semi-crystalline and the nature of the interface between crystalline and amorphous phases in these materials has generated much debate and some rancour, into which the authors have no desire to venture. Concerning the nature of the surface of a crystalline polymer, beyond the results of a few experiments described in section 2.6 almost nothing is yet known.

We have said nothing at all about what is perhaps the most fascinating and important area involving polymers at surfaces and interfaces, the interfacial behaviour of biological macromolecules. The combination of the extra richness and functionality of such molecules with the subtleties of surface phenomena make this a huge and as yet barely developed field, to which we cannot hope to start doing justice. However, it will never prove possible to make progress with these complex systems without first understanding the simpler problems that we describe here.

2

The surface of a simple polymer melt

2.0 Introduction

Everyone is aware of one of the curious properties possessed by the surfaces of fluids; the 'skin' on the surface of water that allows one to 'float' a needle on the surface of pure water and which is used by insects to walk across the surface of ponds. This effect, termed surface tension, is a direct manifestation of the cohesive forces that hold liquids together. Polymer melts are no different from other, simpler liquids; they too exhibit surface tension and to understand and predict this surface tension we need to understand the way in which interatomic forces lead to cohesion in polymers.

The simplest picture of surface energy then, is a static one – near the surface of a fluid some of the contacts between molecules, which in the bulk are energetically favourable, are missing. The energy of these 'missing bonds' may· be equated with the surface energy and from such simple arguments the surface energy may be directly related to the cohesive energy of the liquid. This essentially mechanical picture ignores the Brownian motion of the molecules, which is present at the interface as well as in the bulk. One can picture this Brownian motion producing fluctuations in the mean height of the surface or interface, which as a result becomes diffuse. The resulting disorder has an associated entropy and it is the task of a statistical mechanical theory of surface tension to quantify and predict both the excess energy and entropy associated with the interface. The first theory to attempt this was developed over 100 years ago by van der Waals; it is this treatment that still underlies modern quantitative treatments of the surface tension of polymer melts.

2.1 Surface tension and surface energy

Experiments show that the surface of a fluid behaves as if it were a taut membrane; if one cuts such a membrane one has to apply a given force per unit length of cut to hold the surface together. Thus surface tension has the dimensions of a force per unit length. This must be numerically equal to the energy[*] of the surface; this can be seen from a thought experiment of the type shown in figure 2.1, in which one imagines extending a liquid film and one equates the work done against the force exerted by surface tension to the energies of the new surfaces created. The relationship between the surface energy and the cohesive forces holding the liquid together may be seen by another thought experiment; imagine making a cut in a large piece of liquid and then drawing the two halves of liquid apart. There is a cohesive force between the two halves and we can equate the work done against this force in moving the two halves from contact to infinity (or some distance outside the effective range of the force) with the surface energy of the two new surfaces.

We can make this idea concrete with an elementary model of a liquid. Imagine that each molecule occupies a volume v_0 and interacts with, on average, z other molecules, and that the energy of interaction is ε. This energy can be thought of as a 'bond energy'; its origin may be from dispersion forces, metallic bonds or whatever the source of cohesion is. (There is no need for it to be in any sense directional, like a covalent bond.) At a surface a molecule will interact with fewer of its neighbours than it would in the bulk, so that when we make a cut through the liquid we will

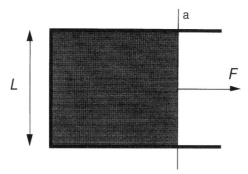

Figure 2.1. A thought experiment illustrating the relation between the surface tension and the surface energy for a liquid film held on a wire frame. The surface tension γ causes a force $F = 2\gamma L$ on the movable wire a; if this wire is moved a distance x the energy associated with the newly created surface is equal to the work, W, done against the force F: $W = 2\gamma x L$. This shows the identity of the surface tension with the surface energy.

---

[*] Strictly speaking this is a free energy; we will make this distinction clearer in section 2.2.

Table 2.1. *The correlation between the latent heat of vaporisation and the surface energy of simple liquids at a temperature of 20 °C*

	Surface energy γ (J m^{-2})	Heat of vaporisation L (J m^{-3})	Molar volume v_0 (m^3)	$\dfrac{\gamma}{Lv_0^{1/3}}$
Water	0.07275	2.25×10^{10}	3.0×10^{-29}	0.10
Methanol	0.0226	9.05×10^{8}	6.6×10^{-29}	0.062
Ethanol	0.0223	6.78×10^{8}	9.7×10^{-29}	0.072
Benzene	0.0289	3.50×10^{8}	1.4×10^{-28}	0.16
Mercury	0.475	3.69×10^{9}	2.5×10^{-29}	0.44

'cut' a number of bonds. If the surface coordination number is z', we have for the surface energy γ

$$\gamma = \frac{z'e}{2v_0^{2/3}}. \tag{2.1.1}$$

We can attempt to relate this to other macroscopic measurements by relating ε to the latent heat of vaporisation of the liquid L. In going from a liquid to a gas we break $z\varepsilon$ bonds per molecule, so the latent heat of vaporisation per unit volume is given by

$$L = \frac{ze}{2v_0}. \tag{2.1.2}$$

Thus the prediction of this simple model is that there is a correlation between the latent heat and surface energy of fluids as follows:

$$\frac{\gamma}{Lv_0^{1/3}} = \frac{z'}{z}. \tag{2.1.3}$$

For an amorphous, liquid, structure, it is not clear what values one should take for the ratio z'/z. For a cubic lattice $z'/z = 1/6$, whereas for a close-packed structure it would be $1/4$. Table 2.1 gives a selection of experimental results; it can be seen that for most materials there seems to be some correlation: liquids with stronger cohesive forces have higher latent heats of vaporisation and higher surface energies.

Of course, in liquids the forces responsible for cohesion are not usually well modelled by a simple pair-wise interaction of the sort assumed here; moreover, this relationship is of little practical use for polymeric materials, because their vapour pressures are negligible for all but the shortest chain lengths and thus the latent heat of vaporisation is experimentally inaccessible. One might ask whether there is any way of calculating the surface tension more directly from the known interatomic forces.

One way is to adopt a continuum approach. Let us revisit the idea of making a new surface in some bulk fluid and separating the two halves to infinity. The cohesive forces in the fluid mean that there will be a force between the two halves of fluid that varies with the separation h; if this force is $F(h)$ per unit area we can write the surface energy as

$$\gamma = \frac{1}{2} \int_a^\infty F(h)\, dh, \qquad (2.1.4)$$

where a is a cut-off distance that represents the distance between the two halves of fluid at equilibrium; we will discuss below what value it is appropriate to ascribe to it.

The form of the force law is particularly simple if the origin of the intermolecular forces is solely the van der Waals, or dispersive, force, as it will be for noble gases, or more practically, pure hydrocarbon fluids such as benzene. If we assume that each molecule interacts through a pair potential depending on the inverse sixth power of separation and we can obtain the total interaction energy by pair-wise summing of the interactions between all molecules, we find that the force $F(h)$ per unit area is given by

$$F(h) = \frac{A}{6\pi h^3}, \qquad (2.1.5)$$

where A is a material constant known as the Hamaker constant (Israelachvili 1991). This may be calculated knowing the dielectric properties of the material in question, or in some circumstances experimental values can be found by the direct measurement of forces between surfaces. Values for hydrocarbon fluids are all rather comparable, with that of benzene, at 5×10^{-20} J, being fairly representative.

Using equations (2.1.4) and (2.1.5) we find the surface energy to be given by

$$\gamma = \frac{A}{24\pi a^2}. \qquad (2.1.6)$$

The cut-off distance a might at first be taken to be half the average intermolecular distance. However, if we use this value we find that predictions of the surface tension are systematically too small. Later in this chapter we discuss fundamental reasons why this static, mechanical approach to surface tension is likely to be in error, but it can also be argued that, even in this framework, one should use a smaller cut-off distance. Israelachvili argues that part of the discrepancy arises from attempting to use an essentially continuum theory at length scales comparable to molecular sizes and he notes that surprisingly good agreement between theory and experiment is achieved if a single, universal value for the cut-off a of 0.165 nm is used (Israelachvili 1991). Agreement to within a few per cent is achieved both for small-molecule and for polymeric hydrocarbons. Although the precise reason for this success is not immediately apparent, it does emphasise the close connection between intermolecular forces and the surface tension.

Equation (2.1.6) fails completely for liquids such as water, methanol and formamide. This is not surprising, because in these liquids hydrogen bonds play a substantial role in providing cohesive forces whose effect is not included in the purely dispersive Hamaker constant. One approach to this problem, initiated by Fowkes (1962), has been to assume that each type of intermolecular force makes an additive contribution to the surface tension. So-called 'group contribution' methods have been developed on the basis of this idea, which are of considerable practical use in classifying and predicting surface tension behaviour in a semi-empirical way.

2.2 Interfacial tensions, contact angle and wetting

2.2.1 Classical thermodynamics of interfaces

Before we proceed to discuss either the experiments that can give information about surface or interfacial tension, or the statistical mechanical theories allowing more precise predictions of these quantities, we should discuss some of the fundamentals of classical interfacial thermodynamics. The interested reader will find a fuller account in the book by Adamson (1990).

Thermodynamically the interfacial tension between two phases may be defined as the increase in Gibbs free energy of the whole system per unit increase in interfacial area, carried out under conditions of constant temperature and pressure, that is

$$\gamma = \left(\frac{\partial G}{\partial A}\right)_{T,p,n}.$$ (2.2.1)

This definition of interfacial tension also encompasses surface tension, in which the two phases are the pure liquid and its coexisting vapour. If we define the interface between two phases, A and B, as a region of finite thickness, t, where the composition changes from that comprising A in the bulk phase to the bulk phase B, then the chemical potential of any species must be equal in the two phases and at the interface;

$$\mu_A^1 = \mu_A^2 = \mu_A^{\text{int}}.$$ (2.2.2)

For each of the phases the most general expression for the Gibbs free energy is

$$dG^i = -S^i \, dT + V^i \, dp + \mu_A \, dn_A^i + \mu_B \, dn_B^i \qquad (2.2.3)$$

and for constant T and p

$$G^i = n_A^i \mu_A + n_B^i \mu_B, \qquad (2.2.4)$$

where the superscript i indicates phase 1 or 2. For the interfacial region a term $\gamma \, dA$ has to be added to (2.2.3) and we eventually obtain

$$G^{\text{int}} = \gamma A + n_A^{\text{int}} \mu_A + n_B^{\text{int}} \mu_B. \qquad (2.2.5)$$

Any transfer either of A or of B to or from the interfacial region necessarily incorporates a change in interfacial area and thus the surface free energy contribution to the total Gibbs free energy must be included. From (2.2.5) we observe that the surface tension is actually the *excess* surface free energy per unit area over that due to the additive free energies of the components in the interfacial region:

$$\gamma = \frac{G^{\text{int}}}{A} - (\Gamma_A \mu_A + \Gamma_B \mu_B), \qquad (2.2.6)$$

where Γ_i is the interfacial concentration of species i (moles per unit area).

Subtracting the general variation of the Gibbs free energy in the interfacial region from the total differential of equation (2.2.5) and dividing by the interfacial area gives

$$\frac{S^{\text{int}}}{A} \, dT - t \, dp + d\gamma + \Gamma_A d\mu_A + \Gamma_B \, d\mu_B = 0, \qquad (2.2.7)$$

which is one form of the Gibbs adsorption isotherm. Rearranging, we find

$$-\frac{d\gamma}{dT} = \frac{S^{\text{int}}}{A} - t \frac{dP}{dT} + \Gamma_A \frac{d\mu_A}{dT} + \Gamma_B \frac{d\mu_B}{dT}. \qquad (2.2.8)$$

We make use of the fundamental equation of chemical thermodynamics to obtain

$$d\mu = \overline{V} \, dp - \overline{S} \, dT,$$

where the bars signify partial molar quantities, and thus

$$-\frac{d\gamma}{dT} = \left(\frac{S^{\text{int}}}{A} - \Gamma_A \overline{S}_A - \Gamma_B \overline{S}_B \right) - (t - \Gamma_A \overline{V}_A - \Gamma_B \overline{V}_B) \frac{dp}{dT}. \qquad (2.2.9)$$

The first term on the right-hand side of equation (2.2.9) is the excess entropy of formation of a surface per unit area and thus is constant. The second term is the associated volume change in the formation of a surface and it is also an entropic term in its nature. Hence the entropy of *interface* formation, ΔS_A^{int}, per unit area is given by

$$\Delta S_A^{\text{int}} = -\left(\frac{\partial \gamma}{\partial T} \right)_{p,n}. \qquad (2.2.10)$$

Using equation (2.2.1) we can define the enthalpy of formation of an interface per unit area:

$$\Delta H_{\mathrm{A}}^{\mathrm{int}} = \gamma - T\left(\frac{\partial \gamma}{\partial T}\right)_{p,n}.$$

(2.2.11)

Note that the simple arguments of section 2.1, being essentially static and mechanical in character, fail to make any prediction about the dependence of the interfacial tension on temperature. We shall see that the determination of surface or interfacial tensions as a function of temperature can give considerable insight into the equilibrium thermodynamics of the interface, therefore the accurate measurement of surface and interfacial tension becomes of some significance.

2.2.2 Measuring surface and interfacial tensions

Many methods of measuring surface and interfacial tensions are based on the use of the Young–Laplace equation. This relates to the surface tension the difference in pressure Δp that is observed across any arbitrarily curved interface. If the curvature is described by two radii R_1 and R_2 (see figure 2.2) and the interfacial tension is γ, then the work of maintaining this interface is balanced by the work needed to displace the interfacial plane along the radius. Thus we find

$$\Delta p = \gamma\left(\frac{1}{R_1} + \frac{1}{R_2}\right).$$

(2.2.12)

If the curved surface is spherical, $R_1 = R_2 = R$ and hence $\Delta p = 2\gamma/R$, demonstrating that smaller bubbles or drops have a greater pressure inside than

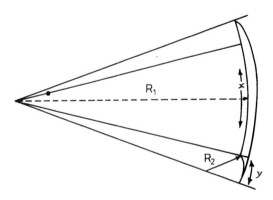

Figure 2.2. A sketch of the differing radii of an arbitrarily curved interface.

do large bubbles or drops. Equally, a plane surface has effectively an infinite radius and there is no pressure difference across the interface.

One of the most familiar consequences of surface tension, capillary rise, may be understood in terms of this equation. If a tube of narrow diameter is dipped into a liquid and the liquid wets the inside of the tube, a curved meniscus will be formed (figure 2.3). Just below the meniscus, the internal pressure inside the liquid will be less than the pressure in the bulk, by virtue of equation (2.2.12). Equilibrium is attained by the liquid rising up the tube until the extra hydrostatic pressure due to gravity balances the discrepancy. At any point a distance z above the minimum of the meniscus this balance is expressed by

$$\gamma\left(\frac{1}{R_1}+\frac{1}{R_2}\right) = \Delta\rho g z + \frac{2\gamma}{b}, \qquad (2.2.13)$$

where $\Delta\rho$ is the density difference either side of the meniscus, g is the acceleration due to gravity and b is the radius of curvature at the apex of the figure of revolution, where $R_1 = R_2$. Numerical solutions to this equation have been obtained and from them one may obtain the value of the surface tension, given experimentally measured values of the height of the capillary rise, the radius of the capillary and the difference in density between the phases. The procedure is outlined in Adamson (1990). The capillary rise method offers a precise and convenient way of measuring surface tensions of simple liquids. Its

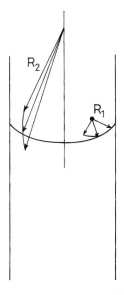

Figure 2.3. The geometry of the meniscus in a capillary rise experiment. $R_1 \neq R_2$.

value for measuring the surface tension of a polymer melt is much more limited, however. Excessively long times are needed to reach the equilibrium capillary height of a polymer melt. In combination with the high temperatures that are usually required, this means that oxidation and degradation are always likely to be a problem.

For polymers interfacial and surface tensions are more practically obtainable from analysing the shapes of pendant or sessile drops or bubbles, all of which are examples of axisymmetrical drops. Bubbles may be used to obtain surface tensions at liquid/vapour interfaces over a range of temperatures and for vapours other than air. Drops can also be used to obtain vapour/liquid surface tensions; but they are particularly suited to determination of liquid/liquid interfacial tensions, for example for polymer/polymer interfaces. All the methods are based on the application of equation (2.2.1). The principles are illustrated in figure 2.4, in which a sessile drop is used as the specific example. Just like for the capillary meniscus, the drop has two principal radii of curvature, R_1 in the plane of the axis of symmetry and R_2 normal to the plane of the paper. At the apex, O, the drop is spherically symmetrical and $R_1 = R_2 = b$ and equation (2.2.12) becomes

$$\Delta p = 2\gamma/b. \tag{2.2.14}$$

At some point P on the surface of revolution a vertical distance z below zero there is an additional contribution to the pressure from the density difference and we recover equation (2.2.13). It is convenient to write this in a dimensionless form; writing $R_2 = x/\sin\phi$ and using equation (2.2.14) we have

$$\frac{1}{R_1/b} + \frac{\sin\phi}{x/b} = 2 + \frac{z}{b}\Delta\rho\frac{gb^2}{\gamma}. \tag{2.2.15}$$

A shape factor β may be defined as

$$\beta = gb^2 \Delta\rho/\gamma \tag{2.2.16}$$

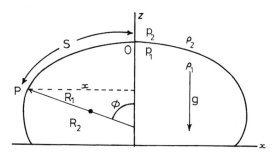

Figure 2.4. A sessile drop of a liquid as an example of a figure of revolution; p_1 and p_2 and ρ_1 and ρ_2 are the pressures and densities either side of the interface.

and thus

$$\frac{1}{R_1/b} + \frac{\sin\phi}{x/b} = 2 + \beta\frac{z}{b}. \qquad (2.2.17)$$

With this approach we can deal with four different types of axisymmetrical drops:

1. Δp, b, $\Delta\rho$, β all positive – a sessile drop;
2. Δp, b negative and $\Delta\rho$, β positive – a captive bubble;
3. Δp, b, $\Delta\rho$, β all negative – a pendant drop; and
4. Δp, b, positive and $\Delta\rho$, β negative – a hanging bubble.

Pendant drops offer an attractive means of determining interfacial tensions for polymers and polymer–polymer combinations. A wide range of temperature can be used, inert atmospheres can be introduced to prevent degradation and only small volumes of polymer melt are needed. The drop shape can be analysed for the interfacial tension in two ways, via the use (in some form) of the tables prepared by Bashforth and Adams of x/b and z/b as functions of β (Bashforth and Adams 1883), cited by Adamson (1990), or by the direct fitting of equation (2.2.16) (as a second-order differential equation) to the drop profile. A full discussion of the methods by which the sessile drops can be analysed to obtain interfacial tensions has been provided by Couper (1993). This author particularly recommends computer fitting of a photographic image of the drop. Nowadays video imaging may be used, with direct transfer of the image into a computer, where subsequent analysis and processing can be carried out. This has been exhaustively described by Anastasiadis *et al.* (1987), who used the technique to determine interfacial tensions for a number of immiscible polymer pairs, as discussed in section 4.3.

Although the use of pendant drops is a robust method and produces interfacial tensions of acceptable accuracy, it is dependent on there being a sufficient difference in density between the two polymers. When the densities become very similar, a true axisymmetric pendant drop may not be obtainable. Figure 2.5 shows an example of a distorted drop of polybutadiene surrounded by poly(2-methyl pentadiene). In some of these cases a sessile drop usually conforms more to the desirable profile than would a pendant one.

A straightforward, simple-to-use method that does not require the use of computerised fitting methods or complicated correction factors that are necessary for the pendant or sessile drop methods is provided by the Wilhelmy plate. A thin plate (generally rectangular but not necessarily so) is immersed into the fluid polymer and withdrawn so that a meniscus is supported by the plate (see

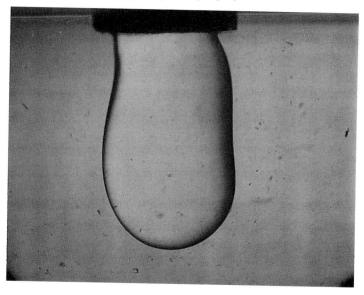

Figure 2.5. A photograph of a distorted pendant drop of polybutadiene.

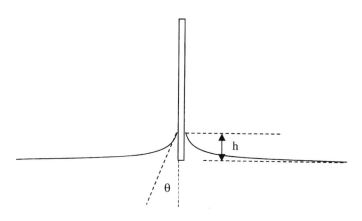

Figure 2.6. A schematic sketch of a Wilhelmy plate at the point of detachment from the liquid surface.

figure 2.6). In this situation then, just at the point of detachment of the plate from the fluid,

$$Mg = mg + \gamma p \cos \theta, \qquad (2.2.18)$$

where M is the total mass of the plate plus the suspended liquid film and m is the mass of the plate alone, which has a perimeter distance p. If the contact angle is zero, which is generally assumed for roughened plates that are wetted by the fluid, then a much simpler equation results:

$$\gamma = \frac{g(M - m)}{P}. \tag{2.2.19}$$

More commonly the liquid height is increased until the surface just touches the plate and the increase in weight is noted. If the polymer melt is non-wetting, then, by using a travelling microscope arrangement or some form of image capture, it may also be possible to measure the contact angle and the distance h, in which case the surface tension can be determined using the relation

$$(h/a)^2 = 1 - \sin\theta \tag{2.2.20}$$

where a is the capillary constant defined by

$$a = [2\gamma/(\Delta\rho\, g)]^{1/2}. \tag{2.2.21}$$

The major problem with the Wilhelmy method for polymer melts is that generally high temperatures are required in order to make the polymers sufficiently fluid. Convection currents from the hot melt then disturb the microbalance used to determine the mass increase and the results can be highly variable. A design that overcomes such problems and greatly reduces the time required to achieve equilibrium has been proposed by Sauer and Dipaolo (1991). A glass fibre replaces the classical Wilhemy plate meaning that very small volumes of molten polymer could be used. Figure 2.7 is a schematic diagram of the arrangement used; the important feature is the baffle tube through which runs the steel wire which connects the glass fibre to the arm of the microbalance. The result is that convection currents generated by the small oven in which the polymer is melted are prevented from swaying the wire or disturbing the microbalance. Other benefits of this method include the zero contact angle of molten polymer on the glass fibre when equilibrium is obtained and the rapidity of attainment of this equilibrium for glass fibre probes; the smaller the fibre diameter the more rapidly is equilibrium estab-lished. Satisfactory values for surface tensions at temperatures up to 400 °C were determined by this instrument; figure 2.8 shows typical results obtained and a comparison with the results of other techniques.

2.2.3 Contact angles

Fluids placed as drops on solids generally have a finite contact angle between the liquid and solid phases, as illustrated by the sessile drop shown in figure 2.9. If the area of the solid covered by the drop is altered by a small amount

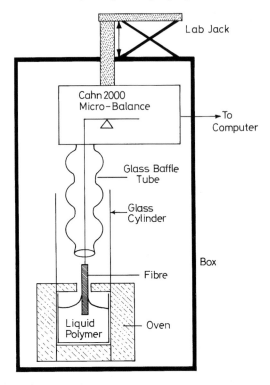

Figure 2.7. A modification to the Wilhelmy plate method developed by Sauer and Dipaolo. After Sauer and Dipaolo (1991).

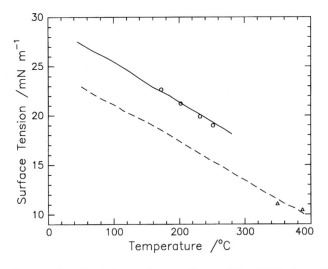

Figure 2.8. Surface tension data obtained using the modified Wilhelmy plate apparatus compared with the results of other techniques (○, PP Wilhelmy; △, PTFE Wilhelmy; ——, PP literature; – –, PTFE literature). After Sauer and Dipaolo (1991).

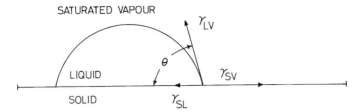

Figure 2.9. The contact angle and relevant surface tensions for a liquid drop on a solid surface.

(ΔA) there will be an accompanying change in the surface Gibbs free energy, ΔG^s, given by

$$\Delta G^s = \Delta A(\gamma_{SL} - \gamma_{SV}) + \Delta A \gamma_{LV} \cos(\theta - \Delta\theta), \qquad (2.2.22)$$

where $\Delta\theta$ is the associated change in contact angle which may accompany the change in area. At equilibrium

$$\lim_{\Delta A \to 0} \frac{\Delta G^s}{\Delta A} = 0 = \gamma_{SL} - \gamma_{SV} + \gamma_{LV} \cos\theta. \qquad (2.2.23)$$

This relation is one statement of the Young–Dupré equation; the various surface tensions refer to the solid/liquid, liquid/vapour and solid/vapour interfaces. In practice contact angle values frequently display hysteresis; the value recorded as the liquid covers fresh solid surface (i.e. as the liquid advances) differs from that obtained as the liquid recedes. Values quoted in the literature are generally advancing contact angles. Various causes have been suggested for this hysteresis, including surface contamination, surface roughness, surface compositional heterogeneity and surface immobility. The influence of surface roughness has been discussed by Wenzel (1936) and surface composition heterogeneity has been considered by Cassie (1948). More recent discussions of these practically important complications may be found in Wu (1982) and Garbassi *et al.* (1994).

For liquids of low relative molar mass the experimental determination of the contact angle is generally straightforward, since a goniometer can be used to measure the angle directly. A commercial instrument has been available for this purpose for some years. Values of contact angles for liquids of low relative molar mass on the surface depend, of course, on the surface tension of the liquid employed. The contact angles for a homologous series of liquids may often be summarised in an empirical relation proposed by Zisman (Jarvis *et al.* 1964):

$$\cos\theta = 1 - \beta(\gamma_{LV} - \gamma_c), \qquad (2.2.24)$$

where γ_c is the critical surface tension, i.e. the surface tension where a contact angle of zero is just obtained. An example of a Zisman plot to evaluate γ_c is shown in figure 2.10; this shows contact angles for a variety of liquids on polyethylene.

Commonly attempts to relate the critical surface tension γ_c to the actual surface energy of the solid γ_s are made. However, from the Young–Dupré equation it is clear that to do this some assumption needs to be made about the solid/liquid interfacial tension γ_{SL}. Values both of β and of γ_c are found to depend on the liquids used (Jarvis *et al.* 1964) and this observation can be rationalised using the semi-empirical Girofalco–Good equation (Girofalco and Good 1957). This equation expresses the interfacial tension between two immiscible phases (A and B) in terms of a geometric mean of their surface tensions as

$$\gamma_{AB} = \gamma_A + \gamma_B - 2\phi(\gamma_A\gamma_B)^{1/2} \tag{2.2.25}$$

where ϕ is an 'interaction' parameter calculated from the molar volumes, polarisabilities, dipolar moments and ionisation potentials of the interacting units. Applying this relation to the Young–Dupré equation one obtains

$$\cos\theta = -1 + 2\phi\left(\frac{\gamma_s}{\gamma_L}\right)^{1/2}. \tag{2.2.26}$$

Using the definition of critical surface tension as

$$\gamma_c = \lim_{\theta \to 0} \gamma_{LV}$$

we have

$$\gamma_c = \phi^2\gamma_s \tag{2.2.27}$$

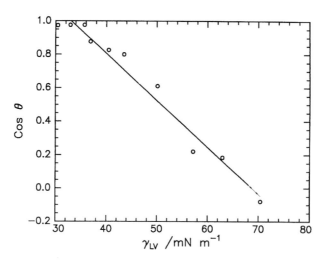

Figure 2.10. A Zisman plot for polyethylene: the contact angles of a variety of liquids on polyethylene are plotted against the surface tension of each liquid.

and thus

$$\cos\theta = -1 + 2\left(\frac{\gamma_c}{\gamma_L}\right)^{1/2}. \tag{2.2.28}$$

Since the gradient of a Zisman plot, β (see equation (2.2.24)) is $\mathrm{d}(\cos\theta)/\mathrm{d}\gamma_{LV}$, we have from equation (2.2.28) $\beta = 2(\gamma_c/\gamma_L^3)^{1/2}$. Thus the slope β in a Zisman plot is not constant; curvature in such a plot is to be expected.

The relation between γ_c and γ_{SV} is clear from equation (2.2.40);

$$\gamma_c = \gamma_{SV} - \gamma_{SL}, \tag{2.2.29}$$

so γ_c is somewhat less than γ_{SV}. The difference is generally quite small, because solid/liquid interfacial tensions are usually substantially smaller than liquid surface tensions.

There is, in addition, a small correction due to the equilibrium spreading pressure, π^0, of the fluid on the polymer. Inclusion of this results in

$$\gamma_c = \gamma_{SV} - \gamma_{SL} - \pi^0. \tag{2.2.30}$$

The value of π^0 depends on the nature of the fluid making up the drop and becomes larger as the contact angle approaches zero. Consequently the critical surface tension will depend on the liquid used, although in practice the contribution of π^0 is generally ignored. Critical surface tensions for polymers are in the range $20-60$ mN m^{-1}, with the majority of values of the order of $30-40$ mN m^{-1}. Higher values are associated with polar polymers, the lowest values being recorded for fluorocarbon polymers.

Hysteresis in contact angle values is also dependent on the drop size and this is attributed to a pseudo-line tension. Line tensions are intimately associated with the equilibrium of three phases and have been discussed extensively by Rowlinson and Widom (1982). When three phases are in equilibrium they will meet in a line of three-phase contact and there will be an excess free energy per unit length, the line tension, associated with this contact line. For a liquid drop on a polymer surface, this contact line is circular and if the line tension is positive a driving force to shrink the drop laterally will be developed. For a drop of radius R and a line tension of σ, the Young–Dupré equation is modified to

$$\gamma_{SV} = \gamma_{LV}\cos\theta + \gamma_{SL} + \sigma/R. \tag{2.2.31}$$

Following Vitt and Shull (1995) we define $\cos\theta_\infty = (\gamma_{SV} - \gamma_{SL})/\gamma_{LV}$; then we have

$$\cos\theta = \cos\theta_\infty - \sigma/(R\gamma_{LV}), \tag{2.2.32}$$

with a positive σ increasing the contact angle. Although theoretical considerations suggest that $\sigma \simeq 10^{-6}$ mN, larger values (up to six orders of magnitude larger) of σ are found when the drop size dependence of contact angles is analysed. These are termed pseudo-line tensions. Smaller drops give smaller values of σ and hence the contact angle for smaller drops should more closely approach the equilibrium value. Vitt and Shull (1995) have attempted to approach this equilibrium value of the contact angle of a polymer on a polymer surface using atomic force microscopy (AFM). Poly(2-vinyl pyridine) was spun as a thin film on to a layer of polystyrene and the bilayers annealed at high temperature to allow the poly(2-vinyl pyridine) to de-wet the polystyrene surface. Equilibrium droplet formation was judged to be complete when the drops were regular in shape with a circular geometry. The profile of the drops was then obtained using AFM and the contact angle calculated from the height and radius of each drop. Equation (2.2.32) suggests that a plot of $\cos\theta$ as a function of $1/R$ should be linear with a slope of σ/γ_{LV}. Data for poly(2-vinyl pyridine) on polystyrene are plotted in this fashion in figure 2.11, with no slope discernible within the resolution of the experiment. Vitt and Shull estimate that the *maximum* value of the pseudo-line tension is 10^{-4} mN, a value that more closely approaches theoretical predictions. Using approximations to the interfacial tension between these two polymers, they suggested that the actual value of the pseudo-line tension was more likely to be in the region of 10^{-6} mN. Consequently, in this experiment, the droplet size range obtained on de-wetting (up to about 2 μm diameter) provides equilibrium contact angle values.

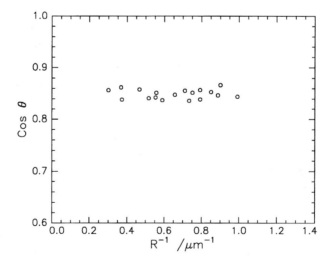

Figure 2.11. The cosine of the contact angle for droplets of poly(2-vinyl-pyridine) of various sizes on the surface of polystyrene. After Vitt and Shull (1995).

2.3 More precise theories of the surface tension of simple fluids

Let us now return to the question of whether we can calculate the surface energies of polymers from first principles. The rough estimates in section 2.1 tell us correctly the order of magnitude of surface tensions and correctly draw attention to the intimate connection between surface energies and the cohesive forces in liquids, but they have a number of drawbacks. Firstly, temperature makes no appearance in these theories, despite the experimental fact that surface tensions depend quite strongly on temperature. Secondly, we have assumed that the density of the liquid near the surface is the same as the bulk density. These shortcomings are seen at their most extreme if we consider a liquid near the liquid–vapour critical point. Here the distinction between liquid and vapour vanishes completely; the surface tension of the liquid approaches zero and the system becomes in effect all interface. An improved theory of surface tension must be able to account for these phenomena, at least qualitatively.

As discussed in section 2.2.1, surface tensions are measured under conditions of constant temperature and pressure – in this case the work done against the surface tension is equal to the increase in the Gibbs free energy. Thus for isothermal changes at finite temperature the surface tension is strictly a surface free energy[†]. The fact that the surface tension at finite temperature has a substantial entropic component is very important, as we shall see.

To proceed, then, let us attempt to calculate the free energy associated with a surface with an arbitrary density profile; the principles of equilibrium will then assert that the equilibrium surface profile will be the one with the lowest value of the free energy. Let us assume, then, that near the surface the density $\rho(z)$ is a function of the depth z. For a uniform system we can write a free energy density as a function of density and temperature, $\psi(\rho, T)$; if we were to assume that the same function describes the free energy density in an inhomogeneous system then we could write the surface tension as follows:

$$\gamma = \int_{-\infty}^{\infty} (\psi(\rho(z)) - \psi(\rho_{L,G})) \, dz, \qquad (2.3.1)$$

where we define $z = 0$ by the relationship

$$\int_{-\infty}^{\infty} (\rho(z) - \rho_{L,G}) \, dz = 0, \qquad (2.3.2)$$

[†] Potential confusion can thus arise about what is meant by the term 'surface energy'. Often it is tacitly assumed that surface energy and surface tension are identical and that by the former one means surface free energy. However 'surface energy' may also be used to refer to the excess surface internal energy. To avoid this confusion from here on we will refer only to surface tension when discussing the surface free energy.

where $\rho_{L,G}$ is the liquid bulk density for $z < 0$ and the vapour bulk density for $z > 0$. This definition is illustrated in figure 2.12.

The function $\psi(\rho, T)$ will typically have the shape illustrated in figure 2.13; below the critical point of the fluid it will have two minima and a double tangent construction defines the two coexisting phases, one liquid and one gas.

It is convenient to define a function $-W(\rho)$, which represents the distance of $\psi(\rho, T)$ above the double tangent line. Now we can rewrite the integral of equation (2.3.1) as

$$\int_{-\infty}^{\infty} (\psi(\rho(z)) - \psi(\rho_{L,G}))\, dz = \int_{-\infty}^{\infty} -W(\rho(z))\, dz. \qquad (2.3.3)$$

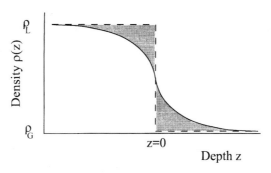

Figure 2.12. A schematic density profile near the surface of a fluid. The position of the origin is defined by the equal area condition of equation (2.3.2).

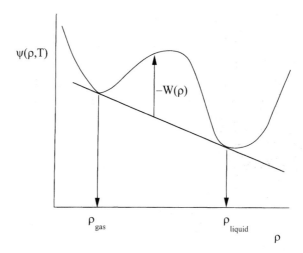

Figure 2.13. The excess free energy density $\psi(\rho, T)$ as a function of the density ρ for a fluid below its critical point. The double tangent construction defines the equilibrium densities of gas and liquid phases.

The surface tension is now expressed as a functional of the density profile near the surface. Unfortunately it now becomes clear that this formulation cannot be correct; the density profile which minimises the functional of equation (2.3.3) is clearly a step function at the surface, which predicts that the surface tension of all liquids is zero! The problem is in our assumption that the free energy density is purely a local function of the density. In fact this cannot be true; the free energy depends both on the local density and on the density of the surroundings. This was first recognised by van der Waals, who introduced a term proportional to the square of the density gradient. Thus augmented we find the surface tension is given by

$$\gamma = \int_{-\infty}^{\infty} \left[-W[\rho(z)] + \tfrac{1}{2}m \left(\frac{\mathrm{d}\rho}{\mathrm{d}z} \right)^2 \right] \mathrm{d}z. \tag{2.3.4}$$

The equilibrium density profile $\rho(z)$ is the one that minimises the surface tension functional; thus by the standard methods of calculus of variations we can write down an Euler–Lagrange equation that the density profile must satisfy:

$$m \frac{\mathrm{d}^2\rho}{\mathrm{d}z^2} = \frac{\mathrm{d}W}{\mathrm{d}\rho}. \tag{2.3.5}$$

So, for a given function $W(\rho)$ and gradient coefficient m, the equilibrium density profile could be found by numerically solving equation (2.3.5) with the boundary conditions that $\rho \to \rho_{\text{liquid}}$ for $z \to -\infty$ and $\rho \to \rho_{\text{gas}}$ for $z \to +\infty$; then the surface tension could be found by evaluating the integral of equation (2.3.4).

In fact we can directly integrate equation (2.3.5) to yield

$$W(\rho) + \tfrac{1}{2}m \left(\frac{\mathrm{d}\rho}{\mathrm{d}z} \right)^2 = 0 \tag{2.3.6}$$

(the constant of integration is zero because $W(\rho_{\text{liquid}}) = W(\rho_{\text{gas}}) = 0$, as is obvious from figure 2.13). Now we can calculate the density profile by the following integral:

$$z = \left(\frac{m}{2} \right)^{1/2} \int_{0}^{\rho} \frac{\mathrm{d}\rho'}{[-W(\rho')]^{1/2}}. \tag{2.3.7}$$

This integral will usually have to be evaluated numerically.

The surface tension can be found once again from equation (2.3.4) if $\rho(z)$ has been evaluated, or from the expression obtained by combining equations (2.3.4) and (2.3.6):

$$\gamma = -2 \int_{-\infty}^{\infty} W[\rho(z)] \, \mathrm{d}z. \tag{2.3.8}$$

However, the most useful form for the surface tension is obtained by changing the variable of integration in equation (2.3.8) using equation (2.3.6);

$$\gamma = \int_{\rho_{gas}}^{\rho_{liquid}} \frac{d\rho}{[-2mW(\rho)]^{1/2}}. \tag{2.3.9}$$

With this equation we can calculate the surface tension by a single (usually numerical) integration without first having to calculate the equilibrium density profile.

Let us make this discussion more concrete by applying this theory to a simple model of a fluid – a lattice fluid model. This is an almost trivial model of a fluid, which does reproduce the transition between a liquid and a gas. It is popular with theoretical physicists because it can be directly mapped on to the much studied Ising model of a magnetic phase transition (Chandler 1987). We will solve it within a mean-field approximation, which allows us to write down simple expressions for the free energy density and gradient coefficient, so we can illustrate the van der Waals theory with explicit calculations. Somewhat surprisingly, perhaps, in view of its great simplicity, it forms the basis for a successful theory for predicting polymer surface tensions, the Sanchez–Lacombe theory.

Again it was van der Waals who identified the key features of the physics of liquids – the atoms or molecules interact by a very strong but short-range repulsion when they are brought close together, a longer range attraction becoming important when they are further apart. The potential has the familiar shape illustrated in figure 2.14a. In the lattice fluid model we regard space as discrete rather than continuous; molecules can occupy cells in space. We specify that no cell can be occupied by more than one molecule and that if two neighbouring cells are occupied there is an associated energy ε (which is negative, indicating that neighbouring molecules are attracted to one another). This amounts to replacing the smooth potential of figure 2.14a by the discrete potential of figure 2.14b, an infinitely high potential barrier preventing two molecules occupying the same cell and a square well of depth ε representing the attraction of two neighbouring molecules. Molecules further apart than nearest neighbours do not interact at all.

To calculate the thermodynamics of this model we can start by writing down the entropy of each site. If the probability that the ith site is occupied is p_i then we can write down the entropy using the Gibbs formula as

$$S = -k \sum_i [p_i \ln(p_i) + (1 - p_i) \ln(1 - p_i)] \tag{2.3.10}$$

because each state has two possible states, occupied or unoccupied, with probabilities p_i and $1 - p_i$, respectively.

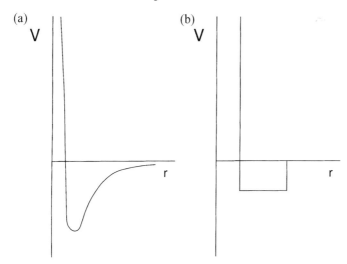

Figure 2.14. (a) A realistic intermolecular pair potential in a liquid, with a hard repulsive core and a more long-ranged intermolecular interaction. (b) The simple potential implicit in the lattice fluid model, with an infinitely steep repulsion defining the size of the lattice cell and an attraction confined to nearest neighbours.

If we can assume that no site is correlated to any other site then the probability that any site is occupied is simply proportional to the overall density ρ:

$$\rho = \frac{nm_{\mathrm{m}}}{\Omega}, \tag{2.3.11}$$

where m_{m} is the molecular mass, Ω is the volume of a cell and $n = p_i$ for all values of i. Thus the entropy per unit volume is given by

$$S = -k[n\ln(n) + (1 - n)\ln(1 - n)]/\Omega. \tag{2.3.12}$$

Using the same approximation – and it is this that constitutes the mean-field approximation – we can calculate the internal energy. Once again assuming that the occupancies of neighbouring sites are uncorrelated, for a coordination number z the energy relative to a fully dense reference state is given by

$$U = \frac{1}{2}\frac{\varepsilon z n(n - 1)}{\Omega}. \tag{2.3.13}$$

Thus we can write the free energy density $\psi(n, T)$ as

$$\psi(n, T) = kT[n\ln(n) + (1 - n)\ln(1 - n)] + \frac{z\varepsilon}{2}n(n - 1) \tag{2.3.14}$$

which typically has the form sketched in figure 2.15.

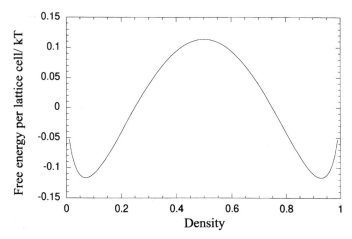

Figure 2.15. The free energy as a function of the density for a lattice gas, assuming a co-ordination number of 12, $\varepsilon = -1$ and $T = 2$.

Note that in this model the free energy curve is symmetrical with respect to the density. Thus the exchange chemical potential is zero for the coexisting liquid and gas phases.

The gradient term can also be calculated explicitly for this model. Consider three layers of cells. We assume that a cell in the central layer i has z' neighbours in layer $i + 1$, z' neighbours in layer $i - 1$ and $z - 2z'$ neighbours in the same layer i. Thus we can write the energy associated with a cell in the ith layer in terms of the densities n_{i+1}, n_i and n_{i-1} in the $(i + 1)$th, ith and $(i - 1)$th layers, respectively, as

$$U_{\text{in}} = \frac{1}{\Omega}\left(\frac{z'\varepsilon}{2} n_i n_{i+1} + \frac{z'\varepsilon}{2} n_i n_{i-1} + \frac{(z - 2z')\varepsilon}{2} n_i^2 - \frac{z\varepsilon}{2} n_i\right). \qquad (2.3.15)$$

If we now assume that the gradients are small we can write

$$n_{i\pm1} = n_i \pm b\left(\frac{dn}{dz}\right)_i + \frac{b^2}{2}\left(\frac{d^2n}{dz^2}\right)_i, \qquad (2.3.16)$$

where b is the size of a cell. If substituted into equation (2.3.15) this yields

$$U_{in} = \frac{1}{\Omega}\left[\frac{z\varepsilon}{2} n_i(n_i - 1) + \frac{z'\varepsilon b^2}{2} n_i\left(\frac{d^2n}{dz^2}\right)_i\right]. \qquad (2.3.17)$$

We note that the configurational entropy depends solely on the local density, so the part of the free energy density that depends on concentration gradients is solely energetic. When we integrate the free energy density to obtain the surface tension we can integrate the term in equation (2.3.17) proportional to the second derivative of the density by parts and in this way we find that the gradient coefficient m is given by

$$m = -\varepsilon z' b^2. \qquad (2.3.18)$$

Now we can use equation (2.3.9) to calculate the surface tension of this lattice gas model as a function of temperature. The results are shown in figure 2.16.

As we expect, as the temperature increases the surface tension decreases. This is because as a density gradient is introduced at the surface there is an extra entropy associated with the surface that reduces the free energy. As expected, at the critical point the surface tension goes to zero. At the critical point the difference between liquid and gas phases disappears, and the system is in a sense all interface. In fact, as one approaches the critical point square gradient theory predicts that the surface tension goes to zero according to a power law:

$$\gamma \sim \left(1 - \frac{T}{T_c}\right)^{\mu} \qquad (2.3.19)$$

where the exponent μ has the value $3/2$. Experimentally, power-law behaviour

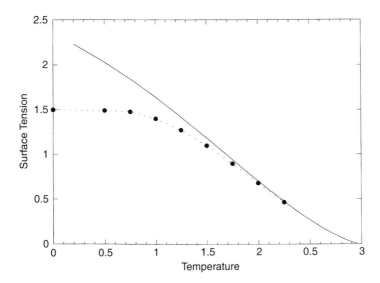

Figure 2.16. The surface tension of a lattice gas as a function of the temperature, assuming that the surface is the (111) face of a FCC lattice, with units chosen so that the lattice spacing is unity. Boltzmann's constant is unity and the interaction energy $\varepsilon = -1$. In these units the critical temperature is 3. The solid line is the prediction of square gradient theory, whereas the points are the predictions of an analogous mean-field theory in which no small-gradient approximation is made.

is observed, but with a different value of μ, closer to 1.26. This discrepancy is an example of the well-known failure of mean-field theories to yield correct values for critical exponents near critical points due to their failure to deal properly with fluctuations (Rowlinson and Widom 1982). However, as the temperature is reduced below the critical point, the fluctuations in density become of much smaller amplitude and mean-field theory can be expected to become more reliable.

However, there is also a problem at low temperatures – using equation (2.1.1) we would predict that for our lattice model at zero temperature the surface tension should be 1.5, in the units used for figure 2.16. In fact, at low temperatures square gradient theory appears to predict a larger value. The origin of this discrepancy is clear if we look at the density profiles at the surface calculated using equation (2.3.7) with the lattice gas free energy density and gradient coefficient. Figure 2.17 shows such a profile; it is well represented by a hyperbolic tangent form

$$n(z) = n_{\text{gas}} + (n_{\text{liquid}} - n_{\text{gas}}) \{\tfrac{1}{2} + \tfrac{1}{2}\tanh[z/(2w)]\}, \tag{2.3.20}$$

where w is a parameter describing the width of the profile. For the profile shown, when the temperature is not too far from the critical temperature, w is large relative to the cell size and we can expect the gradient approximation to work well. However, as the temperature is reduced the profiles become narrower; the gradient approximation breaks down; the profiles increasingly deviate from those calculated without the gradient approximation; and, as a result, the gradient approximation overpredicts the surface tension.

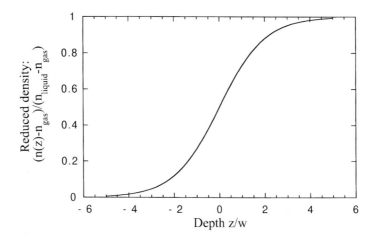

Figure 2.17. The hyperbolic tangent surface density profile.

2.4 Theories of the surface tension of polymers

This simple lattice gas approach to surface tension can be extended to polymers in a relatively straightforward way. The major difference is that the finite length of polymer chains destroys the symmetry between holes and molecules in lattice fluid language. Thus we would expect the equivalent of (2.3.4), function $W(\rho)$, to be strongly asymmetric; the equilibrium fraction of holes in the polymer may well be quite large even when the fraction of polymer molecules in the vapour phase is very small. This can be dealt with at a mean field level by a very simple modification of equation (2.3.14). An additional effect that is not so easily incorporated into the lattice fluid model is that the conformations of polymer chains might be expected to be affected by the presence of a surface, leading to additional contributions to the surface tension. We shall see, perhaps surprisingly, that it is the first factor that seems to be by far the most important and that the surface tensions of polymers, including the dependence of the surface tension on the relative molecular mass, can be predicted remarkably well by using a straightforward extension of the lattice fluid model discussed in the last section.

At the simplest level, we can maintain the same assumption that the local density in the immediate environment of a segment of a chain is everywhere uniform and is the same as the overall density. Thus the internal energy of the polymer system is given by equation (2.3.13) with no modification. The entropy, however, is modified; because of the connections between the various polymer segments of a polymer chain the configurational entropy of the polymer molecules is reduced by a factor of r, the number of lattice cells joined together in a single molecule. Thus the free energy density becomes

$$\psi(n, T) = kT\left(\frac{n}{r}\ln(n) + (1-n)\ln(1-n)\right) + \frac{z\varepsilon}{2}n(n-1). \qquad (2.4.1)$$

The effect of this on the shapes of the free energy curves is shown in figure 2.18; as r increases the curves become increasingly asymmetric, with the density of the coexisting gas phase becoming smaller and smaller. In other words, the vapour pressure of high polymers becomes vanishingly small, as we might intuitively expect.

What effect does the polymeric nature of the fluid have on the gradient term? Recall that in the simple lattice fluid model this term is entirely of energetic origin, with no entropic component; this part of the gradient term will thus be unchanged for a polymer. However we might also expect that concentration gradients in polymeric systems might carry an additional free energy penalty due to the restriction on the number of configurations available to the polymer in

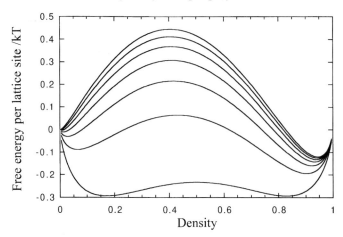

Figure 2.18. The free energy versus the density for various relative molecular masses in the polymer lattice gas model. The temperature is 2.5 and the value of r varies in the range $1-4$ in steps of 0.5.

the concentration gradient; in other words the gradient term should contain both energetic and entropic components. It turns out that, although this is strictly true, for the polymer/air interface the energetic part of the gradient term is dominant. As we shall see in a later chapter the opposite is true for polymer/polymer interfaces. We shall return to this point later, but let us first consider the results of our polymer lattice fluid model assuming a purely energetic gradient term.

Evaluating the integral in equation (2.3.9) with a free energy density given by (2.4.1) and the lattice fluid gradient coefficient (2.3.18) gives us the surface tension as a function both of the temperature and of the relative molar mass, the latter via the parameter r. Figure 2.19 shows the surface tension against r for a temperature of one quarter the critical temperature. For low relative molecular masses the surface tension is an increasing function of increasing relative molecular mass, but for higher relative molecular masses a limiting value is reached. A formula of the form

$$\gamma(r) = \gamma(\infty) - \left(\frac{A}{r}\right)^{z} \tag{2.4.2}$$

represents the trend quite well, with the best value of the exponent $z = 1$. As we shall see, this is close to what is observed for polymer melts of high relative molecular mass.

If we plot the limiting value of the surface tension for a high relative molecular mass as a function of temperature we obtain results such as those plotted in figure 2.20. Just like for the simple lattice fluid, we see a monotonic decrease of surface tension with increasing temperature.

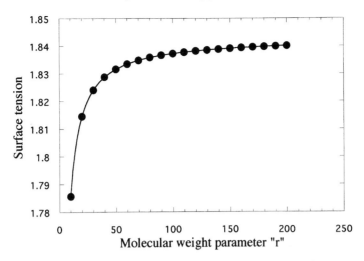

Figure 2.19. The surface tension as a function of the relative molecular mass parameter 'r' for a temperature of 1.5 and an interaction energy of -1, calculated for a (111) surface of a FCC polymer lattice gas model. The solid line is a fit to equation (2.4.2).

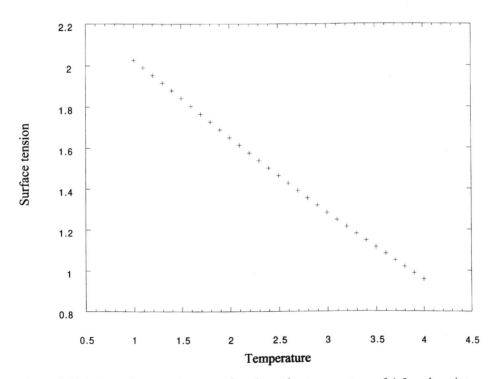

Figure 2.20. The surface tension as a function of a temperature of 1.5 and an interaction energy of -1, calculated for a (111) surface of a FCC polymer lattice gas model in the limit of large values of the relative molecular mass parameter 'r'.

Table 2.2. *Some typical values of surface tensions of polymers*

	150 °C	20 °C
Poly(ethylene terephthalate) ($M_n = 16\,000$, $M_w = 37\,000$)	36.2	44.6
Poly(methyl methacrylate) ($M = 3000$)	31.2	41.1
Polystyrene ($M_n = 9300$)	31.0	39.4
Poly(oxyethylene)-diol ($M = 86–17\,000$)	30.1	42.9
Polyethylene (High M_W)	29.4	36.8
Poly(ethyl acrylate) ($M_w = 28\,000$)	27.0	37.0
Polypropylene (atactic)	22.1	29.4
Poly(tetrafluoroethylene) (High M_W)	16.3	23.9
Poly(dimethyl siloxane) ($M_w = 75\,000$)	14.2	20.4

Table 2.3. *Some typical values of surface tensions of small molecules analogous to the polymers shown in table 2.2*

	γ (mJ m^{-2}) 20 °C
Water	72.75
Di-ethyl phthalate	37.5
Styrene (19 °C)	32.14
Benzene	28.85
Ethanol	22.75
n-Octane	21.8

2.5 Surface tension of polymer melts: the experimental situation and comparison with theory

Using the techniques outlined in section 2.2, values of the surface energies of a variety of polymer melts are now available. Values for some polymer melts of high relative molecular mass are tabulated in table 2.2, with values for some small-molecule liquids in table 2.3, for comparison. Qualitatively, we notice that the polymer melt surface tensions are quite comparable with those of analogous small molecules; non-polar molecules have lower surface tensions than do polar molecules, irrespective of whether they are polymeric.

Perhaps the most extensive and accurate set of data on polymer surface tensions has been gathered by Dee and Sauer using the fibre modification of the Wilhelmy plate method described in section 2.2.4 (Sauer and Dipaolo 1991, Dee and Sauer 1992, Sauer and Dee 1994). This data set permits quite careful testing of the statistical mechanical theories, which consequently can now be considered to have real predictive power.

The surface tension of polymers has a relatively weak dependence on the relative molecular mass. For example, figure 2.21 shows values of surface tension for alkanes at 150 °C (Sauer and Dee 1994). The limiting value for high-relative-molecular-mass polyethylene is about 27 mN m^{-1}, and only for relative molecular masses less than a few thousand is this value significantly reduced. For intermediate relative molecular masses, we find a dependence following equation (2.4.2) with an exponent $z = 1$, just as is predicted by the polymer lattice fluid/square gradient theory outlined above. However, for oligomers with relative molecular masses less than about 200 a value of $z = 2/3$ fits the data better, following an empirical law suggested some years ago by Legrand and Gaines (1969). Very similar results are obtained for the dependence on relative molecular mass of other polymers.

It is found that polymer surface tensions decrease monotonically as a function of increasing temperature. This is illustrated for poly(tetramethylene oxide) in figure 2.22.

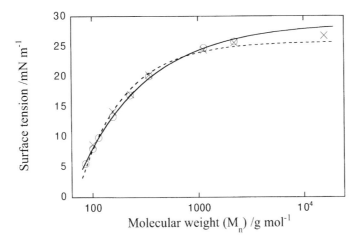

Figure 2.21. The dependence of the surface tension on the relative molecular mass for alkanes at 150 °C. The circles are the data and the crosses give the predictions of gradient theory. The lines are fits to the power law of equation (2.4.2), with the exponent taking the values 2/3 (solid line) and unity (dashed line). Data are from Sauer and Dee (1994).

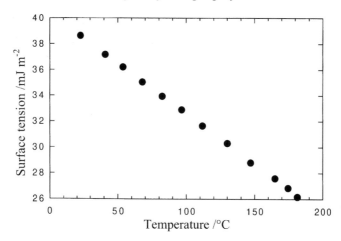

Figure 2.22. The temperature dependence of the surface tension of poly(tetramethylene oxide). After Sauer and Dee (1994).

On comparing the relative molecular mass data of figure 2.21 with the square gradient theory prediction of figure 2.19, and the temperature data of figure 2.22 with the square gradient theory prediction of figure 2.20, we see that the qualitative trends are well described by the simple polymer lattice fluid model. Can the comparison be made any more quantitative? To do so, we need to be able to estimate the parameters of the lattice fluid model, namely the cell volume Ω, the interaction energy parameter ε, and the relationship between the parameter r and the actual relative molecular mass. One way to proceed is to obtain values of these parameters by fitting experimental data on the density of polymers as a function of pressure and temperature to the predictions of the polymer lattice fluid model.

The equation of state for a polymer lattice fluid, developed by Poser and Sanchez (1979), can be written in the form

$$\tilde{\rho} = 1 - \exp\left[-\frac{(\tilde{\rho}^2 + \tilde{P})}{\tilde{T}} - \left(1 - \frac{1}{r}\right)\tilde{\rho} \right], \qquad (2.5.1)$$

where $\tilde{\rho}$ is the reduced, dimensionless density, $\tilde{P}$ is the reduced pressure and $\tilde{T}$ is the reduced temperature. The reduction constants ρ^*, P^* and T^* have definitions in terms of the parameters of the lattice model, but it is probably better to consider them as fitting parameters whose values are tabulated for a variety of polymers by fitting equation (2.5.1) to measured equations of state. Using such best fit parameters and evaluating the surface tension using equation (2.3.9) produces a prediction of the dependences on temperature and relative molecular mass of the surface tension of hydrocarbon liquids and polymer melts with no adjustable parameters. It is found that the theory correctly predicts the observed dependences on relative molecular mass and temperature, but systematically underestimates the actual values by 10% or so.

Attempts to improve on this theory have taken two courses. The first approach is to re-examine the question of the gradient term. As mentioned above, in addition to the gradient term of enthalpic origin included in the simple square gradient theory, one would expect for polymers an additional term of entropic origin reflecting the reduction in number of possible polymer configurations near the interface. At the simplest level, it turns out that such an entropic gradient term has the form

$$F_{\text{config}} = \frac{kTa^2}{36n}\left(\frac{\mathrm{d}n}{\mathrm{d}z}\right)^2 \tag{2.5.2}$$

as we shall see in chapter 4. Including this extra gradient term removes most of the discrepancy between theory and experiment, providing quantitatively accurate predictions for the surface energy of many polymer melts (Sanchez 1992). This is illustrated in figure 2.23, which shows a reduced surface tension versus temperature plot for polymers in the limit of high relative molecular mass, together with experimental data for a number of polymers. Values of the reducing parameters γ^* and T^* are given in table 2.4; these can be used to

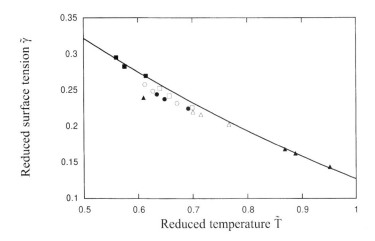

Figure 2.23. The reduced surface tension as a function of the reduced temperature for a number of polymers ($\triangle$, poly(vinyl acetate); ■, polystyrene; □, polyisobutylene; ▲, poly(dimethyl siloxane); ●, linear polyethylene; ○, branched polyethylene), compared with the prediction of a gradient theory using the Poser and Sanchez lattice fluid model and a square gradient term modified to account for loss of polymer configurational entropy near a surface (full line). After Sanchez (1992).

Table 2.4. *Reducing parameters for various polymers, for
use with equation (2.5.3) to predict the temperature
dependence of surface tension (from Poser and Sanchez
(1979))*

	T^*(K)	γ^*(mJ m^{-2})
Poly(dimethyl siloxane)	476	84.3
Poly(vinyl acetate)	590	128
Poly(isobutylene)	643	104
Polyethylene (linear)	649	118
Polyethylene (branched)	673	106
Polystyrene (atactic)	735	109

predict the surface tension of these polymers at any temperature by using the
following parameterisation of the theoretical curve:

$$\tilde{\gamma} = 0.6109 - 0.6725\tilde{T} + 0.1886\tilde{T}^2. \qquad (2.5.3)$$

We should not, however, immediately conclude that the problem of polymer
surface tension is thus completely solved. Firstly, as we will see when we
discuss the origins of the entropic gradient term in chapter 4, in the context of
polymer/polymer interfaces, the form given in equation (2.5.2) is an approximation applicable in a limit of small density gradients which certainly should
not apply here. A more accurate accounting for chain configurations near
interfaces is possible by combining a numerical lattice model with the lattice
fluid equation of state (Theodorou 1989). However, this actually seems to do
worse than the modified square gradient theory at quantitative surface tension
predictions.

On the other hand, Sauer and Dee (1994) have demonstrated that, by refining
the equation of state used in the gradient theory, by using better experimental
data and a different equation of state model, the Flory, Orwoll and Vrij model
(the FOV model) (Flory *et al.* 1964), excellent agreement, at the level of better
than 1%, between theory and experiment is obtained. The square gradient
coefficient still has to be adjusted somewhat from the mean-field value, but it is
still treated as entirely enthalpic in origin, with no temperature dependence.

The FOV model is also a cell model; the polymer molecule is divided into *r*
units, which need not be the monomer units, and each of these units is placed
in a cell. However, in contrast to the lattice fluid approach, each unit may move
within its own cell, but may not leave it. Each unit is considered to have $3c$
degrees of freedom, where the factor *c* accounts for the restriction in number of
degrees of freedom because of the connectivity between the units. This factor

is assumed to be constant and to have a value less than unity. In the bulk this theory leads to the equation of state

$$\tilde{p}\tilde{V}/\tilde{T} = \tilde{V}^{1/3}/(\tilde{V}^{1/3} - 1) - 1/(\tilde{T}\tilde{V}), \tag{2.5.4}$$

where $\tilde{p}$, $\tilde{V}$ and $\tilde{T}$ are the reduced pressure, volume and temperature, respectively. These reduced parameters are defined in terms of fundamental properties of the polymer, V^* the hard core cell volume and ε^* the interaction between units:

$$\tilde{p} = p/p^* = pV^*/\varepsilon^* \tag{2.5.5}$$

$$\tilde{T} = T/T^* = Tck_B/\varepsilon^* \tag{2.5.6}$$

$$\tilde{V} = V/V^*. \tag{2.5.7}$$

The Gibbs free energy for N polymer molecules of r units from this theory is

$$G = Nr\varepsilon^* \left(-\frac{1}{\tilde{V}} + \tilde{p}\tilde{V} - 3\tilde{T}\ln\left(\tilde{V}^{1/3} - 1\right) \right). \tag{2.5.8}$$

The first two terms in the large parentheses are enthalpic contributions and of these the second may be ignored except when pressures are very large and approaching the critical value. The third term is the entropic contribution and has the familiar logarithmic form. More extensive discussions of this equation of state theory (and others) can be found elsewhere (Munk 1989, Sanchez 1992).

To apply this theory to calculate surface tensions one simply uses the free energy density function derived from equation (2.5.8) in conjunction with the square gradient theory of section 2.4. As usual this leads to a smoothly varying density profile between the liquid and vapour phase, which may be visualised in terms of the FOV model as a variation in cell size through the surface, as shown schematically in figure 2.24.

All our discussion of theories of polymer surface tensions up to now has relied ultimately on the van der Waals theory in conjunction with some equation of state. There is, however, another possible approach, which starts out from a principle of corresponding states, as was first applied to polymers by Patterson and Rastogi (1970). We start by assuming spherical molecules interacting via an attractive potential, which is described by the depth of the energy minimum (ε^*) and a hard core or impenetrable radius (r^*). The partition function is written in terms of variables scaled by the characteristic energy and length scale and thus all physical properties can be described in terms of these scaled variables. Clear corollaries with the equation of state

UNIV...
OF SHEFF...
LIBRARY D

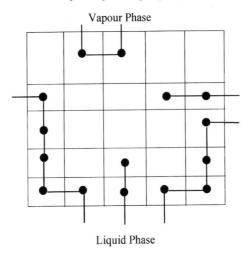

Figure 2.24. A schematic diagram of the Flory lattice model with varying cell sizes near the liquid/vapour interface.

descriptions are evident. The connection with the principle of corresponding states is made by describing the interfacial region as a homogeneous layer with a density intermediate between those of the two equilibrium phases. This interfacial layer is presumed to be of short distance and without the 'tails' of the van der Waals square gradient theory. The relation for the extent and density of this interfacial region provides a further binding with the corresponding states principle by using the potential functions for interaction between the units in the bulk and in the surface layer to define the surface energy. The free energies in the bulk and surface regions thus contain explicitly the interaction potential functions. Additionally, the expression for the contribution of the surface to the free energy contains the ratio of the surface cell free volume to that of a cell in the bulk. It is in the expression for this surface cell free volume that Dee and Sauer (1995) differ from the original approach of Patterson and Rastogi; the expression for the reduced surface tension is

$$\tilde{\gamma} = \{(-m/\tilde{V}) - \tilde{T}\ln[(b\tilde{V}^{1/3})/(\tilde{V}^{1/3} - 1)]\}/\tilde{V}^{2/3}, \qquad (2.5.9)$$

where m is an adjustable parameter that relates the difference between surface and bulk interaction potentials to the bulk potential and b is also an adjustable parameter in the expression for the surface cell free volume as a function of the bulk free volume. We point out here the strong dependence of surface tension on bulk properties; any deviations away from a universal scaled curve of $\tilde{\gamma}$ would be indicative of contributions to the surface tension arising from specifically surface factors such as local modifications in conformation. Using the surface tension and pVT data for the polymers discussed in section 2.2, in combination with the Flory, Orwoll and Vrij equation of state, Dee and Sauer showed that their modified discrete interfacial cell theory produced a universal scaling relation between $\tilde{\gamma}$ and T/T^* which was reasonably well followed by all the polymers over the complete range of relative molecular mass. This was true even for hydroxyl-terminated polyethylene glycols for which the network formation referred to earlier and the high surface energy of hydroxyl units might be expected to lead to depletion of OH groups. Dee and Sauer conclusively demonstrated that any such effects rapidly become negligible as the number of units in the molecule rises above one or two. The general observation of a scaling relation notwithstanding, it was evident that there were differences between different polymers. The basis of the corresponding states principle is that the units are spherical molecules; departures from this are expressed in the value of m which describes the fraction of nearest neighbours to a unit that are lost due to the unit being in the surface layer. Consequently the value of m needs to be adjusted for each polymer type, reflecting the departure from true spherical shape. The adjustment of m to obtain agreement between theory and experiment plays the same role as the adjustment of the square gradient coefficient in the continuum description of the interface. For the range of polymers examined the optimum values of m and b were 0.52 and 2, respectively. In fact, the scaled surface tensions could be collected into two groups of polymers (figure 2.25), those with substituents on the main chain and those without. This grouping again reflects differences in local structure as

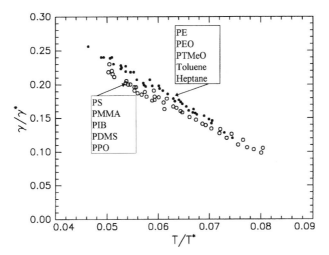

Figure 2.25. The reduced surface tension as a function of the reduced temperature based on the corresponding states principle of Dee and Sauer (1995).

expressed in *m*. The principle of corresponding states thus offers a way of accurately systematising surface tension data; having more adjustable parameters than the square gradient approaches, it may be argued to have less intrinsic predictive power but by the same token it is not limited by potentially unrealistic assumptions.

One conclusion that we can draw from all these various theoretical approaches, irrespective of the differences in detail, is that the surface tensions of polymer melts are not strongly influenced by their polymeric character. Surface tension is largely determined by the bulk *PVT* behaviour of the fluid; the long-chain nature of polymers does not have a strong influence on their bulk compressibility, so it is not surprising that polymer surface tensions are similar in magnitude to the surface tensions of analogous small-molecule liquids and that the dependence on relative molecular mass is weak.

One exception to this conclusion that surface tension is insensitive to questions of surface configuration is provided by situations in which a polymer molecule contains chemical groups which might be expected to orient themselves in a particular way to minimise the surface tension. An example of this is provided by polymers with perfluorinated side-groups, such as poly(hexafluoropropylene oxide) (PHFPO) (Sauer and Dee 1994). This polymer exhibits a huge discrepancy from the predictions of square gradient theory, which is attributed to the fact that the CF_3 groups preferentially locate themselves at the surface, lowering the energy. The same effects are even more important with block copolymers and polymers with different, low-energy end-groups. These inherently surface-active molecules are perhaps more appropriately considered as a separate class from simple homopolymer melts and we shall return to them in chapter 6.

2.6 The polymer surface at a microscopic level

We now turn to the question of what we can say about the structure of the polymer surface at the microscopic level. As we have seen, in addition to predicting the surface tension, square gradient theories also predict the density profile at the surface of the polymer melt. Typically, the density goes smoothly from the melt density to zero (the density of the vapour phase being vanishingly small for high polymers) over a few ångström units, in very much the same way as it does at the surface of a small-molecule liquid.

There have not been many systematic studies of the degree to which the surface of a polymer is diffuse on this scale, although neutron and x-ray reflectivity measurements are sensitive enough to provide this information. What observations exist are consistent with an interface whose width is about 5 or 10 Å. Measurements of the specular reflectivity, as we shall see, cannot distinguish between the intrinsic diffuseness of the surface and its roughness on a small length scale, even if it makes sense to make this distinction at all. At equilibrium the surface of any liquid will be roughened by thermally excited capillary waves; we can estimate their effect in the following way (Buff *et al.* 1965). If we consider a surface whose plane area is A, then a capillary wave of the form

$$z(\mathbf{r}, t) = a(k) \sin(\mathbf{k} \cdot \mathbf{r}), \qquad (2.6.1)$$

where k is the magnitude of the wave-vector $\mathbf{k}$, creates additional surface area with associated surface energy $\frac{1}{4} a(k)^2 k^2 A \gamma$. By the theorem of equipartition of energy, at equilibrium the average value of this extra energy will be $\frac{1}{2} k_B T$, so we can write the mean square value of the amplitude of the capillary wave with wave-vector of magnitude k as

$$\langle a(k)^2 \rangle = \frac{2 k_B T}{A k^2 \gamma}. \qquad (2.6.2)$$

We can now find the total contribution of the capillary waves to the roughness of the interface by integrating the mean squared displacement due to all the possible capillary waves. The number of possible waves is controlled by a quantisation condition; if we consider standing waves in a square of sides L, we know that the x and y components of the wave-vector must be integral multiples of π/L. Thus the density of possible standing wave states in k-space is A/π^2, and the total number of states whose wave-vector has a magnitude between k and $k + dk$ is $A k \, dk / (2\pi)$. Thus we find

$$\langle \Delta z^2 \rangle = \int \frac{1}{2} \langle a(k)^2 \rangle \frac{A k \, dk}{2\pi} = \int \frac{k_B T}{A k^2 \gamma} \frac{A k \, dk}{2\pi}. \qquad (2.6.3)$$

The integral does not converge, so we need to do the integration between maximum and minimum values of k, k_{max} and k_{min}, the choice of which we discuss below. The result, then, is

$$\langle \Delta z^2 \rangle = \frac{k_B T}{2 \gamma \pi} \ln \left(\frac{k_{max}}{k_{min}} \right). \qquad (2.6.4)$$

The divergent nature of this result is, at first, rather disconcerting. The maximum value of the wave-vector clearly has to be set by a microscopic distance; it does not make sense to talk about a capillary wave whose wavelength is much smaller than the intrinsic diffuseness of the interface, so this sets the value of k_{max}. The minimum value of the wave-vector is more puzzling. Ultimately, the wavelength cannot be bigger than the total size of the surface, so we are left with the paradoxical conclusion that the roughness of a liquid surface depends on the size of the container. In practice, however, other factors limit the minimum wave-vector. Often, the technique one uses to measure roughness is not equally sensitive to roughness on all length scales, and the potential energy of long wavelength waves is dominated by gravity rather than surface tension. In the latter situation the long-wavelength cut-off is provided by the capillary length l_c (Rowlinson and Widom 1982):

$$l_c = \left(\frac{\gamma}{\rho g} \right)^{1/2}. \tag{2.6.5}$$

In addition to affecting the observed interfacial width, these capillary waves also have an associated free energy. The magnitude of the correction that this implies for the surface tension is given by the expression

$$\gamma_{exp} = \gamma_{mf} - \frac{3 k_B T k_{max}^2}{16\pi}, \tag{2.6.6}$$

where γ_{exp} is the experimental surface tension and γ_{mf} is the mean-field value calculated by, for example, a square gradient theory. For an interface of order 5 Å wide this leads to a predicted correction of order a few mJ m^{-2}, which is certainly not negligible. However, this factor has not yet been taken into account in theoretical treatments of polymer surface tensions.

These results apply equally well for all interfaces that can be characterised by an interfacial tension; this includes simple liquid surfaces, polymer melt surfaces and interfaces between liquids. Indeed we will refer to capillary waves again when we discuss polymer/polymer interfaces in chapter 4 and again in chapter 8. One important thing to note is that the argument is not in any way altered by the viscosity of the liquid in question. The viscosity greatly affects the dynamics of capillary waves, as we shall discuss in chapter 8 when we consider how surface quasi-elastic light scattering can be used to study these dynamics, but the mean squared amplitude remains the same no matter how heavily damped the capillary waves are. In this respect the situation is exactly the same as the problem of the thermal fluctuations of a mass on a spring; the mean square amplitude of these fluctuations is solely a function of the spring constant and the temperature and remains the same whether the mass is immersed in air, water, a vacuum or treacle.

We now move on to considering the detailed conformations of polymer chains near interfaces. We have already commented that, although it is easy to argue that the conformations of polymers near interfaces must necessarily be strongly perturbed, this does not seem to be the major factor in determining the surface tension. An elegant physical argument about the effect of a surface on polymer configurations was made by Silberberg (1982). He argued that one could account for these modifications by considering an imaginary plane cutting the polymer melt. For every configuration on one side of the plane, one can find an exact mirror image configuration on the other side and in particular each configuration that crosses the plane has an image configuration crossing in the other direction. Now imagine cutting each chain that crosses the plane and joining up the free end with the cut end of the mirror conformation, so that after this procedure there remain no bonds crossing the plane, but the same number of bonds were remade as were broken. From this argument we can deduce that those chains whose centres of gravity lie less than about two radii of gyration from the surface will have their configurations perturbed; in particular the radius of gyration perpendicular to the interface will tend to be reduced, with rather less change occurring in the directions in the plane of the surface.

This picture of chains near the surface adopting rather flattened configurations has been confirmed by computer simulations. This is clearly illustrated in figure 2.26, in which the ratio of the chain end-to-end distances normal to the interface and in the bulk is plotted as a function of the position of the chain (Bitsanis and ten Brinke 1993). This lattice Monte Carlo study was carried out not for a free surface, but for a hard wall; however, in another study, already mentioned above, a lattice model was extended to allow for a variable density by combination with a lattice fluid equation of state theory. This explicitly

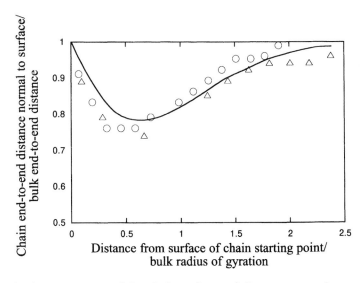

Figure 2.26. The component of the chain end-to-end distance normal to a surface as a function of the distance from the surface at which the chain starts, normalised by the bulk end-to-end distance. Points are from a lattice Monte Carlo simulation of a polymer melt in contact with a neutral wall ($\circ$, $N = 100$; $\triangle$, $N = 50$), whereas the line is the analytical prediction for random walks with a reflecting boundary condition. After Bitsanis and Brinke (1993).

addresses the case of a free surface, with similar results. One quantity that illustrates the surface-induced conformational changes in an illuminating way is the so-called chain-monomer profile. This is defined in a lattice model as the average number of monomers in a layer a given distance z from the surface that belong to the same chain. Figure 2.27 shows this chain-monomer profile in the variable density lattice model; it has a pronounced peak close to the surface, indicating the tendency of chains near the surface to take up flattened configurations.

Finally we should consider the location of the chain ends. On physical grounds it can be argued that a chain with its end near a surface suffers a smaller loss of entropy by virtue of the surface than does a chain with a middle segment at the surface, so that the density of chain ends at the surface should be somewhat enhanced. This picture is supported by simulations; for example in the variable density self-consistent mean field study of PDMS of relative molecular mass 3270 already cited, more than 50% of the population of the outermost layer of the polymer consisted of chain ends, even though chain ends represented only 17% of the total mass. On the other hand experimental evidence for chain-end segregation is much more equivocal. The reason for this is that, in order to detect such segregation, it is necessary experimentally to label the end of the chain in some way so that a surface analysis technique may be used to detect an excess of the ends. However, as we shall see in chapter 6, even a very small substitution at the end of a polymer chain may lead it to be

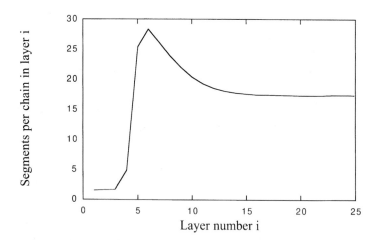

Figure 2.27. Profiles of the average number of monomers belonging to the same chain near a free surface in a variable density lattice model of a poly(dimethyl siloxane) melt. The relative molecular mass is 3720, the layer thickness is 2.79 Å and the temperature is 30 °C. After Theodorou (1989).

rather surface active, hence it is difficult to separate any intrinsic tendency for the ends to segregate driven purely by entropic considerations from a segregation driven by a difference in surface energy or by a strongly unfavourable interaction in the bulk between the end group and the rest of the polymer.

There is no question but that polymers can be prepared with end-groups that do strongly segregate to the surface; an example is provided by a case in which polystyrene with an end-group of perfluorooctyl dimethyl chlorosilane was shown to segregate to the surface strongly (Affrossman *et al.* 1994). Of course this is hardly a minimally perturbing label; a better choice if one wants to try and isolate the intrinsic, entropic driving force for end-group segregation would be to have a short, deuterium-labelled block at the end of the polymer, though even here one should be aware of a small isotopic difference in surface energy between normal and deuterated polymers which itself can drive surface segregation (Jones *et al.* 1989). End-group segregation in a triblock copolymer of normal and deuterated polystyrene was observed in one study (Affrossman *et al.* 1993); another study on a very similar system (Zhao *et al.* 1993) could only put an upper limit on the amount of segregation; their data were consistent with a degree of end-group segregation of no more than a factor of two.

What would clearly be desirable from the experimental point of view are experiments that can probe the chain conformation near the polymer surface, in the same way as neutron scattering experiments so effectively probe bulk conformational properties of polymer chains. At present it is not possible to get surface-sensitive information on the large-scale structure of polymers – that is, on the level of whole chains. However, it is possible to obtain structural information on the smaller scale that would characterise, for example, crystalline packing of chain segments. The technique involved is grazing incidence x-ray diffraction; here x-rays are incident on a sample at less than the angle for total reflection (reflection of x-rays and neutrons from polymer surfaces is discussed more extensively in chapter 3). Thus there is a finite x-ray intensity only for the small distance below the surface of the polymer that an evanescent wave can penetrate, which may be of the order of 50 Å. If crystalline order is present in the sample, it is thus possible to distinguish between the degrees of crystalline order in the bulk and at the surface. Such experiments have been carried out on polyimide surfaces (Factor *et al.* 1991, 1993, 1994). These materials are rather rigid polymers, which are prepared by spin-casting a more soluble precursor polymer that is subsequently converted to the polyimide by a thermal treatment. What was found was that films of poly(pyromellitic dianhydride oxydianiline) (PMDA-ODA) had a markedly enhanced degree of structural order near the surface, compared with that in the bulk, with the enhanced ordering persisting to a depth of about 70 Å. However, this tendency is only observed for polymers that have some tendency to order in the bulk; it is not seen with flexible polymers such as polystyrene.

2.7 Dynamics of the polymer surface

All our discussions have up to now assumed that the polymer segments near a surface are mobile enough to be able to reach full thermodynamic equilibrium. However, we know that, in the case of bulk polymers, this is not always so; as a polymer melt is cooled the segmental mobility decreases until a temperature is reached at which full thermodynamic equilibrium cannot be reached on the experimental time scale. The temperature at which the system drops out of equilibrium is known as the glass transition temperature and is marked by changes in certain easily measurable thermodynamic properties of the polymer; for example, at the transition the heat capacity and the expansivity change discontinuously. In addition, around the glass transition temperature mechanical properties change very rapidly, resulting in the familiar contrast between the rather stiff behaviour of a polymer glass, such as PMMA, and the rubbery properties of a polymer melt of high relative molecular mass, such as unvulcanised polybutadiene.

On first sight, then, one might imagine that we can distinguish between the surface properties of polymer glasses, which we might imagine as static and immobile, and the surfaces of polymer melts, at which individual chains and chain segments are able to move relative to one another. In the latter case, we would expect the surface to be able to respond to changes in its environment. For example, consider a polymer such as poly(hexafluoropropylene oxide), which we discussed in section 2.4, in which there might be a preferential orientation of the side-groups, or a polymer with an end-group that can preferentially locate itself at the surface. Such a polymer may first be equilibrated in one environment, which favours a particular configuration of the side-groups or end-groups; for example, we might immerse the polymer in water, in which case a highly hydrophobic group like a fluorocarbon would be expelled from the surface. On removing the polymer from this watery environment, however, the reverse situation is favoured – the fluorocarbon now is favoured at the surface. If the surface is mobile enough the chains will rearrange in order to achieve this and the change will be reflected in surface properties such as the contact angle, as well as being detectable by surface analysis. This type of behaviour is of considerable technical importance and has been observed in a variety of different contexts, including both polymers with functional groups, as discussed above, and polymers whose surface has been chemically or physically modified; many examples are given by Garbassi *et al.* (1994). An example is shown in figure 2.28, in which the contact angle data are shown for polystyrenes with both hydrophobic and hydrophilic end groups as a function of the time of exposure to water vapour. In the case of a hydrophilic end-group, the contact angle decreases below the value for normal polystyrene as more of the hydrophilic groups find their way to the surface. In contrast, for a hydrophobic end-group the contact angle is initially greater than that of normal polystyrene, but with increasing water exposure times the contact angle drops towards the normal polystyrene value as the hydrophobic groups become hidden.

If the dynamic behaviour of a polymer surface were the same as that of the bulk, we would expect this kind of rearrangement to be possible only for polymers above their bulk glass transition temperature. However, there are some indications that there may be differences in mobility between polymer segments at the surface and those in the bulk. In the example of surface reconstruction given above, the polystyrene is actually below its bulk glass transition temperature, although it may be possible to argue that in this case the exposure to water vapour can produce local plasticisation. Other support for the idea that there may be significant chain mobility near the surface of polymers that in the bulk are glassy comes from computer simulation (Mansfield and Theodorou 1991), whereas patterns of deformation obtained at the surface of polystyrene on interaction with an atomic force microscope have been interpreted as suggesting that, even at room temperature, the surface of polystyrene is effectively rubbery (Meyers *et al.* 1992). More precise measurements using scanning force microscopy to measure the loss tangent $\tan(\delta)$

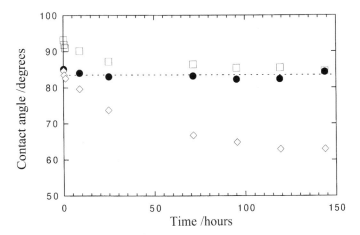

Figure 2.28. The water contact angle for end-terminated polystyrenes (- -●- -, proton terminated; □, fluorosilane terminated; ◇, carboxyl terminated) as a function of the time of exposure to water vapour at 40 °C. After Koberstein (1996).

(Kajiyama *et al.* 1997) led to the conclusion that, for polystyrene of relative molecular mass 26 600 and below, the surface was rubbery at 293 K, even though the bulk glass transition was measured to be at 373 K. On the other hand, polystyrene of relative molecular mass 40 400 (for which the bulk glass transition temperature was measured as 376 K) and higher still had values of $\tan(\delta)$ characteristic of a glassy polymer. On this basis one might estimate that the effective surface glass transition temperature was lowered by about 80 K.

The glass transition behaviour of thin films may be directly measured and there are indications that this is not the same as the bulk behaviour. This is shown by simple de-wetting experiments; a thin polystyrene film will de-wet an untreated silicon wafer substrate at equilibrium; however, a continuous film can be made by spin casting and of course de-wetting cannot take place while the film is in its glassy state. If such a film is now heated, the temperature at which de-wetting does occur is found to decrease as the film is made thinner (Reiter 1993, 1994). Polymer chains in thin films, then, clearly have a higher mobility than do corresponding chains in the bulk; what is not immediately clear is whether this is a confinement effect, possibly due to the fact that the films are thinner than the unperturbed dimensions of the polymer chains, or an effect solely of the free surface.

One can make a direct measurement of the effect of size on the glass transition temperature of a thin polymer film by measuring the film's thickness as the temperature is increased or decreased; the glass transition temperature then is marked by a discontinuity in expansivity. Such measurements may be conveniently carried out by ellipsometry or x-ray or neutron reflectivity. Figure 2.29 shows results obtained for thin polystyrene films heated in air on silicon wafers etched to remove the native oxide (Keddie *et al.* 1994). At thicknesses below about 1000 Å the glass transition temperature drops below its bulk value, with the transition temperature depression exceeding 20 °C for films of a few hundred ångström units. The effect seems to be essentially independent of relative molecular mass, suggesting that the distortion of chain configurations in very thin films is not the major cause of the depression of T_g. Even greater depressions of the glass transition temperature are observed for free-standing films (Forrest *et al.* 1996). On the other hand, experiments on poly(methyl methacrylate) on a substrate of native-oxide-coated silicon reveal no such decrease in T_g; if anything there is a very small increase (Keddie *et al.* 1995, Wu *et al.* 1995). The same polymer on a gold substrate, however, exhibits similar behaviour to the polystyrene on silicon. It would seem that each interface in a thin film has its own effect on the mobility of nearby chains; when there is a strong interaction between polymer segments and the interface this results in a decrease in mobility which can offset the influence of the free

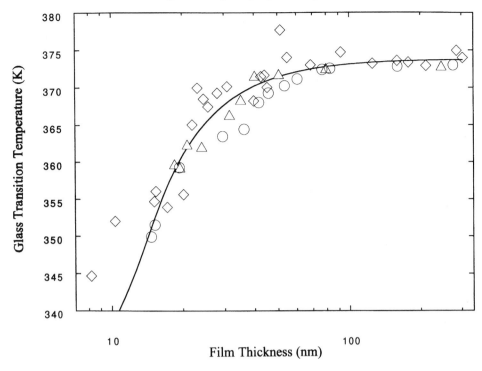

Figure 2.29. Glass transitions of thin polystyrene films of relative molecular masses 120 000 (△), 500 000 (○) and 2 900 000 (◇) as a function of the film thickness. After Keddie *et al.* (1994).

surface in increasing the mobility of nearby segments. The effect of a strongly attractive wall in reducing the mobility of nearby chains has been directly probed by measurements of polymer diffusion near walls (Zheng *et al.* 1995). Another experiment to probe the glass transition at the surface of a polymer was carried out using low-energy positronium annihilation lifetime spectroscopy; this provides a direct, surface-sensitive, probe of the free volume, allowing a direct measure of the glass transition temperature of the top 100 Å of a polystyrene sample (Xie *et al.* 1995). In this experiment, no depression of the glass transition temperature was observed.

Thus the experimental picture does not yet lend itself to a definite conclusion. If it does turn out to be the case that there is a layer of more liquid-like material at the surface of a glassy polymer, there will be important consequences in our understanding of adhesion, friction and fracture. For example, if the extent of the liquid-like layer were large enough to allow significant entanglement to take place if two glassy polymers were brought into contact, then even below the bulk glass transition temperature significant auto-adhesion might be expected to occur; these issues will be discussed more fully in chapter

7. Another intriguing and technologically important finding that has been attributed to the presence of a surface layer of enhanced mobility concerns the alignment of molecules at the surface of rubbed polyimides. Alignment layers for liquid crystals in devices are created by a rather gentle process of rubbing at a temperature well below the glass transition temperature of the polyimide; grazing angle x-ray diffraction reveals that this results in alignment of the chains near the surface (Toney *et al.* 1995).

This evidence for differences in mobility between molecules at the surface and in the bulk is suggestive and intriguing. However, our understanding of this aspect of polymer surfaces is in its infancy, and will not be anything like complete until new experimental techniques – preferably direct, surface-sensitive probes of polymer mobility – and theoretical ideas have been developed.

2.8 References

Adamson, A. W. (1990). *Physical Chemistry of Surfaces*. New York, Wiley.
Affrossman, S., Hartshorne, M. *et al.* (1993). *Macromolecules* **26**, 6251.
Affrossman, S., Hartshorne, M. *et al.* (1994). *Macromolecules* **27**, 1588.
Anastasiadis, S. H., Chen, J-K. *et al.* (1987). *Journal of Colloid and Interface Science* **119**, 55.
Bashforth, F. and Adams, J. C. (1883). *An Attempt to Test the Theories of Capillary Action*. Cambridge, Cambridge University Press.
Bitsanis, I. A. and ten Brinke, G. (1993). *Journal of Chemical Physics* **99**, 3100.
Buff, F. P., Lovett, R. A. *et al.* (1965). *Physical Review Letters* **15**, 621.
Cassie, A. D. B. (1948). *Discussions of the Faraday Society* **3**, 11.
Chandler, D. (1987). *Introduction to Modern Statistical Mechanics*. New York, Oxford University Press.
Couper, A. (1993). Chapter 1 in *Physical Methods of Chemistry* Volume IXA. New York, John Wiley.
Dee, G. T. and Sauer, B. B. (1992). *Journal of Colloid and Interface Science* **152**, 85.
Dee, G. T. and Sauer, B. B. (1995). *Polymer* **36**, 1673.
Factor, B. J., Russell, T. P. *et al.* (1991). *Physical Review Letters* **66**, 1181.
Factor, B. J., Russell, T. P. *et al.* (1993). *Macromolecules* **26**, 2847.
Factor, B. J., Russell, T. P. *et al.* (1994). *Faraday Discussions* **98**, 319.
Flory, P. J., Orwoll, R. A. *et al.* (1964). *Journal of the American Chemical Society* **86**, 3515.
Forrest, J. A., Dalnoki-Veress, K. *et al.* (1996). *Physical Review Letters* **77**, 2002.
Fowkes, F. M. (1962). *Journal of Physical Chemistry* **66**, 382.
Garbassi, F., Morra, M. *et al.* (1994). *Polymer Surfaces: From Physics to Technology*. Chichester, John Wiley & Sons.
Girofalco, L. A. and Good, R. J. (1957). *Journal of Physical Chemistry* **61**, 904.
Israelachvili, J. (1991). *Intermolecular and Surface Forces*. San Diego, Academic Press.
Jarvis, N. L., Fox, R. B. *et al.* (1964). In *Advances in Chemistry Series*. F. M. Fowkes. Washington DC, American Chemical Society. **43**, 317.
Jones, R. A. L., Kramer, E. J. *et al.* (1989). *Physical Review Letters* **62**, 280.

Kajiyama, T., Tanaka, K. *et al.* (1997). *Macromolecules* **30**, 280.

Keddie, J. L., Jones, R. A. L. *et al.* (1994). *Europhysics Letters* **27**, 59.

Keddie, J. L., Jones, R. A. L. *et al.* (1995). *Faraday Discussions* **98**, 219.

Koberstein, J. T., *MRS Bulletin* **21**, 16.

Legrand, D. G. and Gaines, G. L. (1969). *Journal of Colloid and Interface Science* **31**, 162.

Mansfield, K. F. and Theodorou, D. N. (1991). *Macromolecules* **24**, 6283.

Meyers, G. F., DeKoven, B. M. *et al.* (1992). *Langmuir* **8**, 2330.

Munk, P. (1989). *Introduction to Macromolecular Science*. New York, Wiley-Interscience.

Patterson, D. and Rastogi, A. K. (1970). *Journal of Physical Chemistry* **74**, 1076.

Poser, C. I. and Sanchez, I. C. (1979). *Journal of Colloid and Interface Science* **69**, 539.

Reiter, G. (1993). *Europhysics Letters* **23**, 579.

Reiter, G. (1994). *Macromolecules* **27**, 3046.

Rowlinson, J. S. and Widom, B. (1982). *Molecular Theory of Capillarity*. Oxford, Clarendon Press.

Sanchez, I. C. (1992). in *Physics of Polymer Surfaces and Interfaces*. I. C. Sanchez. Boston, Butterworth-Heinemann.

Sauer, B. B. and Dee, G. T. (1994). *Journal of Colloid and Interface Science* **162**, 25.

Sauer, B. B. and Dipaolo, N. V. (1991). *Journal of Colloid and Interface Science* **144**, 527.

Silberberg, A. (1982). *Journal of Colloid and Interface Science* **90**, 86.

Theodorou, D. N. (1989). *Macromolecules* **22**, 4578.

Toney, M. F., Russell, T. P. *et al.* (1995). *Nature* **374**, 709.

Vitt, E. and Shull, K. (1995). *Macromolecules* **28**, 6349.

Wenzel, R. N. (1936). *Industrial and Engineering Chemistry* **28**, 988.

Wu, S. (1982). *Polymer Interface and Adhesion*. New York, Marcel Dekker.

Wu, W., van Zanten, J. H. *et al.* (1995). *Macromolecules* **28**, 771.

Xie, L., DeMaggio, G. B. *et al.* (1995). *Physical Review Letters* **74**, 4947.

Zhao, W., Zhao, X. *et al.* (1993). *Macromolecules* **26**, 561.

Zheng, X., Sauer, B. B. *et al.* (1995). *Physical Review Letters* **74**, 407.

3

Experimental techniques

3.0 Introduction

The recent rapid development in the quantitative description of polymer surfaces and interfaces has, to a large extent, been driven by the development of a variety of powerful experimental techniques. The newcomer to the field might be forgiven for feeling some confusion in the face of this profusion, not least because of the seemingly unstoppable urge of surface scientists to maximise the use of unpronounceable acronyms. Many books and review articles already describe the technicalities of each of these techniques, so our aim in this chapter is to review the physical principles underlying them so that we can distinguish between the different classes of information that each technique is best able to provide.

Broadly speaking, experimental information about the nature of polymer surfaces and interfaces is available in three broad classes, *composition, structure and morphology* and *properties*. Let us discuss each in turn.

Perhaps the most obvious question we can ask about a polymer surface is what it is composed of. Compositional information may be restricted to an elemental analysis, or we may be able to obtain more specific information about the chemical species that are present. Alternatively, it may be possible only to detect the presence of a restricted set of elements or compounds. Such a technique is obviously much less helpful if we are faced with a completely unknown sample (such as an industrial material with some unknown contaminant at the surface), but, on the other hand, if we are doing a closely controlled experiment in which we are able to introduce an element to which we are sensitive as a label (for example by using a deuterium label on a polymer component believed surface active), the technique may be much more appropriate.

The ability of the technique to discriminate between different depths below the surface is also important. We can distinguish between truly surface-sensitive techniques, such as XPS, that analyse the composition within a few tens of ångström units from the surface and near-surface depth-profiling techniques, such as dynamic SIMS, which give one information about the composition with depth in the near-surface region. Other techniques, such as ATR-IR, are neither truly surface sensitive nor depth-profiling, but give some average composition over a region that may be as much as 1 μm from the surface.

Analysis of composition near buried interfaces poses a difficult problem, because direct use of surface-sensitive techniques is obviously ruled out. If one can physically break apart the interface, say by peeling a polymer film off an inorganic substrate, one can use surface analysis tools on the resulting fracture surface, but at the risk of doing substantial and unquantifiable damage to the surface in the process. To study a buried interface *in situ* one has to resort to depth-profiling techniques, in particular those techniques that involve a probe that is able to penetrate substantial amounts of condensed matter, the best example of which is a neutron beam. Thus the technique of neutron reflectivity is unrivalled for the study of solid/liquid interfaces, for which an additional strength is provided by the fact that the analysis does not require a vacuum environment.

When it comes to structure near surfaces or interfaces our choices are much more limited. For the study of the topography of polymer surfaces various types of microscopy are appropriate, such as scanning electron microscopy of a coated surface, transmission electron microscopy of replicas and more recently scanning probe microscopies such as scanning tunnelling microscopy for conducting surfaces and atomic force microscopy (AFM), also known as scanning force microscopy (SFM). Little work has been done on characterising the state of chain order or crystallinity near the surface, one of the few truly surface-sensitive techniques being grazing incidence x-ray diffraction. Morphology in the sense of phase-separated regions from blends or copolymers may be probed by imaging SIMS; more indirectly such morphology may be reflected in differences in local relief, which may be probed by SFM.

Several properties of surfaces have origins that are very sensitive to the surface, but are essentially macroscopic in character. Examples include adhesion, surface tension and the contact angle. Techniques to measure these properties are not discussed in this chapter, but rather are mentioned in connection with discussions of the property involved elsewhere in this book. However, surface properties can also be probed at a microscopic level; for example the forces between surfaces at microscopic separations may be probed using the surface forces apparatus (SFA). The scanning force microscope

(SFM) refines this idea to measure the force between a probe and a surface with fine lateral resolution. Using this basic idea it is now possible to probe a number of properties such as friction and viscoelasticity both at fine lateral resolution and with surface sensitivity.

With these general considerations in mind, we now go on to discuss the techniques in more detail. First we discuss techniques involving reflection of waves of one form or another, then we go on to methods in which the probe consists of fast moving ions. The very important spectroscopic methods are discussed next, before we finish with a look at surface force measurements and the scanning force microscope.

3.1 Reflection of waves from interfaces

3.1.1 Neutron and x-ray reflectivity

When a wave of any kind is incident on an interface between two materials in which the characteristics of propagation of the wave – the refractive indices – are different, a fraction of the wave will be reflected from the interface. For a strictly planar interface, in which the composition varies with the perpendicular distance through the interface but for which in any plane parallel to the interface the composition is uniform, the reflection will be entirely specular; if the wave is incident at a certain angle the reflected wave will make the same angle to the interface (see figure 3.1). The reflectivity is defined as the ratio of the intensities of the reflected and incident beams; it should be carefully distinguished from the reflectance, which is the ratio of the amplitudes of the incident and reflected waves. The reflectance in general is a complex number, because there is usually a change in phase of a wave on reflection, whereas the reflectivity is a real number varying from zero to unity.

If one measures the reflectivity as a function of incident angle and/or

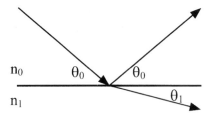

Figure 3.1. A schematic diagram of incident, reflected and refracted beams at the junction between two phases of refractive indices n_0 and n_1. In this case n_1 is smaller than n_0, as is commonly the case for neutrons incident on a material from air, and $\theta_1 < \theta_0$.

wavelength, the resulting function gives one information both about refractive index values in the bulk materials separated by the interface and about the detailed way in which the refractive index changes from one bulk value to the other through the interface. Consequently, since the refractive index is related in a simple manner to the composition, specular reflection can provide information about the composition distribution normal to the surface. The length scale over which a reflectivity measurement is sensitive to gradients in refractive index is defined by the wavelength of the incident wave, or more strictly by the component of its wave-vector perpendicular to the interface. So, for example, visible light from a helium–neon laser, incident on an interface at an angle of 45°, will have a perpendicular component of its wave-vector of about 7×10^{-3} nm^{-1}, so we would expect measurements of light reflectivity to be sensitive to interfacial features at a length scale of about 100 nm. Thus, although light reflectivity is a useful technique for studying processes like polymer adsorption, we cannot usually expect it to be sensitive to the details of an interface at a molecular level, for which we need to be probing distances of order $10-100$ Å. To achieve this, then, we need a wave with a perpendicular component of wave-vector between 0.1 and 0.01 Å^{-1}; this can be achieved by using x-rays or neutrons with wavelengths of around 5 Å at glancing angles of incidence in the range $0.5-5°$. The sensitivity of thermal neutrons and x-rays to these molecular length scales, combined with their power to penetrate into opaque materials, the simplicity of relating the refractive index to the composition and, in the case of neutrons, sensitivity to deuterium labelling, mean that neutron and x-ray reflectivity have played a major role in elucidating the structure of polymer surfaces and interfaces.

Much of the simple theory of reflection is presented in optics textbooks; a classic and clear presentation is to be found in Born and Wolf (1975). The general theory of the reflection of waves, both electromagnetic and particle, is dealt with in more detail in the book by Lekner (1987). Reviews of x-ray and neutron reflectivity, including their application to polymers, are available (Russell 1990, Thomas 1995). In this section we concentrate on neutron reflectivity, primarily because of the advantages offered by deuterium labelling, though as x-ray reflectivity can also provide useful information, contrasts between reflection of neutrons and x-rays will be pointed out where necessary.

We consider a plane wave travelling in medium 0 incident on the smooth surface of medium 1 (figure 3.1); the associated wave-vectors in each medium are k_0 and k_1 and the refractive index at the boundary is given by $n = k_1/k_0$. This refractive index can commonly be written as

$$n = 1 - \lambda^2 A + i\lambda C, \qquad (3.1.1)$$

where the complex term accounts for any absorption in medium 1. For neutron beams

$$A = Nb/(2\pi), \tag{3.1.2}$$

$$C = n\sigma_A/(4\pi), \tag{3.1.3}$$

where N is the atomic number density of medium 1 and b the bound atom coherent scattering length. The term σ_A is the absorption cross section of the atomic species making up medium 1. Values of b and σ_A vary in an unpredictable manner across the periodic table and tabulated values are available (Sears 1992). For the atoms occurring in most polymers, absorption cross sections are either zero or negligible and the absorption term in equation (3.1.1) can generally be ignored. For polymeric species and solvents of low relative molecular mass Nb can be replaced by the scattering length density of the polymer segment or solvent molecule, ρ, and

$$\rho = \frac{N_A d \sum_i b_i}{m}, \tag{3.1.4}$$

where d is the physical density of the polymer (or solvent), m is the relative molecular mass of the segment or solvent molecule and $\sum_i b_i$ is the sum of the bound atom coherent scattering lengths of the atoms making up the unit considered. For x-rays a similar expression is obtained:

$$\rho = \frac{N_A d}{m} \sum_i z_i r_\rho. \tag{3.1.5}$$

Here r_ρ is the electron radius (2.82×10^{-13} cm) and $\sum_i z_i$ is the sum of the electrons in the species. The absorption coefficient, C, for x-rays is $\mu/(4\pi)$, where μ is the mass absorption coefficient for the unit (Creagh and Hubbell 1992). N_A is Avogadro's number in these formulae.

At the interface between the two media the grazing angle of incidence, θ_0, is related to the angle of refraction, θ_1, by

$$n_0 \cos \theta_0 = n_1 \cos \theta_1. \tag{3.1.6}$$

If medium 0 is air, then n_0 is 1 and

$$\cos \theta_0 = n_1 \cos \theta_1. \tag{3.1.7}$$

When n_1 is less than 1 there will exist a critical angle above which the angle of refraction will be real for all incident angles. (Note that, apart from light water and compounds that are heavily hydrogenous, most materials encountered in polymer science have a positive A and thus $n_1 < 1$). At this critical angle of incidence, θ_c, θ_1 is zero and

$$\cos \theta_c = n_1. \tag{3.1.8}$$

When the critical angle is small, as is generally the case for neutron reflectivity, the cosine term can be expanded to give

$$\theta_c = \left(\frac{\lambda^2}{\pi}\rho\right)^{1/2}. \tag{3.1.9}$$

For a smooth polymer surface, the component of the incident beam's wave-vector normal to the surface, k_{z0}, is $(2\pi/\lambda)\sin\theta_0$, while in the polymer the z component of the wave-vector, k_{z1}, is $(k_{z0}^2 - 4\pi\rho)^{1/2}$, which can also be written $(k_{z0}^2 - k_c^2)^{1/2}$, where k_c is the value of the component of the wave-vector normal to the surface at the critical angle. The reflectance of the interface between media 0 and 1 is given by the Fresnel formula

$$r_{01} = \frac{k_{z0} - k_{z1}}{k_{z0} + k_{z1}}. \tag{3.1.10}$$

The reflectivity R is $r_{01}r_{01}^*$, with r_{01}^* the complex conjugate of r_{01}; hence for the smooth interface the Fresnel reflectivity is given by

$$R_F(Q) = \left(\frac{Q - (Q^2 - Q_c^2)^{1/2}}{Q + (Q^2 - Q_c^2)^{1/2}}\right)^2. \tag{3.1.11}$$

Here we have used, instead of wave-vectors, the momentum transfer normal to the surface $Q = (4\pi/\pi)\sin\theta$, a notation preferred by those familiar with small-angle scattering. Replacing for Q_c $(= 4\pi^{1/2}\rho^{1/2})$ we find that when $Q \gg Q_c$ we can write

$$R_F(Q) = 16\pi^2\rho^2/Q^4; \tag{3.1.12}$$

hence, for a smooth, sharp surface at sufficiently high Q values, $Q^4 R_F(Q)$ should become a constant value determined only by the scattering length density of the bulk material.

When a single layer is interposed between the air and the bulk phase (i.e. the bulk material becomes a substrate) as in figure 3.2 there is now reflection at each interface and a transmission coefficient for beam transport through the

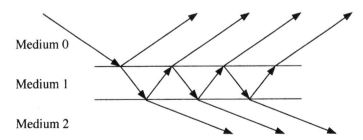

Figure 3.2. Multiple reflection and transmission for a beam incident on a layer of refractive index differing from that of the substrate.

intervening layer. The reflected beam from the interface between media 1 and 2 can undergo reflection as well as refraction at the interface on going from medium 1 to medium 0. This process continues until the transmission losses attenuate the beam to effective extinction. In this circumstance the reflectivity is given by

$$R_F(Q) = \left| \frac{r_{01} + r_{12} \exp{(2\mathrm{i}\beta)}}{1 + r_{01} r_{12} \exp{(2\mathrm{i}\beta)}} \right|^2, \tag{3.1.13}$$

where the r_{ij} terms are the Fresnel reflectances calculated for each interface and β is the phase shift or optical path length in medium 1:

$$\beta = \frac{2\pi}{\lambda} n_1 d \sin{\theta_1}. \tag{3.1.14}$$

The externally reflected beams have different path lengths to the detector and interference between the reflected beams leads to maxima and minima being observed in the reflectivity. The separation of the minima in Q, ΔQ, is related to the layer thickness by

$$\Delta Q = 2\pi/d. \tag{3.1.15}$$

Similar formulae in terms of the Fresnel reflectances may be built up for situations with a small number of discrete layers. A larger number of layers demands the use of a more general method that is easily coded for numerical computation. The optical matrix method used in the reflectivity of light from stratified media (Born and Wolf 1975) is often used. These methods are particularly relevant to polymer surfaces and interfaces for which the interfacial composition variation normal to the interface may change smoothly between limiting values as illustrated in figure 3.3. In such cases the region over which the composition changes can be broken up into a series of discrete strips with a scattering length density approximating to that of the smooth decay. An additional factor is the occurrence of any roughness at the interfaces between such hypothetical strips (or indeed real roughness at the surface). Roughness at any interface reduces the specular reflectivity and is indistinguishable in this respect from diffuseness of the interface. Its effect can be incorporated as a Gaussian smoothing factor on the Fresnel reflectivity coefficients. Such interfacial roughness can be incorporated into the optical matrix method if the method due to Abeles is used in place of that due to Born and Wolf.

For this approach each layer in the interface as shown in figure 3.3 is represented by a characteristic matrix; thus for layer m

$$\mathbf{M}_m = \begin{bmatrix} \exp{(\mathrm{i}\beta_{m-1})} & r_m \exp{(\mathrm{i}\beta_{m-1})} \\ r_m \exp{(-\mathrm{i}\beta_{m-1})} & \exp{(-\mathrm{i}\beta_{m-1})} \end{bmatrix}, \tag{3.1.16}$$

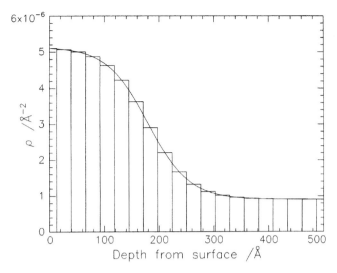

Figure 3.3. The gradual change in scattering length density (continuous line) and its approximation as a series of discrete layers.

where β is the optical path length of layer m as defined before and r_m is the product of the Fresnel reflectance of the mth interface r_m^f and a damping term that accounts for the roughness:

$$r_m = r_m^f \exp\left(-0.5 Q_{m-1} Q_m \langle\sigma\rangle^2\right). \tag{3.1.17}$$

Here $\langle\sigma\rangle^2$ is the mean square roughness of the interface and $Q_i = (4\pi/\lambda)\sin\theta_i$. For a total of n layers a resultant 2 by 2 matrix is obtained by multiplying these n characteristic matrices

$$\mathbf{M}_n = \mathbf{M}_1 \mathbf{M}_2 \mathbf{M}_3 \dots \mathbf{M}_n$$

$$= \begin{bmatrix} M_{11} & M_{21} \\ M_{12} & M_{22} \end{bmatrix}. \tag{3.1.18}$$

The reflectivity of the whole multilayer is then obtained as

$$R(Q) = \frac{M_{21} M_{21}^*}{M_{11} M_{11}^*}. \tag{3.1.19}$$

This scheme lends itself well to computational calculation and, in principle, any number of layers can be incorporated, speed of processing being the limiting factor in simulating the reflectivity.

In practical terms, to observe an effect on the reflectivity due to some particular chemical species, there must be contrast between it and its surroundings. This is easily achieved using neutron beams because the bound atom

Table 3.1. *Scattering length densities for selected polymers and solvents*

Species	Unit chemical formula	$\rho\ (10^{-6}\ \text{Å}^{-2})$
H Polystyrene	C_8H_8	1.4
D Polystyrene	C_8D_8	6.92
H Polymethyl methacrylate	$C_5H_8O_2$	0.90
D Polymethyl methacrylate	$C_5D_8O_2$	6.02
H Polybutadiene	C_4H_6	0.45
D Polybutadiene	C_4D_6	7.1
H Polyethylene oxide	C_2H_4O	0.56
D Polyethylene oxide	C_2D_4O	6.33
Water	H_2O	-0.56
Heavy water	D_2O	6.38
H Cyclohexane	C_6H_{12}	-0.28
D Cyclohexane	C_6D_{12}	6.68

coherent scattering length for deuterium is considerably different from that of hydrogen. Consequently, the scattering length density of the deuterated species is markedly different from that of the hydrogenous isomers (table 3.1 shows some selected values of ρ).

Although deuteration may perturb both the bulk and surface thermodynamics of polymers (see chapters 4 and 5), the perturbation is much less than that engendered by the heavy-atom labelling which may be necessary to obtain sufficient x-ray contrast. Nonetheless, when sufficient x-ray contrast exists naturally, all that has been set out is equally applicable to x-ray reflectometry.

Although this kind of formalism is easily implemented to calculate any profile, it is worth considering some of the qualitative features of the reflectivity curves. Figure 3.4 shows the calculated reflectivity profile for a smooth surface of a pure material, plotted both as $R(Q)$ against Q and as $R(Q)Q^4$. For low Q we have total reflection and the reflectivity is unity. Above the critical value Q_c the reflectivity falls off very quickly; from equation (3.1.12) we know it falls off as Q^{-4} at high Q. For this reason it is convenient to plot the product $R(Q)Q^4$ against Q, which clearly exhibits a constant value at high Q. This kind of plot is very convenient for showing reflectivity data.

The effect of having a thin layer interposed between the air and a semi-infinite substrate is shown in figure 3.5. Pronounced fringes are obtained, whose periodicity is determined by the film thickness via equation (3.1.15) and whose visibility is determined by the difference in scattering length density

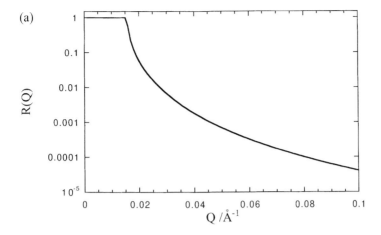

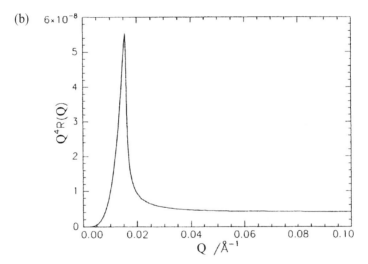

Figure 3.4. Calculated reflectivities for a smooth interface between a semi-infinite medium of scattering length density $\rho = 5 \times 10^{-6}$ Å^{-2} plotted (a) as a semilogarithmic plot of the reflectivity $R(Q)$ against the scattering vector Q and (b) as a plot of the product $Q^4 R(Q)$ against Q.

between the air and the film and the substrate and the film. The effect of roughness at the film/air interface is also shown; this leads to a loss of visibility of fringes at high Q, as a result of the decrease of the reflectance at high Q indicated by equation (3.1.17). Even quite modest values of root mean square roughness produce distinctive effects.

When repeated layer structures appear in polymer specimens, Bragg peaks are observed in the reflectivity due to this regular pattern of layers. Figure 3.6a shows the reflectivity calculated for four alternating layers of deutero-poly-

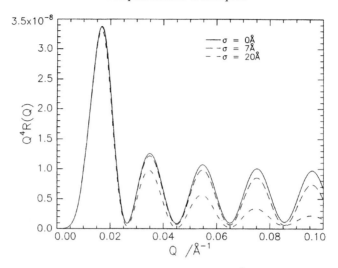

Figure 3.5. Calculated reflectivity curves for a 300 Å film of a material with a scattering length density $\rho = 5 \times 10^{-6}$ Å^{-2} on a silicon substrate (scattering length density 2.08×10^{-6} Å^{-2}) plotted as $Q^4 R(Q)$ against Q. The influence of increasing root mean square roughness of the air/substance interface is shown.

styrene and hydrogenous polymethyl methacrylate each of thickness 150 Å; the Bragg peaks are clearly evident in this simulation. This kind of structure is exactly what is found experimentally in thin films of linear block copolymers, as discussed in chapter 6. However in the block copolymer case, rather than having an integral number of layers of identical thickness, 'half layers' are found at the top and bottom surfaces of the annealed films. Incorporating such half layers produces the simulated reflectivity of figure 3.6b, clearly different from that of figure 3.6a. For comparison actual reflectivity data for such a block copolymer are shown in figure 3.7, which incorporates the scattering length density profile normal to the surface actually used to fit the data.

As our final example, the reflectivity is shown in figure 3.8 for a polymer film of which the surface has become enriched by some means in deuterated polymer of the same chemical type, producing a scattering length density profile at the surface of the form shown in figure 3.3. The major effect of this is visible close to the critical scattering vector; above Q_c the reflectivity drops off much more slowly than it does in the absence of enrichment. This effect can be understood by analogy to the optical phenomenon of frustrated total reflection. At a scattering vector above Q_c, but below the critical value for a semi-infinite material with the same scattering length density as the surface, an evanescent wave will penetrate the sample. The intensity in the evanescent wave decays exponentially, so if the scattering density remains constant over a distance

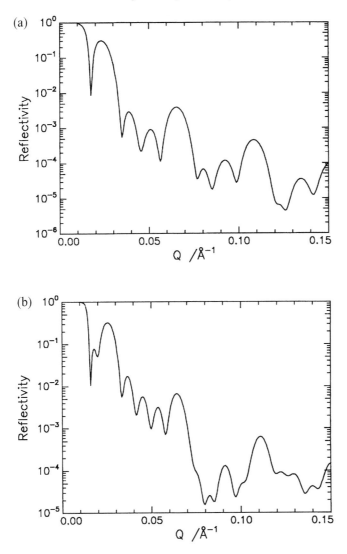

Figure 3.6. (a) Simulated reflectivity curves for samples with four layers of deutero-polystyrene and polymethyl methacrylate each of thickness 150 Å. (b) The reflectivity of the same system incorporating half layers at the top and bottom surfaces.

much greater than the decay length then no wave is transmitted and the reflectivity will be unity. If, however, within a few decay lengths of the surface the scattering length density drops down to a value lower than the value for total reflection at that scattering vector, then some intensity will be transmitted and the reflectivity will be less than unity. The way in which the reflectivity falls from unity in this frustrated total reflection regime is a very sensitive function of the shape of the scattering length density profile.

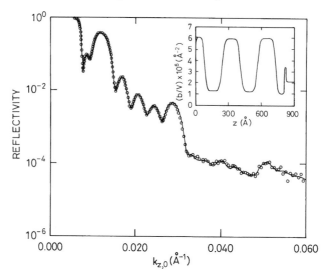

Figure 3.7. Neutron reflectivity data from an annealed sample of a deuterostyrene–methyl methacrylate linear diblock copolymer, with a polystyrene volume fraction of 0.52 and a total relative molecular mass of 80×10^3. After Mayes *et al.* (1994).

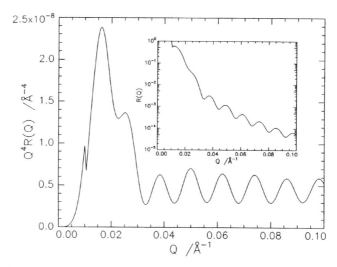

Figure 3.8. The calculated reflectivity curve for the volume fraction distribution of deuteropolystyrene in hydrogenous polystyrene shown in figure 3.3 plotted as $Q^4 R(Q)$ against Q (in the main plot) and $R(Q)$ against Q (in the inset).

The large differences in reflectivity that occur on the segregation of a deuterium-rich material to the surface are very clearly shown in figure 3.9, showing the effect of annealing on a mixture of normal and deuterated polystyrene of high relative molecular mass; before annealing the reflectivity

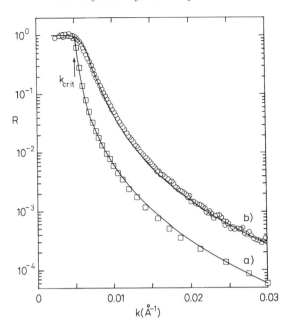

Figure 3.9. Neutron reflectivity from a 15% blend of d-PS (relative molecular mass 1.03×10^6) with h-PS (relative molecular mass 1.8×10^6), (a) before annealing and (b) after annealing. The solid line through the data is the Fresnel law prediction in the unannealed case and the prediction of mean-field theory for the annealed case. See section 5.1 for more details. After Jones *et al.* (1990).

falls off abruptly around the critical edge following the Fresnel law, whereas after annealing the deuterated polystyrene segregated at the surface presents a potential barrier, which manifests itself in the reflectivity by a much more gradual fall off from total reflection. Note that these same data are plotted as RQ^4 against Q as figure 5.3 later, where the details of the physics underlying the segregation phenomena are discussed.

Inherent to the optical matrix description of the reflectivity of neutron beams is the assumption that the neutron beam is a wave. Alternatively and entirely equivalently, the neutron could be viewed as a quantum mechnical particle with the reflection process the result of the interaction of the neutron with a potential energy barrier, V. The reflectivity is then the modulus squared of the reflected wave amplitude, the latter quantity being related to the potential energy barrier by the time-independent Schrödinger wave equation,

$$-\frac{\hbar^2}{2m}\frac{\mathrm{d}^2\psi}{\mathrm{d}z^2} + V\psi = E\psi, \qquad (3.1.21)$$

where E is the effective neutron energy $\hbar^2 k^2/(2m)$ and k is the perpendicular component of the neutron wave-vector $2\pi \sin\theta/\lambda$. Because the neutron wavelengths used are large compared with the range of the nuclear force, the potential energy barrier may be considered to be continuous (this is the Fermi pseudopotential approximation) and related to the scattering length density ρ by

$$V = \frac{2\pi\hbar^2}{m}\rho. \qquad (3.1.21)$$

A thin layer on a substrate now can be viewed as a potential barrier (see figure 3.10). To overcome the potential energy barrier the effective neutron kinetic energy normal to the barrier must be greater than the height of the potential barrier; since E is dependent on the angle of incidence, the range of angles over which total reflectivity will be observed is finite. This kind of one-dimensional scattering problem is treated in elementary quantum mechanics textbooks; solution of the Schrödinger equation gives the reflectivity as

$$R = 1 - \frac{4\alpha_0\alpha_2}{(\alpha_0 + \alpha_2)^2 \cos^2(\alpha_1 d) + (\alpha - \alpha_0\alpha_2/\alpha_1)^2 \sin^2(\alpha_1 d)}, \qquad (3.1.22)$$

where d is the potential energy barrier thickness and the α_i terms are related to the neutron effective kinetic energy and the potential energies in figure 3.7 by

$$\alpha_i^2 = \frac{2m}{\hbar^2}(E - V_i). \qquad (3.1.23)$$

Inside the uniform layer the perpendicular component of the neutron wave-vector is given by

$$k_i = (k^2 - 4\pi\rho)^{1/2}. \qquad (3.1.24)$$

The critical wave-vector for total reflection is $k_c = (4\pi\rho)^{1/2}$, so inside the layer k_i is imaginary and the neutron wave is evanescent. When the potential energy barrier thickness, d, is sufficiently thin, the neutron beam can tunnel through the barrier.

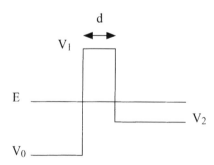

Figure 3.10. The potential energy profile for neutrons with an effective kinetic energy, E, from a region of potential energy V_0, passing through a thin barrier of potential energy V_1 to a region where the potential energy is V_2.

A particularly interesting situation arises if, on the other side of the barrier, there is a potential well, for example by having a substrate with an effective potential larger than V_2. For certain values of the incident wave-vector reflections from the barrier and the substrate will be in phase and there will be a resonance, leading to an enhancement of the neutron intensity inside the potential well. If this potential well is formed by a hydrogenous material, for example a polymer, neutrons inside the well are scattered incoherently (i.e. isotropically) by the proton-containing material and thus do not contribute to the specular reflectivity. In this case, even below the critical wave-vector corresponding to the substrate the reflectivity will fall below unity at well-defined values of the incident wave-vector. These characteristic resonant dips have been observed by Norton *et al.* (1994) and are shown in figure 3.11; they can be used as aids in developing the correct model for the distribution of material in thin polymer films.

Optical matrix calculations of the reflectivity from any system are exact under all circumstances. Such calculations may become lengthy when the interface is divided into many layers and in these circumstances more rapid approximate methods may provide as much, if not more, insight into the interfacial structure. Of the approximate methods proposed the kinematic approximation first described by Als-Nielsen (1984) (and elaborated on by Crowley *et al.* (1991a, b)) is the most useful. This is equivalent to a first Born

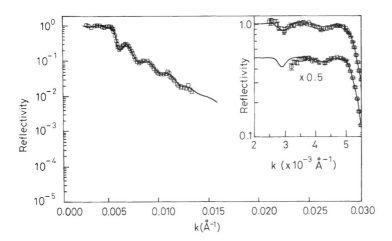

Figure 3.11. The reflectivity of a mixture of deuteropolyethylene propylene (d-PEP) at a volume fraction of 0.137, in its hydrogenous equivalent. Segregation of the d-PEP at the surface results in the creation of a potential well for the neutrons, leading to an enhancement of incoherent scattering under resonant conditions. The result is that the reflectivity drops below unity at resonant k vector values below the critical value. After Norton *et al.* (1994).

approximation in scattering theory; in optical terms it amounts to a neglect of refraction effects within the material. It is valid in the limit of large values of the scattering vector Q compared with the critical value Q_c; in these circumstances the reflectivity can be written as

$$R(Q) = \frac{16\pi^2}{Q^4} |\rho'(Q)|^2, \qquad (3.1.25)$$

where $\rho'(Q)$ is the one-dimensional Fourier transform of the gradient of the scattering length density normal to the interface:

$$\rho'(Q) = \int_{-\infty}^{\infty} \exp(iQz) \frac{d\rho}{dz} \, dz. \qquad (3.1.26)$$

Denoting the number density of species i by n_i then $\rho = \sum_i n_i b_i$. On substituting this into (3.1.25) we find

$$R(Q) = \frac{16\pi^2}{Q^4} \sum_{i,j} b_i b_j h'_{ij}(Q), \qquad (3.1.27)$$

where $h'_{ij}(Q)$ is the gradient in the partial structure factor. In cases in which i and j refer to the same species then

$$h'_{ii} = |n'_i(Q)|^2 \qquad (3.1.28)$$

and $h'_{ii}(Q)$ is related to the partial structure factor by

$$h_{ii}(Q) = Q^2 h'_{ii}(Q) = (n_i(Q))^2, \qquad (3.1.29)$$

where $n_i(Q)$ is the one-dimensional Fourier transform of the number density distribution of species i normal to the surface. When i and j are different the gradient of the cross partial structure factor is obtained:

$$h'_{ij}(Q) = \text{Re}\,(n'_i(Q)n'_j(Q)). \qquad (3.1.30)$$

The possibilities become clearer if we consider a polymer, with components A and B at an interface between two media. For the sake of simplicity one of these media is considered to be air, the second medium is signified by S (substrate). The components A and B are distributed in some fashion relative to each other and the interface with the scattering length density at any distance, z, from the interface given by

$$\rho(z) = b_a n_a(z) + b_b n_b(z) + b_s n_s(z). \qquad (3.1.31)$$

Replacing this into equation (3.1.27) then

$$R(Q) = \frac{16\pi^2}{Q^4} b_a^2 h'_{aa}(Q) + b_a^2 h'_{aa}(Q) + b_a^2 h'_{aa}(Q)$$

$$+ 2b_a b_b h'_{ab}(Q) + 2b_a b_s h'_{as}(Q) + 2b_b b_s h'_{bs}(Q). \qquad (3.1.32)$$

Hence, for this type of system, if six reflectivity profiles were measured for analogous systems with different contrasts and thus different values of b_i, then the six simultaneous equations could be solved to provide the six partial structure factors $h_{ij}(Q)$ or $h'_{ij}(Q)$. Fourier transformation of the partial structure factors provides the 'Patterson' function of the number density in real space (Thomas 1995); however, the reflectivity data usually do not extend sufficiently far in Q to justify Fourier transformation and recourse is made to models in the interpretation of the partial structure factors.

Explicit examples of the use of models will be given in chapter 8. It has been pointed out that, even in the absence both of Fourier transformation and of model fitting, important information about the arrangement of the molecule at the interface can be obtained by this kind of approach, namely the centre-to-centre separation between the components (Simister *et al.* 1992). The ability to do this stems from the properties of Fourier transforms and some observations concerning the nature of the number distribution of each species. Thus components A and B are probably symmetrical about the centres of their distributions since when $|z|$ is large the number densities are zero, i.e. they are even functions;

$$h_{ab}(Q) = \pm(h_{aa}(Q)h_{bb}(Q))^{1/2} \cos{(Q\delta_{ab})}, \qquad (3.1.33)$$

where δ_{ab} is the separation between the centres of the two distributions. On the other hand, for the subphase, the number density in the interfacial region will differ from that of the bulk and is an asymmetric distribution and is thus an odd function, hence

$$h_{as}(Q) = \pm(h_{aa}(Q)h_{ss}(Q))^{1/2} \sin{(Q\delta_{as})}. \qquad (3.1.34)$$

The $\pm$ operator results from uncertainty about the phase and introduces some inevitable ambiguity. The loss of phase information in neutron reflectivity can be a problem which may lead to non-unique solutions. The problem is alleviated by the use of many different contrasts, or by arguments about which scattering length density profiles are physically reasonable. The latter type of argument is, of course, hugely strengthened if data from other, complementary analysis techniques are available. The real space but relatively low-resolution ion beam techniques discussed in section 3.2 are ideal for this purpose for solid samples (Kramer *et al.* 1990), but of course they are much more difficult to use for studying interfaces involving liquids.

The kinematic approximation was originally applied to the analysis of the reflectivity from liquid surfaces, for which there is a smooth change in scattering length density at the liquid/vapour interface. The approximation has also been shown to be valid for an isolated layer bounded on both sides by media of the same scattering length density and for an adsorbed layer at an interface. In all of these cases, the kinematic approximation only applies in regions where $Q \gg Q_c$; notably the approximation fails to predict total reflection. Major contributions to the development of the kinematic approximation to describe neutron reflectometry data have been made by Thomas and colleagues (Crowley *et al.* 1991a, b, Lu *et al.* 1992, 1993a, b, Cooke *et al.* 1996) in their work on surfactant surface excess layers of low relative molecular mass. However, the principles have been extended to polymers and the results will be discussed in chapter 8. Examples of the variation in reflectivity profiles obtained by exploiting the range of scattering lengths available are shown in figure 3.12.

The essence of a reflectometry experiment is the determination of the specularly reflected intensity as a function of Q, the reflectivity then being given by the ratio of this reflected intensity to the incident intensity having subtracted any background contributions. Variation in Q can be achieved in two ways: (i) by altering the angle between the incident beam and the surface by movements of sample, beam or both (the detector then has to move through an angle 2θ to collect the reflected beam) and (ii) by using a broad-wavelength beam incident on the sample at a fixed angle and subsequently analysing the reflected beam either by time of flight (neutrons) or by energy dispersion (x-rays).

X-ray reflectometers are available commercially either as stand-alone instruments or as additional attachments to conventional diffractometers. Rotating anode sources are generally not necessary since the specularly reflected intensity is usually high over a sufficiently wide range of Q. An advantage of x-ray reflectometers is that the changes, ΔQ, between measuring points can be made very small and thus details in the reflectometry profile can be well resolved. The major drawback, referred to earlier, is the limited possibility of changing the scattering length density without a concomitant change in the thermodynamic properties and hence organisation of the system.

Neutron reflectometers require either a nuclear reactor or a pulsed spallation source to provide the neutron beam. There is no decided advantage for one source or the other for solid samples. A pulsed source with its 'white' neutron beam does have the advantage that the outline of the reflectivity profile over the whole range of Q available becomes evident at an early stage. The stepwise manner of collection necessitated at a reactor source delays appreciation of the full profile; moreover, the necessity to change angles and wavelength means that the resolution may vary with Q. No such problems apply to pulsed sources and certainly for experiments at liquid interfaces pulsed sources are to be preferred for the ease of coverage of Q. There are now a number of neutron reflectometers around the world, some of which are optimised for liquid surfaces but most are used on solid samples.

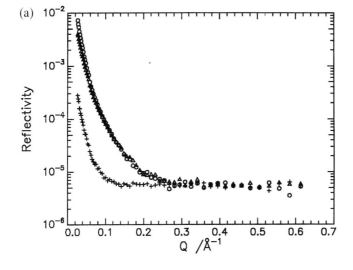

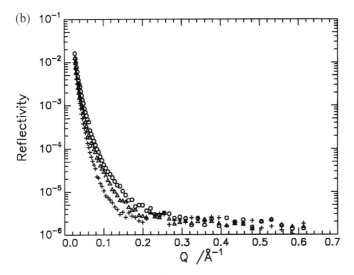

Figure 3.12. The reflectivity of $1\,\mathrm{mg\,m^{-2}}$ polymethyl methacrylate (PMMA) spread on a 0.1% wt/vol polyethylene oxide (PEO) solution in water: (a) for a null reflecting subphase ($\circ$, dPMMA/dPEO; $\triangle$, dPMMA/hPEO; and +, hPMMA/dPEO) and (b) lower set for a subphase of D_2O ($\circ$, dPMMA/hPEO; $\triangle$, hPMMA/dPEO; and +, hPMMA/hPEO). (S. K. Peace, PhD thesis, University of Durham, 1996.)

An important practical feature of experimental reflectivity data is that contributions to the measured reflectivity can arise from background signal. Background can arise from several sources; examples are non-specular reflection from a rough surface, small-angle scattering from the bulk sample and incoherent scattering by the sample. The latter contribution may be significant for polymers since the incoherent scattering cross section of hydrogen is very large. Although this background can be subtracted to leave the reflectivity, its effect is to limit the Q region over which the reflectivity provides information about the surface or interfacial organisation. In practice for solid polymers the maximum Q value used is rarely greater than 0.1 Å^{-1} since background scattering overwhelms the reflectivity at this stage. This is what in practice limits the resolution of neutron reflection to very fine-scale details of the interface.

3.1.2 Optical reflectivity and ellipsometry

As we stressed at the beginning of the last section, the theory of reflection applies to any kind of wave, so in principle measurement of the reflectivity of light from an interface as a function of angle or wavelength should yield information about the refractive index profile through an interface. However, as we explained above, the wavelength of light is such that one could only really hope to discriminate the structural details of interfaces on length scales of 100 nm or bigger, which is too large to resolve details on the molecular level. Nonetheless, it is possible, for example, to measure the total amount of polymer adsorbed at a solid/liquid interface. The advantages of the technique are that it is cheap to set up and very fast in acquiring data. These advantages have been put to good advantage to follow the kinetics of polymer adsorption (Dijt *et al.* 1990, Gast 1992).

An optical reflection technique in more widespread use is ellipsometry. In this technique one does not measure the reflection coefficient directly; instead one effectively measures the (complex) ratio of the reflectances of light polarised in a direction parallel to the plane of the interface (the 's-wave') and the light polarised in a direction in the plane defined by the incident and reflected beam (the 'p-wave'). The geometry is illustrated in figure 3.13.

Conventionally the ratio of the reflectances of the p-wave r_p and the s-wave r_s is written

$$\frac{r_p}{r_s} = \tan{(\psi)}\exp{(i\delta)}, \tag{3.1.35}$$

where ψ and δ are known as the ellipsometric angles. Originally ellipsometry was done by measuring the ellipticity of polarisation of reflected circularly polarised light, hence the name. Modern ellipsometers use a variety of different schemes to measure the ratio of equation (3.1.35), but in all cases the quantities that result from the measurement, sometimes as a function of the wavelength and angle of incidence, are ψ and δ. Expressions for the s and p reflectances

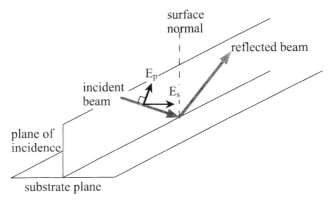

Figure 3.13. The geometry of an ellipsometry experiment, showing the direction of the electric vector for the s-polarisation E_s and the p-polarisation E_p.

can be found for the simplest situation of a simple interface between two semi-infinite media of refractive indices n_0 and n_1 (Born and Wolf 1975); the most prominent feature is that, as is well known from elementary optics, there exists an angle of incidence for which the p-wave reflectance is zero – the Brewster angle θ_B, defined by

$$\tan(\theta_B) = \frac{n_1}{n_0}. \tag{3.1.36}$$

At the Brewster angle, for a truly sharp interface, ψ is zero and δ changes discontinuously from π to zero. In the presence either of roughness at the interface or of a thin layer of some third medium, ψ does not go all the way to zero and the change in δ as a function of angle of incidence becomes less abrupt. Under these conditions ellipsometry is very sensitive to the presence of thin adsorbed layers, and provides an accurate measure of the amount adsorbed.

Rather clumsy expressions for r_s and r_p can be written down for the case of a thicker third layer between the medium from which the light is incident and the substrate. These can be solved in terms of the refractive index and the thickness of the intervening layer. This is perhaps how ellipsometry is most often used; as a quick, convenient and accurate way of measuring the thickness of films of polymers or other materials in the range of thickness of tens of ångström units up to 1 μm or so. We have already met in chapter 2 an example of how far this approach can be taken; if the thickness of a polymer film is measured as a function of temperature then the glass transition temperature of films down to tens of ångström units in thickness may be measured by detecting the temperature at which there is a break in the expansivity versus temperature curve (Keddie *et al.* 1994). Ellipsometry is very often used to study adsorbed polymer layers which are then analysed as if the layer were one of uniform refractive index, thus yielding some sort of effective thickness. These measurements can be done quite fast, allowing the kinetics of adsorption to be conveniently measured. A recent example of this kind of approach is provided by work on the adsorption of polymers with ionic end-groups from solvent to a solid/liquid interface (Pankewitsch *et al.* 1995).

In principle matrix methods analogous to the ones discussed in the section about neutron reflectivity can be applied to calculate ellipsometric angles for an arbitrary refractive index profile (Lekner 1987) and analytical approximations have also been developed (Charmet and de Gennes 1983). In practice the use of ellipsometry to obtain fine details of the structure of interfaces at the level of tens of ångström units is likely to be difficult and to require extreme care.

3.1.3 Surface quasi-elastic light scattering

Here we discuss a technique that is also based on the reflection of light waves, but in other respects is rather different from the methods discussed above in that it obtains information about the dynamics of an interface by detecting the time correlations of the reflected light. Fluid surfaces are in a continual state of disturbance, these disturbances being driven by thermal fluctuations of the fluid interface. These fluctuations can be Fourier decomposed into a set of surface waves and it was Kelvin (1871) who showed that the properties of these capillary waves, as they are known, are mainly determined by the surface tension. When a thin layer of a different material is present at the fluid surface, either as a spread film or as a surface excess, then the properties of these capillary waves are modified. *A priori* we expect that because of the visco-elastic properties of the surface film the capillary waves will be damped more rapidly due to increased resistance to changes in the surface shape. These visco-elastic properties of fluid interfaces may be characterised in terms of a number of parameters; suitable analysis of the capillary waves provides values for each of these parameters, that in turn may be related to the surface organisation at the interfacial region. In particular we are concerned with the dynamic elastic moduli represented by γ, the surface tension and ε, the surface elasticity or dilational modulus. Although each of these is accessible by a number of other techniques, these are restricted by a number of limitations; the surface may be perturbed by the technique used or the frequency range accessible may be limited. Surface quasi-elastic light scattering (SQELS) suffers from neither of these restrictions and in principle is able to provide *all* the surface visco-elastic parameters. The theoretical basis of the method is outlined here and experimental details together with methods of data analysis are provided. Experimental data on a variety of polymer systems are considered in chapter 8. Detailed treatments of the properties of capillary waves and the scattering of light by liquid surfaces are available in the literature (Earnshaw 1986); we restrict ourselves to a summary of these aspects here.

We consider an interface defined by the Gibbs dividing surface in the $x-y$ plane that is disturbed by a wave propagating in the x direction (figure 3.14a). The transverse disturbance from the mean position of the interface is given by

$$\zeta(x,\ t) = \zeta_0 \exp[i(qx + \omega t)], \qquad (3.1.37)$$

where q is the interfacial wave-vector and ω the complex wave frequency. The interfacial wave-vector is related to the capillary wave length, Λ, by $q = 2\pi/\Lambda$. The wave acts as a diffraction grating and will diffract light incident on the

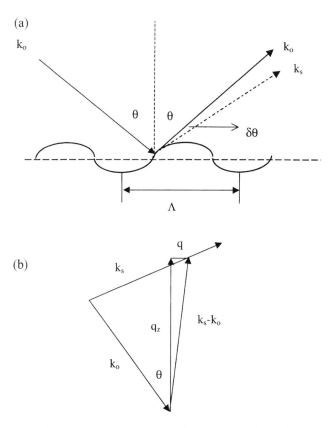

Figure 3.14. (a) A schematic diagram of incident, specularly reflected and scattered light at a liquid surface. In (b) the vector diagram defines q, the scattering vector in the plane of the liquid surface.

surface. When light is incident on the surface a specularly reflected beam will be produced, accompanied by a scattered beam at a small angle $\delta\theta$ to the specular beam (figure 3.14b).

The surface wave-vector is defined by

$$q = 2k_0 \sin\left(\frac{\delta\theta}{2}\right)\cos\theta, \qquad (3.1.38)$$

where k_0 is the wavenumber of the incident light beam. The capillary wave frequency is composed of a propagation frequency ω_0 and a damping coefficient Γ,

$$\omega = \omega_0 + i\Gamma. \qquad (3.1.39)$$

Both the frequency and capillary wave-vector are related to the material properties of the fluid via the dispersion relation. For pure liquid surfaces the dispersion relation has been derived independently by Lamb (1945) and Levich (1962), the frequency and damping of the capillary waves being given by

$$\omega_0 = \left(\frac{\gamma_0 q^3}{\rho}\right)^{1/2}\left(1 - \frac{1}{2y^{3/4}}\right), \qquad (3.1.40)$$

$$\Gamma = \frac{2\eta q^2}{\rho}\left(1 - \frac{1}{2y^{1/4}}\right), \qquad (3.1.41)$$

where γ_0, ρ and η are the surface tension, density and viscosity of the pure fluid, respectively, and

$$y = \frac{\gamma_0 \rho}{4\eta^2 q}. \qquad (3.1.45)$$

For when a surface layer is present on the fluid, Goodrich (1981) has defined five types of motion (figure 3.15) to which the molecules may be subjected, each of which is characterised by an independent visco-elastic modulus, which

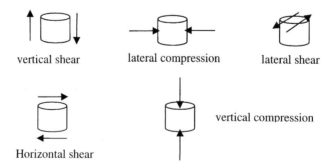

vertical shear   lateral compression   lateral shear

Horizontal shear   vertical compression

Figure 3.15. An element of the fluid surface showing the five possible surface motions defined by Goodrich (1981).

itself can be expanded as a linear response function including an elastic component and a viscosity component.

Of these five types of motion, one is the classical lateral shearing frequently discussed for monolayers of low relative molecular mass and this is of no relevance here because this motion does not scatter light. Two of the remaining four have no quantitative theory developed for them and they are not relevant for polymers at a fluid interface. It is the remaining two, vertical shearing and lateral compression, that engage our interest. For vertical shearing the pertinent visco-elastic modulus is the surface tension γ; lateral compression is parameterised by the dilation modulus, ε. Upon incorporating the additional energy dissipation viscosity terms, these become

$$\gamma = \gamma_0 + i\omega_0\gamma', \tag{3.1.43}$$

$$\varepsilon = \varepsilon_0 + i\omega_0\varepsilon', \tag{3.1.44}$$

where γ' and ε' are the transverse shear viscosity and the dilational viscosity, respectively, and γ_0 and ε_0 are the surface tension and dilational modulus at the capillary wave frequency ω_0.

Lucassen-Reynders and Lucassen (Lucassen 1968, Lucassen-Reynders and Lucassen 1969) have derived the dispersion relation for a liquid surface in the presence of a surface film. They showed that periodic disturbance of such a film-covered surface results in a surface tension that varies from point to point on the surface because of the fluctuations in surface concentration. Consequently, in addition to a transverse stress being developed, a finite tangential surface stress is also present. The solution to this dispersion equation has two roots, one of which corresponds to the capillary waves (transverse motion) and one of which corresponds to longitudinal or dilational waves derived from the transverse stress. The dispersion relation ($D(\omega)$) obtained for a film at the interface between two media is

$$[\eta(q - m) - \eta'(q - m')]^2 + \{\varepsilon q^2/\omega_0 + i[\eta(q + m) + \eta'(q + m')]\}$$

$$\times \{\gamma q^2/\omega_0 + g(\rho - \rho')/\omega_0 - \omega_0(\rho' - \rho)/q$$

$$+ i[\eta(q + m) + \eta'(q + m')]\} = 0 \tag{3.1.45}$$

and $m = (q^2 + i\omega_0\rho/\eta)^{1/2}$ for $\mathrm{Re}\,(m) > 0$, where the primed terms relate to the second fluid and in most cases the gravitational term can be neglected. For a free liquid surface the second fluid is air and both η' and ρ' approximate to zero. Loudon (1984) has a single dispersion relation from which equations relating to a fluid plus a surface film and a film between two fluids can be obtained as special cases.

The spectrum of light scattered by the capillary waves is just the power

spectrum of the waves; in the presence of a surface film this power spectrum is
(Langevin and Bouchiat 1971)

$$P(\omega) = \frac{k_B T}{\pi \omega} \, \text{Im} \left(\frac{i\omega\eta(m+q) + \varepsilon q^2}{D(\omega)} \right). \qquad (3.1.46)$$

In principle, because the only change in dipole moment is transverse to the
surface, observation of the scattered light should only lead to evaluation of the
surface tension and transverse shear viscosity. However, because the transverse
and dilational modes are coupled, dilational parameters are extractable in
principle from the spectrum of the scattered light. The coupling between
capillary and dilational waves is implicit in the treatment of Lucassen-Reynders
and Lucassen and has been made explicit in a consideration of surface waves
by Earnshaw and McLaughlin (1991, 1993), the coupling coefficient being
$\eta(q - m) - \eta'(q - m')$. Consequently, when a free liquid surface is being
investigated, the coupling is at a maximum and dilational visco-elastic parameters should be accessible. Furthermore, the existence of dilational waves and
this coupling give rise to resonance phenomena enhancing the energy dissipation, resulting in a maximum in the damping coefficient and capillary wave
frequency at a particular value of ε_0/γ_0 (≈ 0.16).

The shape of the power spectrum of the scattered light from a liquid surface
is shown in figure 3.16.

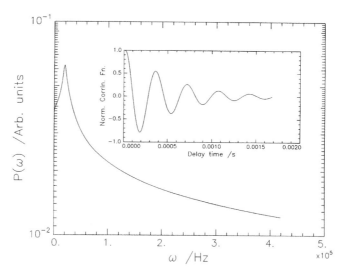

Figure 3.16. The power spectrum of capillary waves on a liquid surface with the
following properties: $\gamma_0 = 50 \, \text{mN m}^{-2}$, $\gamma' = 2 \times 10^{-4} \, \text{mN s m}^{-1}$, $\varepsilon_0 = 10 \, \text{mN m}^{-1}$,
$\varepsilon' = 0 \, \text{mN s m}^{-1}$. The inset is the heterodyne correlation function of this power
spectrum.

The capillary wave frequency is observed as the location of the maximum in $P(\omega)$ and is shifted by ω_0 from the incident light frequency (taken as zero in figure 3.16). Capillary wave damping influences the width of the spectrum. All four surface properties influence the experimental observables ω_0 and Γ in a manner that is not describable by a unique relation. Broadly, the surface tension influences the frequency according to the leading term in equation (3.1.41), the damping increasing with γ' until it becomes so large that wave propagation ceases. Note that, for large values of ε_0 and low q values, both ω_0 and Γ become insensitive to ε_0. Using a higher q value can restore this sensitivity somewhat. The maxima in ω_0 and Γ often observed result from resonance between capillary and dilational modes and the optimum conditions for obtaining ε_0 and ε' is the resonance condition under which the capillary waves (the directly observable property) are most influenced by the dilational waves.

In practice the frequency shifts, ω_0, due to the capillary wave frequency are small and optical mixing (or heterodyne) spectroscopy methods must be used. In the detection of the scattered light one can use frequency-domain or time-domain-based methods. The former results in obtaining the power spectrum whereas the latter detects the Fourier transform of the power spectrum, i.e. its correlation function. We concentrate solely on the latter technique here, but the optical train preceeding detection is identical both for spectrum analysis and for photon correlation spectroscopy. The design of suitable instrumentation allowing ease of control of several factors with good resolution and sensitivity to changes in capillary wave frequency is due to Hård and Neumann (1981). Figure 3.17 shows the essential features.

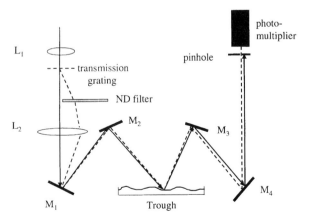

Figure 3.17. A schematic diagram of the SQELS apparatus. L and M represent lenses and mirrors, respectively.

Spatially filtered light from a laser source passes through a transmission grating where a proportion of the beam is diffracted, forming a series of weaker reference beams of much lower intensity than the main beam. Reference beams on one side of the main beam pass through a neutral density filter where their intensity is reduced further. The main beam and the reference beams are focused onto the liquid surface by the two lenses L_1 and L_2 with the aid of the periscope arrangement of mirrors. Each of the separate diffraction orders that form the reference beams is incident at a different angle on the surface and thus selecting different reference beams allows a range of surface wave-vectors, q, to be investigated. Specularly reflected and reference beams from the liquid surface, the latter now containing a proportion of scattered light, are directed to a photomultiplier by a second arrangement of mirrors. The specularly reflected beam is collected in a light trap and discarded, but each weak reference beam (and its associated scattered light) is directed on to the photocathode of a photomultiplier via a pinhole. By changing the orientation of mirror M_4, each reference beam can be directed on to the photomultiplier tube. The output from the photomultiplier tube is conditioned by its associated amplifier and discriminator and passed to a digital correlator. By this means the heterodyne correlation function is observed as a damped oscillatory signal. The signal observed is not the pure Fourier transform of the power spectrum but rather its convolution with various instrumental and adventitious exterior factors; we discuss this later when we consider the analysis of such correlation functions.

Representing the intensities of the scattered light and the reference beam as I_S and I_R, respectively, the heterodyne correlation function is

$$G(\tau) = (I_S + I_R)^2 + I_S I_R g^1(\tau) + I_S^2[g^2(\tau) - 1], \qquad (3.1.47)$$

where $g^1(\tau)$ and $g^2(\tau)$ are the field and intensity correlation functions respectively of the scattered light and τ is the delay time. It is $g^1(\tau)$ that is the Fourier transform of the power spectrum referred to earlier, so measures must be adopted to ensure that I_R greatly exceeds I_S in order to ensure that the homodyne contribution (I_S^2 term) does not contribute. A value of $I_R/I_S > 35$ is sufficient but significantly larger values are usually used in practice; this control can be achieved by the use of appropriate neutral density filters as q is varied.

Additionally it is important that the liquid surface is isolated from vibration and shielded from adventitious draughts or acoustic waves. It is generally sufficient to place the liquid container on an active vibration-isolation table that is placed on a heavy optical table to which all other optical elements are fixed. The whole should be placed in a draught-proof enclosure.

The form of the power spectrum given by equation (3.1.46) and shown in figure 3.16 is approximately Lorentzian, but deviations from Lorentzian behaviour become very apparent in the high-frequency shift wing of the spectrum. Consequently it is not appropriate to use a simple Lorentzian function to analyse SQELS data (although much data analysis has been done in this way in the past). Several instrumental factors also influence the observed signal. These include instrumental line broadening due to more than one capillary wave-vector being illuminated by the finite diameter of the light beam incident on the surface, fast processes at short times due to after-pulsing in the photomultiplier tube and additional background contributions due to very-long-wavelength, slow vibrations (building vibrations). Each of these factors and the non-Lorentzian form can be incorporated into the analysis of hetero-dyne correlation functions.

Fundamental parameters of the capillary waves, i.e. their frequency and damping, can be obtained by writing the correlation function as

$$G(\tau) = B + Af(\tau)\exp\left(-\beta^2\tau^2/4\right), \tag{3.1.48}$$

where B is the background contribution, A is an amplitude factor and the Gaussian term is the instrumental line broadening referred to above, all other instrumental factors being omitted here for clarity. The function $f(\tau)$ can be represented by an exponentially damped cosine function incorporating a phase term, ϕ, that accounts for the deviations of the power spectrum from a Lorentzian form:

$$f(\tau) = \cos\left(\omega_0\tau + \phi\right)\exp\left(-\Gamma\tau\right). \tag{3.1.49}$$

Fitting this function to the experimental data provides unbiased values for ω_0 and Γ. Such values are important in two ways. Firstly, ω_0 and Γ respond to the variations in surface visco-elastic parameters; however, estimation of these parameters requires a value of q, the surface wave-vector. Calibration of the optical train in terms of q is frequently done by obtaining ω_0 and Γ for pure liquids of known viscosity and density and then using equations (3.1.40) and (3.1.41) to obtain values of q. More precise values of q can be obtained by numerical evaluation of the relevant dispersion equation for the range of ω_0 and Γ values investigated.

The frequency and damping of the capillary waves are the only two experimental observables in a SQELS experiment. Obtaining the four surface visco-elastic parameters can be a more demanding process. In many cases some of these parameters have been assumed either to be zero (γ' generally) or to have their classical zero frequency value (γ_0 usually); in some cases such assumptions have been compounded by the use of a Lorentzian to fit the SQELS data. Since there is no *a priori* basis for assuming that *any* of the visco-elastic moduli should either be zero or have a zero frequency value, one should be very cautious about approaching the data from this viewpoint. A *direct* analysis of the SQELS data has been pioneered by Earnshaw and McGivern (1990) and is best described as spectral fitting. Equation (3.1.48) is again used to describe the correlation function, but $f(\tau)$ is now replaced by the numerical Fourier transform of the power spectrum, equation (3.1.46), using γ_0, γ', ε_0 and ε' as adjustable parameters of the fitting procedure. This approach can be successful provided due care is taken in its use. Fitting should be repeated from different starting points and it should be ensured that the resonance condition is passed through during the fitting process, since this is where the surface visco-elastic parameters are most sensitive to the form of the correlation function. A typical fit for a water surface covered by a spread polymer monolayer is shown in figure 3.18, together with the residuals of the fit. Generally, values of γ_0 and γ' obtained by this method are both precise and accurate; as remarked earlier ε_0 and ε' are less sensitive parameters (particularly ε'). The smaller the noise in the correlation functions the smaller the errors in the dilational parameters, particularly ε'. Discussion of results obtained using SQELS on a variety of polymer systems is given in chapter 8.

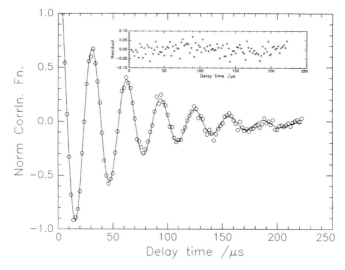

Figure 3.18. The correlation function obtained for a surface excess layer in a 0.1% wt/wt solution of polyethylene oxide ($M_r = 100\,000$) in water at $q = 871$ cm^{-1}. The solid line is the fit using the complete power spectrum expression; the inset shows the residuals of the fitting.

3.2 Ion beam methods

3.2.1 The interaction of energetic ions with matter – introducing the techniques

In this section we discuss two groups of surface analysis methods in which the surface is probed by a beam of energetic ions. Surface sensitivity is achieved by virtue of the fact that ions interact strongly with matter and thus they have only limited penetrating power. This penetrating power is energy dependent and ranges from a few ångström units for ions with energies in the range of tens of kilo-electron-volts, up to micrometres for ions with energies of a few mega-electron-volts. Thus rather low-energy ion beams are used to achieve good surface sensitivity, whereas mega-electron-volt ion beams allow one to acquire information about the near-surface region. In addition, rather different detection modes are used for these two energy ranges, resulting in two separate classes of techniques. In secondary ion mass spectrometry (SIMS), low-energy ions are incident on the surface and, as a result of complex interactions between the ions and the material, fragments of the sample are physically removed from the surface, often themselves ionised, by processes collectively termed sputtering. These secondary ions are detected and analysed to give a very sensitive picture of the surface composition of the material. Thus SIMS is essentially a surface-sensitive technique; however, by using high fluxes of ions, one can remove enough material that a composition–depth profile may be obtained. This constitutes the technique of dynamic secondary ion mass spectrometry. On the other hand, in Rutherford backscattering, forward recoil scattering and nuclear reaction analysis, the incident ions penetrate up to micrometres into the sample. The product of the interaction is itself a high-energy ion whose energy carries information about the depth at which the interaction took place; thus these techniques intrinsically yield depth-profile information and are not very surface sensitive. The technique of low-energy ion scattering spectrometry (LEISS) combines some features of both these classes; low-energy ions are used, conferring surface sensitivity, as in SIMS, but it is the energy of the backscattered ions that is detected, as in RBS, giving extremely surface-sensitive information on the sample's elemental composition.

3.2.2 Secondary ion mass spectrometry

In secondary ion mass spectrometry (SIMS) ions with energies in the kilo-electron-volt range consitute the probe. The penetrating power of these ions is rather small and the majority of the ion–matter interaction takes place in the outermost few ångström units of the material. However, these interactions are very strong and result in the physical removal from the surface of charged, molecular fragments, some of which may be quite large. These fragments are collected from above the surface, collected and mass analysed, either in a quadrupole system or by time-of-flight techniques. In general one can expect a beam of energetic ions to create highly reactive products of radiation chemistry that initiate a complex sequence of reactions causing rather severe chemical changes in the affected surface layer; however, if very low fluxes of ions are used damage to the surface is minimised and rather large fragments are sputtered from the surface. Analysis of the pattern of reaction fragments allows specific chemical identification to be made. This mode of operation is referred to as static secondary ion mass spectrometry.

In general use positive noble gas ions, usually of argon or xenon, are produced at an energy of up to around 4 keV, with a current density of about 1 nA cm^{-2}. The ion beam may be defocused over the whole analysis area or focused to a spot which is rastered across the area; the latter allows the possibility of imaging. The sputtered ions are extracted from the sample region by the application of an electric field and then analysed. Both positive and negative ions can be analysed, though it is found that for most polymer analysis problems the positive-ion spectra are more informative. Perhaps the most common type of mass spectrometer used for static SIMS work is the quadrupole, though the advantages of time-of-flight methods (TOF-SIMS) are now being seen as increasingly compelling. We will discuss the special features of TOF-SIMS below.

An important practical problem in the analysis of non-conducting materials such as polymers is the build-up of charge on the surface. The predominant cause of the charging is the emission of large numbers of secondary electrons by the surface; thus, even for an incident beam of positive ions, the surface will usually acquire a positive charge. The result of this is that positive ions sputtered from the surface will be accelerated to higher energies that may be larger than the maximum energy that can be detected by a quadrupole mass analyser. To get round this problem two approaches can be used. By far the most common is to apply a flood of low-energy electrons to neutralise the surface. With careful choice of energy and current this seems to work well, but there is the danger that the electrons themselves can stimulate emission of ions from the surface. Less commonly, one can neutralise the incident ion beam by passing it through a low-pressure gas, giving the technique of fast-atom bombardment mass spectrometry (FABMS).

In a quadrupole SIMS instrument one can typically detect ions up to 1000 in relative molecular mass. The detailed mechanisms that underly sputtering are extremely complex and as yet one cannot expect quantitative predictions of the relative intensities and relative molecular mass of the sputtered fragments for a given chemical structure. Instead the approach that has been taken has been to study well-characterised surfaces to check that the ion spectra can be understood in terms of the known chemical structure, making analogies, if appropriate, with more conventional mass spectrometry. The ion spectrum of such a well-characterised material may be regarded as its 'finger-print' and the ion spectrum for an unknown sample can be compared with library spectra to yield a confident identification of the chemical species present at the surface of the sample. A substantial compendium of SIMS spectra for many organic materials is now available (Briggs *et al.* 1989).

An example of a SIMS spectrum is shown in figure 3.19. This is the positive-ion spectrum obtained from poly(ethylene terepthalate) and is reasonably typical of the SIMS spectra of homopolymers. What is striking is the rich structure of the data; many quite large molecular fragments are being sputtered and the careful analysis of their distribution and intensities allows an unambiguous identification of the surface. For example, the main repeat unit gives a peak at 193 amu, whereas the other major features are obvious

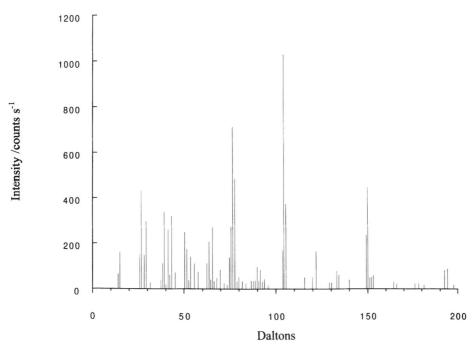

Figure 3.19. The positive secondary ion mass spectrum of poly(ethylene terephthalate) using 4 keV Ar$^+$. After Briggs (1987).

fragments of this unit, for example HOOC—C_6H_4—CO^+ at 149 amu and $C_6H_5^+$ at 77 amu.

Despite the power of the static SIMS technique using a quadrupole mass analyser, there are some difficulties that can be overcome by using the more recently developed time-of-flight technique for mass analysis. The major problem with the quadrupole analyser is that it has an intrinsically low ratio of ions transmitted to ions introduced; this transmission is typically in the range 0.01–0.1. Moreover, the device works by scanning through the mass range, detecting only one ion mass at a time. A time-of-flight mass analyser overcomes both problems; it has a higher intrinsic transmission, that may exceed 0.5 and it detects all masses in parallel. The combination of these two factors gives an overall increase in sensitivity by a factor of order 10^4. In addition the accessible mass range is much greater than that of the quadrupole analyser; in principle it is unlimited, but in practice molecular masses of up to 10 000 may be detected, in contrast to the upper limit for quadrupole analysers of about 1000. Its mass resolution is also greatly superior.

The drawback is extra complexity and cost. A pulsed ion beam is used, with the sputtered ions extracted by a rather higher field than for a quadrupole instrument and detected by an electron multiplier after passage through a drift tube. The pulsed nature of the beam leads to greatly prolonged analysis times; typically one might use a pulse 1 ns long with a repetition rate of 10 kHz, with the consequence that the beam is only incident on the sample for a fraction of 10^{-5} of the total analysis time. Of course the sensitivity is greatly improved but one can still expect analysis times about ten times longer than for quadrupole systems.

Figure 3.20 shows the information that can be obtained in much higher mass ranges using time-of-flight SIMS. The spectrum is for a thin layer of oligomeric polystyrene on a silver support (Bletsos *et al.* 1987); what is seen as a series of strong peaks are ions corresponding to complete oligomers cationised with Ag^+. Remarkably, the distribution of oligomers seems to correspond reasonably well to the relative molecular mass distribution obtained by gel permeation chromatography.

During a typical SIMS experiment the position of the beam spot is rastered across the surface; thus, it is possible to imagine gating the acquisition of the ion spectra in order to achieve a spatial image of the secondary ion spectrum. The limit of possible spatial resolution is effectively set by the problem of achieving enough signal intensity without damaging the sample surface and clearly in this regard the time-of-flight approach, with its great intrinsic sensitivity, offers huge advantages.

As mentioned above, the determining factor that allows one to obtain

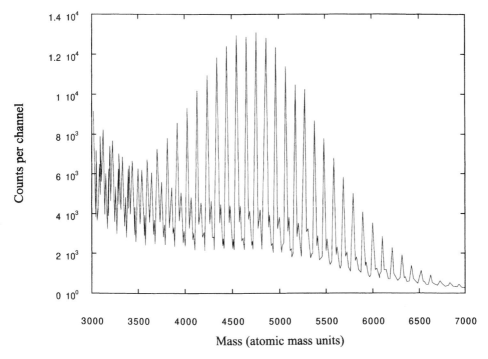

Figure 3.20. The time-of-flight SIMS spectrum of a polystyrene oligomer whose number average relative molecular mass is 4964, deposited as a thin film on a silver substrate. After Bletsos *et al.* (1987).

informative SIMS spectra with a rich pattern of fragments of high relative molecular mass is the beam current density. A 1 nA beam provides about 6×10^9 singly charged ions per second, so if this is delivered to a 1 cm^2 area for a couple of minutes one accumulates a dose of about 10^{12} ions cm^{-2}. At this dose, when on average the distance of ion–surface interactions is of order 100 Å, one is analysing an essentially undamaged surface, because the zone of damage resulting from an ion–surface interaction spreads out a distance of order 10 Å from the impact site. However, if the dose is increased by a couple of orders of magnitude, then no part of the surface is unaffected by collisions. Thus, for doses of greater than about 10^{14} ions cm^{-2}, the high-mass-unit fractions that are of such value in doing chemical analysis are lost and instead the spectrum is composed only of single atoms or very small clusters such as CH or CH$_2$. However, in exchange for the loss of chemical information, one acquires the possibility of depth-profiling, because at these higher doses material is removed at a big enough rate to make depth-profile measurements feasible. This constitutes the technique of dynamic secondary ion mass spectrometry.

3.2.3 *Dynamic secondary ion mass spectrometry*

Dynamic SIMS is an extremely well-established and widely used depth-profiling technique for the study of inorganic materials and in particular semiconductors (Feldman and Mayer 1986, Vickerman *et al.* 1989). Its strengths include relatively good depth resolution combined with very high sensitivity and the ability to profile all elements simultaneously. It has been much less widely used for polymers, even though with some care all these advantages can apply to the polymer case too.

In typical recent experiments (Schwarz *et al.* 1992) a 20 nA 2 keV Ar^+ beam was incident on the sample at an angle 30° off normal. The beam was rastered over a 0.5 mm square, but in order to avoid edge effects from the sputtered crater, ions are only accepted from a 0.15 mm square in the centre of the sputtered region. Analysing polystyrene under these conditions, material is sputtered at a rate of about 1 nm min^{-1}. This sputtering rate is a complex function both of the beam conditions and the material being analysed; in practice it is determined empirically. For dynamic SIMS on inorganic materials it is usual to measure the depth of a crater after sputtering using a stylus profilometer and then to assume a constant sputtering rate. For polymers this procedure may lead to errors due to the danger of the surface being penetrated by the stylus; it is more satisfactory to prepare a standard sample whose thickness has been independently measured, say by ellipsometry, and analyse this under the same conditions. If the standard sample is prepared as a multilayer, one can also deduce the effective depth-resolution function from the observed broadening of interfaces that are known to be sharp. This broadening arises from a variety of physical processes known collectively as ion beam mixing. For the conditions mentioned above, this resulted in an instrumental depth resolution of approximately Gaussian form with a full width at half maximum of 11 nm. This figure would increase if the incident beam energy were increased.

Another potential difficulty with dynamic SIMS is that, in multicomponent systems, various elements may sputter at different rates. This will lead to segregation of the more slowly sputtering component at the surface. At steady state this surface segregation must exactly compensate for the lower sputtering rate, so that the ratio of sputtered ions will accurately reflect the composition of the material, but transient effects will occur before this steady state is reached. This is a problem if one is interested in the composition profile immediately beneath the surface. What can be done is to coat the sample with a thin, sacrificial layer of polymer (typically about 50 nm thick); steady state is then reached by the time etching has reached the sample proper. A final specimen preparation requirement is to coat the sample with gold, to alleviate charging problems (though it may still be necessary to flood the sample with electrons).

Figure 3.21 shows a typical depth profile achieved using dynamic SIMS; in this case the sample is a mixture of two polymers that differ only in their isotopic labelling. Thus the possibility of their having differential sputtering rates is not a problem. Note that ion beam mixing has led to broadening of what is a physically sharp interface between polymer and substrate at distance 0 Å; this corresponds to a Gaussian resolution function with a full width at half maximum of 50 Å. This profile illustrates well one strength of the dynamic SIMS technique; its ability to probe composition–depth profiles not only near polymer surfaces but also at buried interfaces.

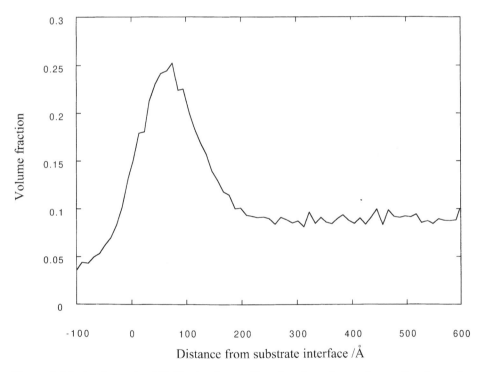

Figure 3.21. A dynamic SIMS depth profile showing the volume fraction of a polystyrene of relative molecular mass 29 000, with both ends carboxy terminated, in a matrix of deuterium-labelled, normally terminated polystyrene of relative molecular mass 500 000, after annealing at 160 °C for 36 h. The volume fraction, derived from the intensity of the CH^- ion, is measured as a function of the distance from the substrate, which is a native-oxide-coated silicon wafer. After Zhao *et al.* (1991).

3.2.4 Rutherford backscattering spectrometry

An alternative way to get depth-profile information relies on the increased penetrating power of higher energy ions. At incident ion energies in the mega-electron-volt range, interactions with the electrons of material are relatively weak and simply have the effect of gradually slowing down the ion. However, as Rutherford discovered in his classic experiment, the ion has a small probability of a direct collision with a nucleus, with which it will interact very strongly, suffering a large change of momentum. A relatively light ion, such as a helium ion, can be completely reversed in its direction if it hits a more massive nucleus. As long as the energy of the incident ion is too small to overcome the Coulomb barrier, preventing it from getting close enough to the target nucleus to take part in nuclear reactions, the collision will be elastic. Thus the energy with which the ion is backscattered is determined by classical considerations of conservation of energy and momentum and, for a given incident ion energy and scattering angle, depends solely on the mass of the nucleus from which it has been scattered. Thus, if one directed a monoenergetic beam of ions at a thin foil containing a mixture of elements and analysed the energies of the ions that were scattered back, one would see a series of peaks, each one corresponding to an element (or more strictly, since the separation is by mass, an isotope) present in the foil. If, instead of a foil, one used a bulk target, scattering events would also take place in the bulk of the sample. However, an ion would have lost some energy before the scattering event by virtue of electronic collisions and, on being scattered, it will lose some more energy as it returns to the sample surface. Thus ions scattered from below the surface will emerge at a lower energy than those scattered from the surface so knowing the stopping power of the material we are able to convert this energy decrement into a depth. In this way we can convert our spectrum of scattered energies into depth-profiles of the heavy nuclei in the sample. This technique is known as Rutherford backscattering (RBS). For more details the reader is referred to the monograph by Chu *et al.* (1978), while a good introduction is provided in the textbook by Feldman and Mayer (1986).

The strengths of RBS are in determining the concentration–depth profiles of relatively heavy elements in a polymer matrix largely composed of elements of low atomic number. This is because the sensitivity of the technique is determined by the Rutherford scattering cross sections, which are approximately proportional to the square of the atomic number of the target nuclei. The depth resolution of the technique is generally limited by the energy resolution of the detection system and typically is of the order of 300 Å for polymer materials. One of the particularly attractive aspects of the technique is the reliability and ease with which the data can be quantified; the ionic stopping powers required for conversion of energy to depth are well known and tabulated, while the scattering cross-sections are, with a few well-known exceptions, accurately predicted by the Rutherford formula. The ease of quantification to some extent compensates for the only moderate depth resolution and sensitivity compared with those of dynamic SIMS. Another drawback is the fact that the presence of heavy elements can hinder the analysis of lighter elements, because back-scattering events from light elements close to the surface can yield detected ions at the same energy from ions scattered from heavier nuclei deeper in the sample.

As an example of RBS applied to a polymer system, figure 3.22 shows the spectrum obtained from a sample of the conjugated polymer poly(phenylene vinylene) after exposure to arsenic pentafluoride vapour (Masse *et al.* 1990). This dopant diffuses into the polymer and reacts with it to form an electrically conducting complex and RBS is well suited to following the kinetics of the doping process by providing concentration–depth profiles of the elemental components of the dopant as a function of time. The peaks in the spectrum

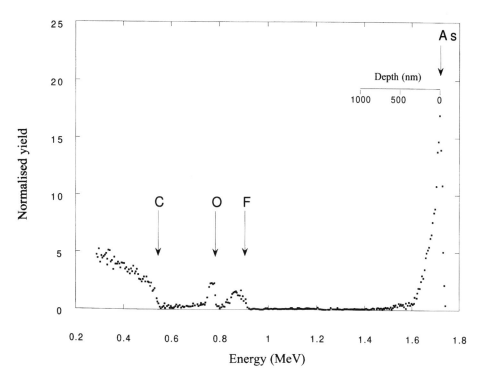

Figure 3.22. The RBS spectrum of a poly(phenylene vinylene) sample after it had been exposed to arsenic pentafluoride vapour for 5 days. After Masse *et al.* (1990).

correspond to different elements present near the surface of the sample and these elements may be identified using the kinematic formula for the fraction of the incident energy retained by the backscattered ion as a function of the mass of the target nucleus. Ions retain a higher proportion of their energy when rebounding from a heavier nucleus, so the peak at highest energy corresponds to the arsenic. Decreasing in energy, the next peak corresponds to fluorine, while at slightly lower energies another peak indicates that oxygen is present in the surface layer, presumably in the form of a layer of arsenic oxide. More detailed analysis, utilising the Rutherford formula for the scattering cross section, would yield the stoichiometry of the doped layer. Finally, at around 0.5 MeV and below, one sees an edge marking the onset of scattering from carbon atoms at the surface and below. Note that, even though the majority element in the sample is the carbon of the polymer, the scattering yield from the carbon is lower than that from the arsenic near the surface, because of the Z^2 dependence of the scattering cross section. Finally, we note that the arsenic peak is highly asymmetric, decaying rather gradually on the low-energy side. This indicates that there is a concentration profile of arsenic going into the film; using the known stopping powers we can calculate the superposed depth scale, which shows that the surface concentration of arsenic decays as one penetrates into the bulk over a distance of order 1000 Å.

In all techniques using mega-electron-volt ion beams as the probe particles, one needs to be alert to the potential difficulties that beam damage to the sample can cause. At the doses used for ion beam analysis, some permanent damage to a polymeric sample is inevitable and is usually visible to the naked eye in the form of a brown or black beam spot. However, the degree to which this damage affects the validity of the analysis depends on the sample. In some polymers (for example polystyrene) the major beam-induced reaction is cross-linking, which is relatively benign. However other polymers, such as poly-(methyl methacrylate) undergo chain scission and depolymerisation, resulting in gross mass loss. The effects of beam damage should always be monitored by the examination of spectra following sequential exposure. In some cases the problems can be greatly alleviated by holding the sample at liquid nitrogen, or lower, temperatures.

It is usual to use incident ions for RBS with energies roughly in the range 1–3 MeV. One can use much lower energy ions, in the kilo-electron-volt energy range, such as one might use for SIMS, but instead of detecting sputtered ions as one would do in a SIMS experiment, one can detect and analyse the energy of backscattered incident ions. This constitutes the technique of low-energy ion scattering spectrometry, or LEISS. Just as in RBS, the energy of the back-scattered ions depends on the mass of the target nuclei with which the collision has taken place. Unlike RBS, however, the penetrating power of the ions is very small, so LEISS is highly surface sensitive. Also, at these low energies the effect of the electron clouds around nuclei at screening the nuclear charge is much more important and the scattering cross sections are no longer given by the Rutherford formula, making quantification more difficult.

LEISS has found some use for polymers in situations in which its extreme surface sensitivity is valuable. In fact, so great is its surface sensitivity that spectra can give some structural information about the surface via shadowing and orientation effects. Recent work has been reviewed by Vargo *et al.* (1993).

3.2.5 Forward recoil spectrometry

The major drawback to RBS, that has greatly limited its application to polymers, is its insensitivity to light elements. A variant of the RBS technique, forward recoil spectrometry, or FRES (sometimes the alternative acronym ERD, for elastic recoil detection, is used), overcomes this problem, at least to the extent that it permits the profiling of hydrogen and deuterium concentrations. FRES has proved particularly useful in fundamental surface and interface studies, because its discrimination between hydrogen and deuterium allows the powerful strategy of deuterium labelling to be used; reviews of its application to polymers include those by Green and Doyle (1990), Shull (1992) and Jones (1993). In the nuclear collisions that occur when ions are incident on the sample, not only are the bombarding ions scattered, but the target nuclei themselves also must recoil in order to conserve momentum. If one detects not the scattered incident ions but the recoiling target nuclei, typically by arranging grazing angles of incidence and detection for the ion beam and recoiling nuclei, one can build up depth-profiles of the light atoms in the sample.

Figure 3.23 shows a typical sample geometry for a FRES experiment. The beam is incident on the sample at a glancing angle of 15°. Emerging from the sample at a similar glancing angle of exit are both recoiling hydrogen and deuterium nuclei and forward scattered helium ions that have undergone Rutherford scattering from heavy nuclei in the sample. These forward scattered helium ions vastly outnumber the recoiling light nuclei, so they have to be removed from the detection system. This is most easily done using a stopper foil, a 10 μm thick film of poly(ethylene terephthalate) that completely stops the helium ions while letting the lighter and more penetrating hydrogen and deuterium ions through. In passing through the stopper foil the H and D ions lose an amount of energy that has some statistical spread (known as 'straggling'). This effectively degrades the depth resolution to a value of about 800 Å.

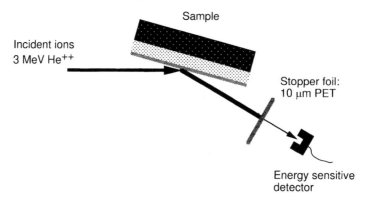

Figure 3.23. The schematic geometry of a typical forward recoil spectrometry experiment.

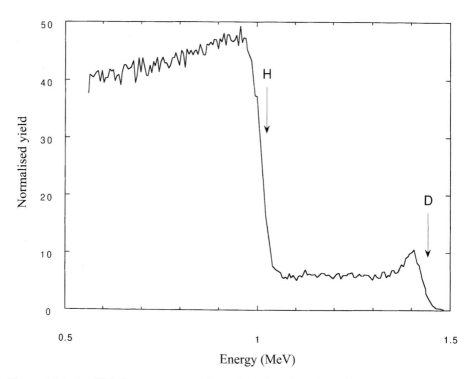

Figure 3.24. A FRES spectrum of a blend of 15% deuterated polystyrene (M_r 1.03×10^6) and 85% normal polystyrene, after annealing for 5 days at 184 °C.

Figure 3.24 shows a typical FRES spectrum, in this case from a blend of normal and deuterated polystyrene that after annealing shows some surface segregation of the deuterated component. The signal at the highest energy consists of deuterons recoiling from collisions at the surface of the film. Ions detected at lower energies correspond to deuterons recoiling from collisions below the surface, until at around 1 MeV one sees a sharp rise in the signal, to which hydrogen nuclei recoiling from the surface of the sample start to contribute. As in the case of RBS, such spectra can be converted into quantitative depth-profiles using the known stopping powers and collision cross sections; however, for the case of FRES the cross sections are not accurately given by the Rutherford formula and have to be determined empirically.

Figure 3.25 shows the depth profile of the deuterated component of the mixture, as deduced from the spectrum shown in figure 3.24. The data clearly show that substantial segregation of the deuterated component has occurred and one could extract from the data a quantitative value of the total surface excess. However, it is clear that virtually all information about the detailed shape of the concentration profile has been washed out by the relatively broad resolution function. Two approaches have been taken for improving the resolution of the technique. Firstly, one can improve the resolution substantially by simply reducing the beam energy. At lower incident beam energies, a thinner stopper foil may be used, reducing the effect of straggling and the

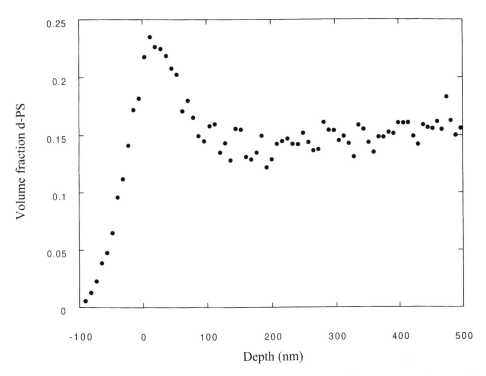

Figure 3.25. The depth profile of deuterated polystyrene deduced from the FRES spectrum shown in figure 3.24. After Jones *et al.* (1989).

stopping power of the target is increased, leading to an improvement of resolution from 800 Å for an incident energy of 3 MeV to 250 Å for an incident energy of 1.3 MeV (Genzer *et al.* 1994). The cost of this improvement in resolution is that the depth for which a deuterium depth profile is accessible is roughly halved. Even greater improvements in resolution can be made by using a more glancing angle of incidence; in this way one can achieve a resolution at the surface of 140 Å, though at the cost of reducing the probing depth to 1500 Å.

The second method provides improved depth resolution by doing away entirely with the need for a stopper foil, by using a time-of-flight method to separate the contributions from each type of ion electronically. In this way one can achieve resolutions of order 300 Å with no loss of probing depth (Sokolov *et al.* 1989), though at the cost of a more expensive and complex detection system. In fact, in principle the time-of-flight method can be combined with the use of heavier incident ions to allow the possibility of simultaneous depth-profiling, without interference, of a number of light elements. This kind of approach has not yet found wide application to polymers, possibly because problems associated with beam damage become more severe when heavier ions are used as the probe beam.

3.2.6 Nuclear reaction analysis

Rutherford backscattering and forward recoil scattering rely on essentially elastic collisions between the incident ions and the target nucleus. However, in some circumstances the interaction of ion and target may lead to a nuclear reaction in which energy is released and reaction products different from the original particles are emitted. For example, ^{3}He ions of around 1 MeV kinetic energy can undergo a fusion reaction if they encounter a deuterium ion, releasing in the process 18.35 MeV of energy and producing a proton and a ^{4}He nucleus. This reaction may be used to depth-profile deuterium by using an incident ^{3}He beam; either the proton or the ^{4}He ion may be detected and, as in RBS and FRES, the energy spectrum of the detected ion may be converted into a deuterium depth-profile using the known stopping powers for ions in the material.

If the exiting ^{4}He ions are detected, it is possible to achieve depth resolutions at the surface as high as 140 Å (Chaturvedi *et al.* 1990) with accessible depths of up to 1 μm. Although the resolution is degraded at greater depth due to straggling, this technique still allows the characterisation of buried polymer/polymer interfaces with resolutions of around 200 Å. Even greater depths are accessible if the protons from the reaction are detected (Payne *et al.* 1989), but this is at the cost of worse depth resolution, which is of order 300 Å at the surface.

A slightly different approach uses a resonant nuclear reaction. For example, ^{15}N incident on hydrogen produces a high-energy γ-particle via the following reaction when the nitrogen beam is at the resonance energy, 6.385 MeV:

$$^{15}\text{N} + {^1}\text{H} \rightarrow {^{12}}\text{C} + {^4}\text{He} + \gamma(4.43 \text{ MeV}).$$

This reaction can be used to profile hydrogen in polymers (Endisch *et al.* 1992) in the following way. Exactly at the resonance energy, the yield of γ-particles is proportional to the concentration at the surface. If the beam energy is raised slightly, then the incident ions will be slowed down to the resonance energy as they penetrate into the material. By measuring the yield as a function of the incident beam energy, a depth-profile can be built up whose depth resolution depends on the width of the resonance, modified by the effect of energy straggling in the nitrogen beam for reactions taking place below the surface. Depth resolutions as small as 35 Å at the surface are possible. A drawback of the technique, however, is the need to use rather high-energy beams of a higher mass ion than helium, meaning that careful precautions have to be taken to alleviate the effects of beam damage. On the other hand the detection of γ-particles can take place with high efficiency over a very wide range of solid angles.

A wide range of other nuclear reactions have been used in material science for the depth-profiling analysis of a number of different elements, some of which are tabulated by Feldman and Mayer (1986). Whether these techniques will in the future prove useful for polymers remains to be seen.

3.3 Surface spectroscopy

3.3.1 Surface sensitive chemical analysis

An important advance in the study of polymer surfaces was made in the midseventies, when it was realised that the ultra-high-vacuum surface analysis techniques that had been developed for the study of clean metal and semiconductor surfaces could be usefully applied to polymers. Clark's group at Durham was particularly successful at applying x-ray photoelectron spectroscopy (XPS, sometimes also known as ESCA, standing for electron spectroscopy for chemical analysis), that rapidly became one of the most popular methods of analysing the chemical composition of the topmost few tens of ångström units of a polymer sample. The technique rapidly became extremely popular both in academic and in industrial laboratories, where its flexibility and versatility has proved invaluable in practical problem-solving situations. XPS relies on the ionisation of core electrons by an incident beam of x-rays; the resulting emitted photoelectrons are detected and their energy analysed. Surface sensitivity arises from the fact that the energies of the emitted electrons (typically up to of order 1 keV) are such that they strongly interact with matter. This leads to an elastic mean free path of order 10 Å for polymers; thus almost all the observed signal in a polymer XPS analysis arises from the top 30 Å of the sample.

At first sight one might think that a technique that in effect measures the energy levels of core electrons would be useful only for determining the atomic composition of surfaces. However, it turns out that the energies of the core levels are slightly affected by their local environments. Thus a sample containing a given element in more than one chemical environment will produce a series of core level peaks at energies that are measurably different in an XPS spectrometer. These 'chemical shifts' provide the basis for the use of XPS for chemical as well as atomic analysis. XPS measurements also provide information about electrons with much lower binding energies – the valence electrons. This information is much more difficult to interpret simply, but the shape of the valence band spectra can be regarded as a finger-print of a given polymer.

Spectroscopies that detect transitions between different vibrational energy levels provide an extremely powerful method of chemical analysis. For bulk materials, of course, this underlies the widespread use of infra-red and Raman spectroscopy. Neither of these techniques can be made truly surface sensitive for polymer systems. It is, however, possible to use infra-red spectroscopy using the evanescent wave that penetrates slightly beyond a totally internally reflecting interface as the exciting radiation. The evanescent wave typically penetrates a distance of order 1 μm, so the resulting technique, attenuated total reflection infra-red spectroscopy, cannot be considered truly surface sensitive. The technique's low cost and relatively simple data analysis, combined with the intrinsic power of vibrational spectroscopy makes it extremely useful, however. Surface-enhanced Raman spectroscopy is truly surface sensitive, but because it relies on a slightly mysterious huge enhancement of the Raman signal very close to a rough metal surface it will be useful only in a few rare situations. Potentially more useful is high-resolution electron energy loss spectroscopy (HREELS). Here low-energy electrons are used as the incident probe; as in XPS the low penetrating power of low-energy electrons provides the source of surface sensitivity. The energy spectrum of the scattered electrons contains features corresponding to the loss of energy to vibrational excitations in the surface. Thus in principle the HREELS experiment provides the same kind of chemical information as infra-red spectroscopy but with true surface sensitivity. In practice, however, with current instrumentation the resolution obtainable with HREELS is rather less than that for infra-red spectroscopy.

Our purpose here is to provide an overview of these techniques and the information they can provide. The reader who needs to know in more depth about the use of the techniques is referred to one of the many books and review articles on the subject, for example Briggs (1990) and Sabbatini and Zambonin (1993).

3.3.2 X-ray photoelectron spectroscopy

A commercial x-ray photoelectron spectrometer uses a fixed anode x-ray source, typically producing magnesium $K\alpha$ or aluminium $K\alpha$ radiation, which directs a beam at the sample surface. The sample chamber is held at ultra-high vacuum and the resulting photoelectrons are collected and their energy analysed in an electrostatic analyser such as a cylindrical mirror analyser.

The basis of the x-ray spectroscopy technique is of course the photoelectric effect; an incident photon of frequency v can result in the photoemission of an electron with kinetic energy E_K that is related to the binding energy of the electron E_B by

$$E_K = hv - E_B, \qquad (3.3.1)$$

where h is Planck's constant. X-ray photons have energies of order 1 keV, so even the tightly bound core electrons of atoms will be emitted.

A low resolution XPS spectrum shows a number of typical overall features: at very low binding energies (that is photoelectrons detected with kinetic energies close to the excitation energy) one sees the complex and not very prominent structure due to valence electrons, while at higher binding energies one sees strong and well-defined peaks corresponding to atomic core-level transitions characteristic of the various elements making up the surface composition. Surface analysis at this apparently crude level is often invaluable in trouble-shooting situations, for additives and contaminants that may well affect the practical surface properties of a material, such as adhesion and wettability, often contain elements such as silicone and fluorine that are absent from a pure polymer surface and thus may easily be spotted in this kind of experiment. In addition, modern instruments have the capability of detecting spectra from rather small illuminated areas, allowing composition mapping of the surface with resolution of order a few tens of microns or better.

A higher resolution spectrum concentrating on one of the core-level peaks shows more detail. Figure 3.26a and figure 3.26b show respectively the carbon 1s and the oxygen 1s peaks obtained from a poly(methyl methacrylate) sample. These show clearly that these peaks have internal structure; curve fitting allows one to resolve sub-peaks, each corresponding to an atom in a different chemical

(a)

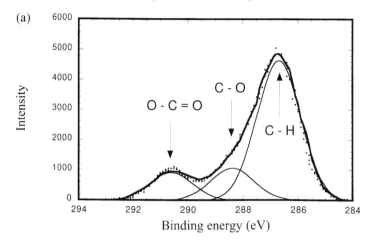

(b)

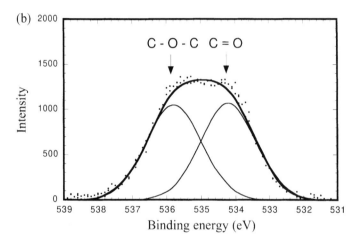

Figure 3.26. Details of the XPS spectrum from a poly(methyl methacrylate) surface, showing (a) the carbon 1s peak and (b) the oxygen 1s peak. In each case internal structure corresponding to atoms in different chemical environments may be resolved. After Green *et al.* (1990).

environment. These chemical shifts have been tabulated, so it is possible to assign each component to different functional groups, allowing a considerable amount of information about the chemistry of the surface to be extracted. In addition to these main peaks resulting from the ionisation of a single electron, secondary peaks resulting from multi-electron processes can also be found. For example, in association with a core electron ionisation event a valence electron may be promoted to a higher level. This process will occur in systems containing unsaturation, which is particularly important for aromatic systems, whereby a promotion of an electron from a π-orbital to a π^*-orbital results in a so-called 'shake-up' satellite about $6-7$ keV lower in energy.

The surface chemical analysis made possible by XPS has proved useful in a number of areas. Simple detection of surface contamination has already been mentioned; another technically important area is the detection of changes in surface functionality introduced by treatments such as plasma and flame modification and chemical etching. These treatments are extensively used in practice to modify surface properties such as adhesion and wettability and the use of XPS and other surface analysis techniques permits one to associate these changes with the introduction of specific chemical functionality at the surface. Excellent entries into the extensive literature in this area may be found in the monograph by Garbassi *et al.* (1994) and the review by Briggs (1990).

The surface composition of mixed polymer systems is also conveniently studied using XPS. For example, the surface of a polymer blend will generally have a different composition to the bulk and, if the chemical differences between the two polymers are sufficient, such a surface enrichment will be detectable using XPS. Such blend surface segregation will be discussed more extensively in chapter 5. Even stronger surface segregation effects can be expected in the case of systems containing block copolymers and here again XPS has been used with success.

One feature of such surface segregation problems is that the surface composition often decays to bulk values over length scales of order tens of ångström units. Although XPS cannot be considered a true normal depth-profiling technique in the way that dynamic SIMS and neutron reflectivity are, some information about the depth profile is available by systematically varying the sampling angle (angle-resolved XPS). As mentioned above, the source of surface sensitivity of XPS is the rather small value of the elastic mean free path of the emitted photoelectrons. The intensity of a given peak $I(i)$ is related to the number density of the corresponding element $N(i)$ and the elastic mean free path λ (which itself will depend on the peak energy) by

$$I(i) \propto \int_0^\infty N(i) \exp\left(\frac{-\zeta}{\lambda}\right) d\zeta, \qquad (3.3.2)$$

where ζ is the path length through the material taken by the emitted electrons. If the electrons are emitted at an angle ϕ to the normal then the path length is related to the depth by $\zeta = z/\cos(\phi)$. Thus, by obtaining spectra for a number of different take-off angles, it is possible to build up depth-profile information that is effectively in the form of a Laplace transform. It is difficult to make such depth-profile information completely quantitative, however, in part because of the intrinsic difficulties of inverting Laplace transforms with noisy and incomplete data, and also because of uncertainties surrounding the precise values of the elastic mean free path (which itself will be composition dependent). An example of angle-resolved XPS in use to study surface segregation in a polystyrene/poly(vinyl methyl ether) blend (Bhatia *et al.* 1988) is discussed in more detail in section 5.1.

All our discussion so far relies on the use of the core energy levels. However, the structure at small binding energies due to transitions among the valence electrons conveys valuable information about the electronic structure of conducting and semi-conducting polymers near their surfaces. Illustrations of the use of XPS and the analagous technique using lower energy photons, ultra-violet photoelectron spectroscopy (UPS), to study this area of growing importance are given in a review by Malitesta *et al.* (1993) and the monograph by Salaneck *et al.* (1996).

3.3.3 Vibrational spectroscopy – HREELS and IR

Although, as we have seen, XPS does carry chemical information by way of the chemical shifts in the core levels, the chemical information is conveyed essentially by rather small perturbations on signals that are fundamentally associated with sensitivity to atomic species. An ideal surface spectroscopic technique would combine the surface sensitivity of XPS with the chemical specificity of vibrational spectroscopies such as infra-red and Raman. For looking at thin adsorbed layers on metal surfaces, infra-red reflection–absorption spectroscopy provides just such a tool, but unfortunately it relies for its operation on the interaction of electromagnetic radiation with a metal and thus cannot be used for examining polymer surfaces, except for very thin polymer films on a conducting substrate. Surface-enhanced Raman spectroscopy also relies on the presence of a metal surface; for reasons that do not seem fully understood at present Raman cross sections are enhanced by orders of magnitude very close to rough metal (typically silver) surfaces.

A genuinely surface-sensitive vibrational spectroscopy is provided by high-resolution electron energy loss spectroscopy (Pireaux 1993). In this ultra-high-vacuum technique a rather low-energy beam of electrons, typically in the range 1–10 eV, impinges on the sample and the electrons scattered from the surface are analysed and detected in an electrostatic analyser. Electrons detected at lower energies than the incident beam have lost energy by exciting various quantum transitions in the material. At loss energies around 5–10 eV it is electronic transitions that are excited and spectroscopy in this range is complementary to the valence band structures studied at very low binding energies in XPS and UPS. However, at smaller values of the energy loss, 0–

400 meV, the electrons lose energy to vibrational excitations of exactly the same class as those that are observed in infra-red and Raman spectroscopy. However, the energy resolution in a HREELS experiment is typically an order of magnitude worse than that in an infra-red spectrometer, so the vibrational bands are much less sharp.

Figure 3.27 shows a series of HREELS spectra obtained from samples of normal, partially deuterated and fully deuterated polystyrene. The region between 500–2000 cm^{-1} shows a complex structure arising from incompletely resolved main chain and phenyl group vibrational modes. As the level of deuteration is increased we see that the C—H stretch peak around 3000 cm^{-1} is replaced by the corresponding C—D stretch at an energy a factor of $\sqrt{2}$ lower, around 2200 cm^{-1}.

HREELS has up to now been relatively little used for polymer surfaces. However, its combination of great surface sensitivity with excellent chemical specificity seems to offer great promise for the future, particularly if future instrumental developments lead to improved resolution.

3.4 Direct measurement of forces between surfaces

3.4.1 Introduction

When two surfaces covered with polymer and surrounded by a solvent for the polymer approach each other a repulsive force will be generated when the outer segments of the adsorbed polymers begin to overlap. This is caused by a

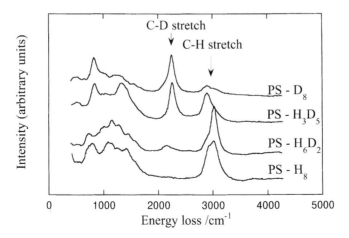

Figure 3.27. HREELS spectra of a series of selectively deuterated polystyrenes for an incident energy of 7 eV and specular geometry. The units on the intensity axis are arbitrary. After Vilar *et al.* (1987).

repulsive osmotic pressure being developed from the unfavourable entropy changes due to the conformational restrictions of confining the adsorbed polymer layers between two surfaces. This effect is made use of in the steric stabilisation of colloids (Napper 1983) and evidently the force–distance curve for two such surfaces is related to the volume fraction profile of the polymer normal to the surface. These forces can be directly measured using the surface force apparatus (SFA) originally designed by Tabor (Tabor and Winterton 1969) and subsequently developed by Israelachvili (Israelachvili and Adams 1978) so that forces between two surfaces immersed in a fluid can be measured. Since the development of the original surface forces apparatus the technique has been developed to allow shearing as well as normal forces to be applied. A hybrid apparatus has been described that should allow the simultaneous measurement of surface forces and the polymer volume fraction profile between the surfaces, by neutron reflectivity. Finally, the related technique of scanning force microscopy holds the promise of being able to measure forces with much more lateral resolution than can the SFA.

Forces between polymer-covered surfaces are a subset of the general subject of intermolecular forces, which are more fully discussed elsewhere (Maitland *et al.* 1981, Israelachvili 1991). Before the SFA and the results obtained from it are discussed, a brief consideration of intersurface forces is needed. If the intermolecular potential energy function between two *individual* molecules is generated solely by van der Waals forces it is purely attractive and of the form

$$U(r) = -K/r^n, \qquad (3.4.1)$$

where r is the intermolecular separation, K is a constant for the molecules and the exponent n takes the value 6 when the interaction is over a short enough range that retardation effects become important. To find the interactions between macroscopic bodies, the simplest approach is to assume that the intermolecular forces are pair-wise additive (for an introduction to the approaches that go beyond this crude approximation see Israelachvili (1991)). The resulting potential is found by an integration over volume elements in each body. This results in a potential energy of interaction (and hence a force between macroscopic bodies) that decays much more slowly with separation than does that predicted by equation (3.4.1). For example, if a molecule approaches a flat surface made from identical molecules, the energy of the interaction between them is

$$U(D) = \frac{-2\pi K\rho}{(n-2)(n-3)D^{n-3}}, \qquad (3.4.2)$$

where D is the distance from the surface and ρ the number density of particles. If $n = 6$ then

$$U(D) = \frac{-\pi K \rho}{6D^3};\tag{3.4.3}$$

the force generated by this potential is obtained by differentiating with respect to D.

If a large sphere of radius R made up of these molecules approaches the flat surface, the interaction potential is obtained by integration of equation (3.4.3) over the sphere's diameter; we obtain

$$U(D) = \frac{-\pi^2 K R \rho^2}{6D}.\tag{3.4.4}$$

If both surfaces are flat and the two bodies have infinite thickness, then on close approach the nett interaction energy per unit surface area for $n = 6$ is

$$U(D) = \frac{-\pi K \rho^2}{6D^2}.\tag{3.4.5}$$

Note that, for macroscopic particles, the interaction potential decays much more slowly than does that between molecules ($\propto r^{-6}$), decaying only as the reciprocal separation for a sphere and as $1/D^2$ for two flat surfaces. The combination $K\pi^2\rho^2$ is a material constant, with dimensions of energy, known as the *Hamaker constant*. It typically takes values of order 10^{-19} J for interactions between identical solids across a vacuum.

The interaction energy potential is generally easier to derive for flat surfaces; on the other hand, it is usually more accurate to measure the force–distance relation ($F(D)$) between two curved surfaces, because the interaction area is precisely definable. The relationship between $U(D)$ per unit area for two flat surfaces and $F(D)$ between two curved surfaces is given by the Derjaguin approximation (Derjaguin 1934) which is derived for the close approach of two spheres of radii R_1 and R_2:

$$F(D) \cong 2\pi \left(\frac{R_1 R_2}{R_1 + R_2}\right) U(D),\tag{3.4.6}$$

where $U(D)$ is the interaction energy per unit area as a function of the separation of two flat surfaces. As long as the separation distance D and the potential energy interaction range are much smaller than R_1 or R_2, then this approximation is applicable to any type of force–distance relation. When one sphere has a much larger radius than the other ($R_2 \gg R_1$) then

$$F(D) = 2\pi R_1 U(D).\tag{3.4.7}$$

Of particular relevance to use of the SFA is the force–distance equation for two cylinders crossed at some angle θ to each other:

$$F(D) = 2\pi (R_1 R_2)^{1/2} U(D)/\sin\theta.\tag{3.4.8}$$

3.4.2 Surface forces apparatus

The surface force apparatus (SFA) was developed to measure weak forces between two surfaces whose separation may be as small as 1 Å, but the value of which also needs to be known accurately. Figure 3.28 shows an SFA of the type used to determine the force–distance curves for adsorbed polymer layers. The solid surfaces consist of two cylinders crossed at right angles to each other on which are glued mica sheets that have been coated with silver to make them semi-reflecting. The distance between the two cylinders is measured from the spacing and shape of the fringes of equal chromatic order obtained by focusing white light passed through the cylinders on to a spectro-

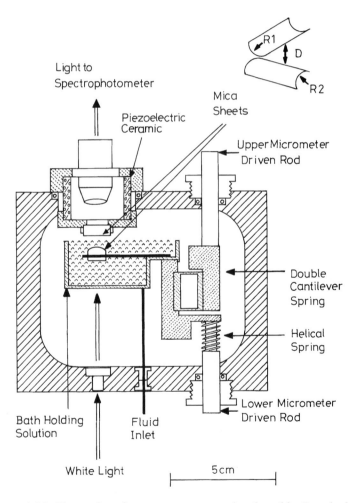

Figure 3.28. The surface forces apparatus, as developed by Israelachvili.

graph grating. The surfaces are brought close to each other by use of the upper and lower control rods, small displacements being effected by applying a voltage to the ceramic piezoelectric tube allowing movements as small as 1 Å to be obtained. The piezoelectric element is also central to the measurement of the force. This element is expanded or contracted a known amount by an applied voltage and the distance the cylindrical surfaces actually move is measured. Multiplication of the difference between the two distances (i.e. distance actually moved – distance associated with the applied voltage) by the spring constant of the force-measuring spring provides the difference in the force between the initial and final positions. If the cylinders are crossed at 90° to each other, then $\sin \theta$ in equation (3.4.8) is 1 and effectively one of the cylinders has an infinite radius and the force–distance relation simplifies to equation (3.4.7).

The use of the SFA to study force–distance profiles between polymer-coated surfaces was pioneered by Klein and co-workers (Klein 1983, Klein and Luckham 1984, Luckham and Klein 1990). Much of the work reported concerns 'brush'-like layers in which polymers are anchored to the surface by one end. This anchoring may be achieved by the use of a strongly adsorbing end-group (Taunton *et al.* 1988) or by using a copolymer with one short block, that acts as an anchor on the surface of the mica half cylinders, the longer block forming a 'buoy' solvated by the solvent (Hadzioannou *et al.* 1986, Patel *et al.* 1988). A review of the earlier work using the SFA to determine force–distance profiles has been published by Patel and Tirrell (1989). Results using SFA to measure forces between polymer brushes will be discussed in detail in chapter 6. Here, to illustrate the technique, we describe its application to uniformly adsorbed polymer layers.

Polyethylene oxide (PEO) absorbs 'uniformly' onto mica and experiments using the SFA on solutions of this polymer in 0.1 M potassium nitrate were reported by Klein and Luckham (1984). Figure 3.29 shows force–distance curves obtained for PEO of relative molecular mass 1.6×10^5 g mol^{-1} at an initial concentration of 10^{-3} g l^{-1}.

Initial compression of the two layers followed curve A, whereas rapid decompression followed by rapid compression followed curve B; however, if, after rapid decompression, the system was left for 1 h, the force–distance curve followed curve A again. Whilst compression was under way Klein and Luckham monitored the refractive index in between the two cylinders; at the distance of closest approach ($\simeq$ 100 Å), the refractive index was approximately that of pure PEO and, from the refractive index, the calculated adsorbed amount was about 4 mg m^{-2}. Removing the solution and replacing it with pure salt solution and repeating the force–distance determination produced identical

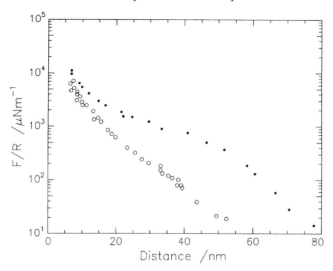

Figure 3.29. Force–distance curves for two surfaces in an SFA immersed in 10^{-3} g l^{-1} PEO in 0.1 M KNO_3 solution: filled points, compression; and open points, rapid decompression following compresssion. After Klein and Luckham (1984).

data; all the polymer had been adsorbed and none was desorbed when pure solvent was present. Interaction between the two layers is first detected at about three times the unperturbed radius of gyration; a rather extended layer is to be expected since the aqueous salt solution is a good solvent for PEO. To understand the force curves in a more quantitative way, it is necessary to have some theory of the properties of adsorbed polymer layers, of the kind discussed in chapter 5. Some understanding can be reached with simpler arguments, by assuming that on compression the volume fraction of polymer adsorbed on the surfaces does not alter and making use of the relation for the osmotic pressure between the two surfaces to the intersurface potential energy function, $U(D)$, i.e.

$$\pi = \partial U(D)/\partial D = \partial[F(D)/(2\pi R)]/\partial D. \qquad (3.4.9)$$

Klein and Luckham were able to estimate the slopes of the force–distance curves in the two limiting regions. Just when interaction between the two polymer-coated surfaces begins, the slope should be large and negative, while at close approach (< 100 Å separation) they obtain

$$\frac{F(D)}{2\pi R} \sim D^{-1}. \qquad (3.4.10)$$

These predictions seem to be supported by the data in figure 3.30. Subsequently, de Gennes stated (de Gennes 1987) on the basis of his self-similar

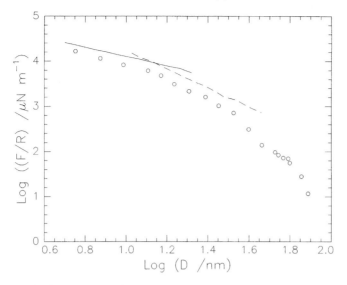

Figure 3.30. A double-logarithmic plot of force–distance curves for forces between surfaces after incubation in 150×10^{-3} g l^{-1} PEO in 0.1 M KNO$_3$. The solid line has a slope of -1; the broken line has a slope of -2. After Klein and Luckham (1984).

description of adsorbed polymer layers (de Gennes 1981) that, for a fixed amount adsorbed, the repulsive force would be due to the osmotic pressure at the mid-point between the two surfaces and, since the correlation length $\xi \sim D/2$ at this point, then

$$\frac{F(D)}{2\pi R} \sim D^{-2}. \qquad (3.4.11)$$

Although it is evident that the *effective* concentration in the intervening space changes continuously (as monitored by the refractive index), there is a region in figure 3.30 for which there does appear to be a force dependence which scales as D^{-2}.

The hysteresis observed on rapid decompression was attributed to the volume fraction of polymer near the surface being higher than the equilibrium value commensurate with surface separation. As a result the concentration between the surfaces is lower and therefore the repulsive force is reduced. This hysteresis is characteristic of situations in which polymers are uniformly adsorbed; it is absent from experiments in which polymers are grafted by one end to the mica. These experiments – on the forces between polymer 'brushes' – are thus somewhat easier to interpret.

Taunton *et al.* (1990) compared the behaviour of polystyrene adsorbed on to mica for cases in which the adsorbing moiety was either a small zwitterionic end group or a polyethylene oxide block with a degree of polymerisation of 50 or 170. The adsorbed polymers were surrounded by toluene or xylene, both being good solvents for polystyrene and the relative molecular mass range of the zwitterion-terminated polystyrene covered was from 26 500–660 000. No hysteresis was observed on compression and decompression of layers adsorbed on both surfaces. By noting the distance at which the onset of interaction was observed, the relation between effective layer thickness, L_0, and relative molecular mass obtained was

$$L_0 \propto M^{0.6}. \tag{3.4.12}$$

Figure 3.31 shows the 'master curve' force–distance profile obtained by normalising the separation distance by the value of $2L_0$ and shifting the force values by amounts that vary with the relative molecular mass.

They remarked that these shift factors are directly related to the extent of surface coverage. It also seems reasonable that the shift factors should also be related to the means of attachment, i.e. the space occupied by the absorbing group on the mica surface. However, the shapes of the force–distance curves were identical whatever the adsorbing group. This suggests that the possible complicating effects of using a polymeric anchor are negligible. More details of how such force curves can be analysed in terms of theories of grafted polymer layers are given in chapter 5.

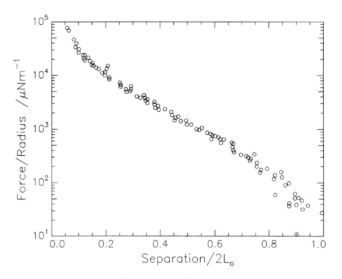

Figure 3.31. The master curve of force profiles for polystyrene brushes attached to mica surfaces and surrounded by a good solvent for the polymer. Relative molecular masses are in the range $26\,500 \leqslant M_r \leqslant 660\,000$. After Taunton *et al.* (1990).

3.4.3 Composition profiles of adsorbed polymers under compression

We remarked above that the force–distance curves do not appear to be particularly sensitive to the shape of the concentration profile normal to the surface used in the theoretical calculations. Evidently, it would be of value to obtain the concentration profile simultaneously with determination of the force–distance data. Cosgrove *et al.* (1995) have described a modified surface force apparatus that allows neutron reflectometry data to be collected; figure 3.32 is a schematic diagram of the apparatus.

The first results obtained were for polystyrene of high relative molecular mass adsorbed from cyclohexane at 55 °C on to the quartz plates that composed the opposing faces of the apparatus. The plates were subsequently dried and reflectivity experiments performed at ambient temperature with the surrounding cyclohexane 'solvent' being contrast matched to the quartz. Because of the design necessary for the neutron reflectivity experiments, the distance of closest approach of the two plates was about 100 Å, but force–distance data indicated that the onset of interaction between the two polymer layers was at about $30R_g$, which, given the poorer than theta conditions of the measurement, seems a large distance. Volume fraction profiles deduced from the neutron reflectometry data by the optical matrix method (see section 3.1) are shown in figure 3.33.

The volume fraction of polymer up to 500 Å from the quartz plate through

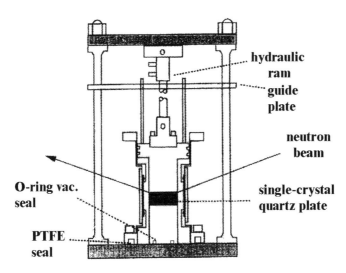

Figure 3.32. A schematic diagram of an apparatus for the simultaneous measurement of the force between polymer-bearing plates and the volume fraction profiles of the adsorbed polymer. After Cosgrove *et al.* (1995).

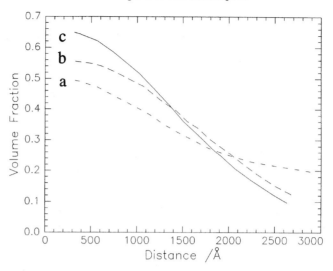

Figure 3.33. Volume fraction profiles for polystyrenes adsorbed onto quartz plates in the modified SFA apparatus of figure 3.32: (a) infinite plate separation, (b) a separation of about 4800 Å and (c) a separation of about 4000 Å. After Cosgrove *et al.* (1995).

which the neutron beam is incident is quite large, being 0.5 and greater. At these concentrations and at the temperature used, cyclohexane solutions of polystyrenes are extremely viscous. This would be especially true for the relative molecular masses of polystyrene used (20 000 000 and 4 000 000). In later experiments (Cosgrove *et al.* 1995) a linear diblock copolymer of polystyrene with a short block of polyethylene oxide was used. The surrounding solvent was toluene and in this situation the polyethylene oxide block acts as the anchor on the quartz surface. Volume fraction profiles obtained indicated that there was a 50 Å thick layer of pure polyethylene oxide adjacent to the quartz with the volume fraction of polystyrene being small and approximately constant over about 100 Å for polystyrene block relative molecular masses of 35 000 and 74 000. Surprisingly, the maximum distance from the surface was not significantly different for each of these two copolymers and, in proportion to the radius of gyration, the extension of the block of higher relative molecular mass was less than that of the small block. This seems somewhat contrary to expectations on the basis of excluded-volume behaviour and its dependence on relative molecular mass. This type of experiment has considerable potential for ascertaining the correct relation between force–distance curve and the concentration profile of polymer, be it adsorbed or end-tethered. The experiments are extraordinarily difficult to carry out even before the data analysis is attempted, but continued efforts should eventually provide rewarding insight.

3.4.4 Rheological aspects of surface forces measurements

An understanding of the rheology of thin films between solid surfaces is clearly relevant to lubrication and the SFA with its precisely determined and controllable separations between the two surfaces is ideal for such studies, provided that one of the surfaces can be moved relative to the other. Modifications to the SFA to provide a shearing motion and descriptions of early experiments on simple liquids have been published by Israelachvili (1992) and Bhushan and Israelachvili (1995). For fluids of low relative molecular mass shearing of a thin film produces a 'stick–slip' motion. Granick has discussed such data in terms of real and imaginary moduli (Granick *et al.* 1995). Determination of the rheology of polymer-bearing surfaces using the SFA is a rapidly developing field and we restrict discussion to two types of experiments that can be done.

The two experiments are exemplified by the sketches in figure 3.34. In the first, one surface is moved normally to the other in a sinusoidal manner, whereas in the second one surface is moved parallel to the second surface. For the first case, solvent is either squeezed from or drawn into the space between the surfaces until the load applied is balanced by steric repulsion from the increase (decrease) in osmotic pressure due to increase (decrease) in segment concentration in the intersurface volume. Hydrodynamic interactions between the two surfaces may become dominant and interactions associated with the flow of solvent between the two surfaces are termed lubrication forces. For the geometry of the SFA the hydrodynamic force, F_H, is related to the velocity of motion of the surfaces normal to each other by

$$\frac{6\pi R^2 \dot{D}}{F_H} = \frac{D}{\eta_0},\qquad(3.4.13)$$

where R is the radius of the two surfaces, D the distance between them, with $\dot{D} = dD/dt$, and η_0 is the viscosity of the bulk fluid. The left-hand side of equation (3.4.13), denoted G (with units of $m^2\,s\,kg^{-1}$), should thus have a

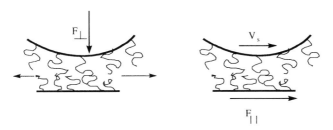

Figure 3.34. A schematic diagram of experiments investigating the rheology of adsorbed polymer layers using SFA.

linear variation with the separation D, with gradient $1/\eta_0$. Klein *et al.* (1993) adsorbed a zwitterionically ended polystyrene on to the toluene-surrounded mica surfaces of an SFA and then oscillated the upper surface with respect to the lower surface over a range of amplitudes for various separations. Their results are shown in figure 3.35, which also includes the line obtained for pure solvent and for a solution of polystyrene with no functional groups.

An extended brush-like layer is formed for the zwitterionically terminated polystyrene and comparison of force–distance curves before and after oscillatory motion showed that they were not displaced by the lubrication forces. These force–distance curves showed that the two brush-like layers interact at distances of about 2500 Å. Above this separation, values of G scale with D in precisely the same manner both in the presence and in the absence of tethered polymer. The slopes of these lines, moreover, gave a viscosity (η_0) that agreed with the bulk viscosity of toluene. For the tethered polymer the line through these large separation data has an x axis intercept that corresponds closely to twice the polymer layer's thickness. Hence, in this region the shear plane in the system has shifted by a distance of $2L_H$ (L_H is hydrodynamic layer thickness $\approx$ equilibrium layer thickness) and thus equation (3.4.13) becomes

$$G = (D - 2L_H)/\eta_0. \qquad (3.4.14)$$

The observation that the η_0 value obtained was that of the pure solvent suggests

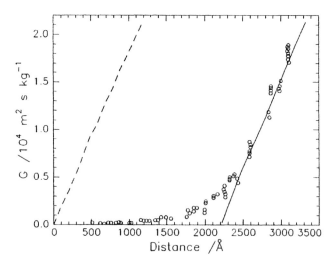

Figure 3.35. G as a function of the surface separation for two surfaces with brush-like polystyrene layers immersed in toluene. The dashed line corresponds to data in the absence of the polymer (and also after incubation in the presence of a polymer with no functional ends). The solid line has a slope of η_0^{-1}. After Klein *et al.* (1993).

that the depth of penetration of *flowing* toluene into the brush is small; Klein *et al.* calculated this to be about 100 Å.

When $D < 2L_H$, the brushes apparently interpenetrate (but see below when shearing motion is discussed), relative motion of polymer and solvent occurs and solvent molecules experience increased drag. Assuming that for these cases the concentration of polymer is uniform, the flow of solvent was analysed by the treatment used for flow in porous media (Brinckman 1947). In this approximation η_0 in equation (3.4.13) is replaced by an effective viscosity, η_{eff}, for which a scaling relation was obtained:

$$\eta_{eff} \sim \eta_0 a^{-2} \Gamma^{2\alpha} D^{2(1-\alpha)}, \qquad (3.4.15)$$

with a the statistical step length and Γ the surface concentration. The exponent α is the scaling exponent and thus has a value of $3/4$ in good solvents and 1 in theta solvents. When the comparison between the theoretical prediction and data in the small-D region is made, there is no agreement whatever exponent is used. The major discrepancy is that the experimental data approach $G = 0$ more rapidly than does the prediction. A possible source for this behaviour is that the brush-like layers become solid-like at small values of D; a rationale for this view based on reptation arguments was made by Klein *et al.* (1993).

Shearing of two polymer-bearing surfaces past each other can be done in the SFA if a *sectored* piezoelectric element is used. Proper interpretation of the data necessitates that the two surfaces be kept parallel to each other. In a second series of experiments, Klein and coworkers (Klein *et al.* 1995) achieved this by coupling of applied potentials to various surfaces of the piezo element. For pure liquids between the two surfaces classical stick–slip behaviour was observed. When each surface was coated with a polymer brush and a good solvent used, the shear force measured was many orders of magnitude less than that for the pure solvent even though the normal loads applied to the two surfaces were comparable. Experiments were also made on polystyrene adsorbed under conditions which approached theta state. The shear forces observed were now much larger; figure 3.36 shows both normal and shear forces obtained as a function of separation for brush-like and adsorbed layers (the brush polymer's relative molecular mass was 1.4×10^5; that of the adsorbed polymer was 1.8×10^6). The large difference in behaviour for these two cases was attributed to the increased penetration of the polymer layers for the adsorbed polymer case.

For the brushes, the degree of interpenetration is small. Within this interpenetration zone the concentration can be viewed as semi-dilute, and the effective viscosity is that of chains which have the dimensions of the correlation length (ξ). As a result the additional shear force over that of the pure

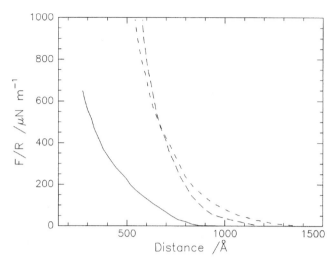

Figure 3.36. Normal and shear forces for brush layers (in toluene) and adsorbed layers (in cyclopentane) of polystyrene. The solid line is the shear force for an adsorbed layer; the shear force for a brush was zero. The long dashed line is the normal force for an adsorbed layer and the short dashed line is the normal force for a brush. After Klein *et al.* (1995).

solvent is very small. (All of this refers to the region $D \approx 2L_H$.) For the same normal load, the adsorbed polymer layers interpenetrate strongly (there is no osmotic repulsion under theta conditions) and become entangled. Sliding the two layers past each other requires an increased shear force. Evidently these types of experiments are somewhat ahead of current theory. For surveys of the use of the SFA for investigation of lubrication and shearing see the reviews by Dan (1996) and Luckham (1996).

3.4.5 *Scanning force microscopy*

The surface forces apparatus measures the forces between surfaces of macroscopic dimensions. The requirement that large areas be brought to very small separations imposes demanding requirements on the flatness of the surfaces used and by far the majority of experiments with the SFA have used freshly cleaved mica as a surface that can be relied upon to be atomically smooth over macroscopic areas. The technique of scanning force microscopy allows the force between a tip of small radius of curvature and virtually any kind of surface to be measured. In addition, the tip can be scanned over the surface to provide an image, which can be of very high resolution, of the force as a function of position. The drawbacks of the technique are that it is difficult to obtain an absolute measure of the tip–surface separation and that the details of the nature of the interaction between the tip and the surface are not always well known.

The scanning force microscope (SFM) – or atomic force microscope (AFM), as it is also commonly known – operates on similar basic principles to the surface forces apparatus. A comprehensive account of the technical details of SFM can be found in the book by Sarid (1994), while the application of the technique to polymers has been reviewed by Kajiyama *et al.* (1996). In a typical SFM (see figure 3.37) a small, pyramid-shaped tip is mounted on a cantilever which acts as a spring, with a spring constant typically of order $0.1\ \mathrm{N\,m^{-1}}$. In many cases the tip and cantilever are micro-fabricated as a whole from silicon, silicon oxide or silicon nitride. The cantilever is mounted on a piezo-electric drive that moves the tip in the vertical direction, while the resulting displacement is measured by reflecting a laser beam from the cantilever onto a quadrant photodiode. The tip can be scanned in the horizontal plane using two further piezo-electric drives. The instrument may be operated in a number of different modes. If the tip is kept stationary in the horizontal direction and moved vertically towards the surface, a force–distance profile analogous to the ones obtained from the surface forces apparatus is obtained. To obtain an image of the surface the tip may be scanned across the surface with a feedback mechanism maintaining a constant displacement of the tip – constant-force mode, or with the tip maintained at constant height – constant-

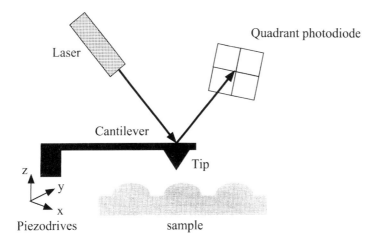

Figure 3.37. A schematic diagram of a scanning force microscope. The tip–cantilever assembly can be driven in the *x*, *y* or *z* direction by piezo-electric drives. The deflection of the tip can be detected both in the vertical and in the lateral direction by measuring the deflection of an incident laser beam with a quadrant photodiode.

height mode. A third imaging method is particularly useful for soft materials like polymers – it is the tapping mode, in which the cantilever is made to oscillate vertically. Further contrast modes are possible if the lateral force that develops as the sample is scanned is observed, which can yield information about the local mechanical and frictional response of the surface.

The scanning force microscope has been used to study grafted polymers at the solid/liquid interface by O'Shea *et al.* (1993) and Overney *et al.* (1996). In the first study, images of the surface at low coverage were recorded in constant-force mode and force measurements were made in good solvent conditions. In the second study, a force–displacement curve was obtained for a brush in good solvent conditions. Figure 3.38 shows the results, together with the prediction of self-consistent field theory (see chapter 6). The theory provides a good fit for separations of more than 10 nm, but when the brush is compressed beyond this point the force systematically deviates from the prediction, which the authors attribute to increasing penetration of the brush by the tip.

The most common use of SFM for polymers has been to characterise the surface topography of polymer thin films. An example of the use of SFM to measure contact angles for polymers was given in chapter 2; another example of this approach can be found in a study in which the effect of block copolymers on the contact angle between immiscible polymers was studied (Israels *et al.* 1995). The mechanisms underlying the kinetics of polymer de-wetting can also be elucidated by SFM studies of the local topography near the

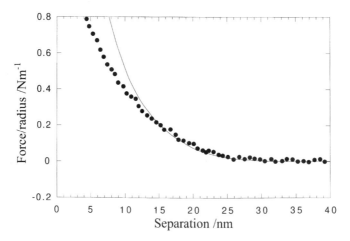

Figure 3.38. Force–displacement curves measured for interactions between a scanning force microscope's tip and a layer of styrene-4-vinyl pyridine block copolymer (the degrees of polymerisation of the styrene and 4-vinyl pyridine blocks were 200 and 20, respectively) in toluene. The solid curve is the prediction of self-consistent field theory. After Overney *et al.* (1996).

contact line (Lambooy *et al.* 1996), as discussed in chapter 5. Finally, SFM is a convenient way of studying the interesting surface morphologies of thin block copolymer films (Carvalho and Thomas 1994).

In addition to measuring the vertical force on the tip, it is also possible to measure the lateral force, thus obtaining information about the frictional properties of the surface being imaged. This can be used to provide some degree of contrast to distinguish between chemically different materials (Krausch *et al.* 1995). A challenge is to use this kind of technique to yield surface-specific information about the dynamic and mechanical properties of a surface. For example, Kajiyama and coworkers have studied polymer surfaces using lateral force microscopy and scanning viscoelastic microscopy (Kajiyama *et al.* 1997). In the latter technique a vertical sinusoidal motion is applied to the sample and the response of the tip measured, from which a local complex viscoelastic modulus may be determined, as discussed in chapter 2.

3.5 References

Als-Nielsen, J. (1984). *Physica* B&C **126**, 145.

Bhatia, Q. S., Pan, D. H. *et al.* (1988). *Macromolecules* **21**, 2166.

Bhushan, B. and Israelachvili, J. N. (1995). *Nature* **374**, 607.

Bletsos, I. V., Hercules, D. M. *et al.* (1987). *Macromolecules* **20**, 407.

Born, M. and Wolf, E. (1975). *Principles of Optics*. Oxford, Pergamon Press.

Briggs, D. (1987). Analysis and chemical imaging of polymer surfaces by SIMS. *Polymer Surfaces and Interfaces*. W. J. Feast and H. S. Munro. Chichester, John Wiley & Sons.

Briggs, D. (1990). Applications of XPS in Polymer Technology. *Practical Surface Analysis. Volume 1 – Auger and X-ray Photoelectron Spectroscopy*. D. Briggs and M. P. Seah. Chichester, Wiley. **1**, 657.

Briggs, D., Brown, A. *et al.* (1989). *Handbook of Static Secondary Ion Mass Spectrometry*. Chichester, Wiley.

Briggs, D. and Seah, M. P. Eds. (1990). *Practical Surface Analysis. Volume 1 – Auger and X-ray Photoelectron Spectroscopy*. Chichester, Wiley.

Brinckman, H. C. (1947). *Applied Surface Research* A**1**, 27.

Carvalho, B. L. and Thomas, E. L. (1994). *Physical Review Letters* **73**, 3321.

Charmet, J. C. and de Gennes, P. G. (1983). *Journal of the Optical Society of America* **73**, 1777.

Chaturvedi, U. K., Steiner, U. *et al.* (1990). *Applied Physics Letters* **56**, 1228.

Chu, W.-K., Mayer, J. W. *et al.* (1978). *Backscattering Spectrometry*. New York, Academic Press.

Cooke, D. J., Lu, J. R. *et al.* (1996). *Journal of Physical Chemistry* **100**, 10298.

Cosgrove, T., Zarbakhsh, A. *et al.* (1995). *Faraday Discussions* **98**, 189.

Creagh, D. C. and Hubbell, J. H. (1992). X-ray absorption coefficients. *International Tables for Crystallography C*. A. J. C. Wilson. Dordrecht, Kluwer.

Crowley, T. L., Lee, E. M. *et al.* (1991a). *Physica B* **173**, 143.

Crowley, T. L., Lee, E. M. *et al.* (1991b). *Colloids and Surfaces* **52**, 85.

Dan, N. (1996). *Current Opinion in Colloid and Interface Science* **1**, 48.

de Gennes, P. G. (1981). *Macromolecules* **14**, 1637.

de Gennes, P. G. (1987). *Advances in Colloid and Interface Science* **27**, 189.

Derjaguin, B. V. (1934). *Kolloid Zeitschrift* **69**, 155.

Dijt, J. C., Stuart, M. A. C. *et al.* (1990). *Colloids and Surfaces* **51**, 141.

Earnshaw, J. C. (1986). Light scattering at the fluid interface. *Fluid Interfacial Phenomena*. C. A. Croxton. New York, Wiley.

Earnshaw, J. C. and McGivern, R. C. (1990). *Langmuir* **6**, 649.

Earnshaw, J. C. and McLaughlin, A. C. (1991). *Proceedings of the Royal Society* A **433**, 663.

Earnshaw, J. C. and McLaughlin, A. C. (1993). *Proceedings of the Royal Society* A **440**, 519.

Endisch, D., Rauch, F. *et al.* (1992). *Nuclear Instruments and Methods in Physics Research* B **62**.

Feldman, L. C. and Mayer, J. W. (1986). *Fundamentals of Surface and Thin Film Analysis*. Englewood Cliffs, NJ, Prentice Hall.

Garbassi, F., Morra, M. *et al.* (1994). *Polymer Surfaces: from Physics to Technology*. Chichester, John Wiley & Sons.

Gast, A. P. (1992). Scanning angle reflectometry. *Physics of Polymer Surfaces and Interfaces*. I. C. Sanchez. Boston, Butterworth-Heinemann.

Genzer, J., Rothman, J. B. *et al.* (1994). *Nuclear Instruments & Methods In Physics Research Section B – Beam Interactions With Materials And Atoms* **86**, 345.

Goodrich, F. C. (1981). *Proceedings of the Royal Society* A **374**, 341.

Granick, S., Demirel, A. L. *et al.* (1995). *Israel Journal of Chemistry* **35**, 75.

Green, P. F., Christensen, T. M. *et al.* (1990). *Journal of Chemical Physics* **92**, 1478.

Green, P. F. and Doyle, B. L. (1990). Ion beam analysis of thin polymer films. *New Characterisation Techniques for Thin Polymer Films*. H.-M. Tong and L. T. Nguyen. New York, John Wiley & Sons 139.

Hadzioannou, G., Patel, S. *et al.* (1986). *Journal of the American Chemical Society* **108**, 2869.

Hård, S. and Neumann, R. D. (1981). *Journal of Colloid and Interface Science* **83**, 313.

Israelachvili, J. (1991). *Intermolecular and Surface Forces*. San Diego, Academic Press.

Israelachvili, J. N. (1992). *Surface Science Reports* **14**, 109.

Israelachvili, J. N. and Adams, G. E. (1978). *Journal of the Chemical Society Faraday Transactions I* **74**, 975.

Israels, R., Jasnow, D. *et al.* (1995). *Journal of Chemical Physics* **102**, 8149.

Jones, R. A. L. (1993). Ion beam analysis of composition profiles near polymer surfaces and interfaces. *Polymer Surfaces and Interfaces II*. W. J. Feast, H. S. Munroe and R. W. Richards. New York, John Wiley.

Jones, R. A. L., Kramer, E. J. *et al.* (1989). *Physical Review Letters* **62**, 280.

Jones, R. A. L., Norton, L. J. *et al.* (1990). *Europhysics Letters* **12**, 41.

Kajiyama, T., Tanaka, K. *et al.* (1996). *Progress in Surface Science* **52**, 1.

Kajiyama, T., Tanaka, K. *et al.* (1997). *Macromolecules* **30**, 280.

Keddie, J. L., Jones, R. A. L. *et al.* (1994). *Europhysics Letters* **27**, 59.

Kelvin, L. (1871). *Philosophical Magazine* **42**, 370.

Klein, J. (1983). *Journal of the Chemical Society Faraday Transactions I* **79**, 99.

Klein, J., Kamiyama, Y. *et al.* (1993). *Macromolecules* **26**, 5552.

Klein, J., Kumacheva, E. *et al.* (1995). *Faraday Discussions* **98**, 173.

Klein, J. and Luckham, P. F. (1984). *Macromolecules* **17**, 1041.

Kramer, E. J., Jones, R. A. L. *et al.* (1990). *Polymer Preprints* **31**, 75.

Krausch, G., Hipp, M. *et al.* (1995). *Macromolecules* **28**, 260.

Lamb, H. (1945). *Hydrodynamics.* New York, Dover.

Lambooy, P., Phelan, K. C. *et al.* (1996). *Physical Review Letters* **76**, 1110.

Langevin, D. and Bouchiat M. (1971). *Comptes Rendus des Académies des Sciences* **B272**, 1422.

Lekner, J. (1987). *Theory of Reflection.* Dordrecht, Martinus Nijhoff.

Levich, V. G. (1962). *Physicochemical Hydrodynamics.* Englewood Cliffs, NJ, Prentice-Hall.

Loudon, R. (1984). Surface excitations. *Surface Excitations.* V. A. Agronovich and R. Loudon. Amsterdam, North Holland.

Lu, J. R., Li, Z. X. *et al.* (1993a). *Langmuir* **9**, 2408.

Lu, J. R., Simister, E. A. *et al.* (1992). *Langmuir* **8**, 1837.

Lu, J. R., Thomas, R. K. *et al.* (1993b). *Langmuir* **9**, 1352.

Lucassen, J. (1968). *Transactions of the Faraday Society* **64**, 2221.

Lucassen-Reynders, E. H. and Lucassen, J. (1969). *Advances in Colloid and Interface Science* **2**, 347.

Luckham, P. F. (1996). *Current Opinion in Colloid and Interface Science* **1**, 39.

Luckham, P. F. and Klein J. (1990). *Journal of the Chemical Society Faraday Transactions* I **86**, 1363.

Maitland, G. C., Rigby, M. *et al.* (1981). *Molecular Forces: Their Origin and Determination.* Oxford, Oxford University Press.

Malitesta, C., Morea, G. *et al.* (1993). X-Ray photoelectron spectroscopy analysis of conducting polymers. *Surface Characterisation of Advanced Polymers.* L. Sabbatini and P. G. Zambonin. Weinheim, VCH.

Masse, M. A., Composto, R. J. *et al.* (1990). *Macromolecules* **23**, 3675.

Mayes, A. M., Russell, T. P. *et al.* (1994). *Macromolecules* **27**, 749.

Napper, D. H. (1983). *Polymeric Stabilisation of Colloidal Dispersions.* London, Academic Press.

Norton, L. J., Kramer, E. J. *et al.* (1994). *Journal de Physique* II **4**, 2534.

O'Shea, S. J., Welland, M. E. *et al.* (1993). *Langmuir* **9**, 1826.

Overney, R. M., Leta, D. P. *et al.* (1996). *Physical Review Letters* **76**, 1272.

Pankewitsch, T., Vanhoorne, P. *et al.* (1995). *Macromolecules* **28**, 6986.

Patel, S. and Tirrell M. (1989). *Annual Review of Physical Chemistry* **40**, 597.

Patel, S., Tirrell, M. *et al.* (1988). *Colloids and Surfaces* **31**, 157.

Payne, R. S., Clough, A. S. *et al.* (1989). *Nuclear Instruments and Methods in Physics Research* B**42**, 130.

Pireaux, J. J. (1993). Electron induced vibrational spectroscopy. *Surface Characterisation of Advanced Polymers.* L. Sabbatini and P. G. Zambonin. Weinheim, VCH.

Russell, T. P. (1990). *Materials Science Reports* **5**, 171.

Sabbatini, L. and Zambonin, P. G. Eds. (1993). *Surface Characterisation of Advanced Polymers.* Weinheim, VCH.

Salaneck, W. R., Stafström, S. *et al.* (1996). *Conjugated Polymer Surfaces and Interfaces.* Cambridge, Cambridge University Press.

Sarid, D. (1994). *Scanning Force Microscopy.* New York, Oxford University Press.

Schwarz, S. A., Wilkens, B. J. *et al.* (1992). *Molecular Physics* **76**, 937.

Sears, V. F. (1992). Scattering lengths for neutrons. *International Tables for Crystallography C.* A. J. C. Wilson. Dordrecht, Kluwer.

Shull, K. R. (1992). Forward recoil spectrometry of polymer interfaces. *Physics of Polymer Surfaces and Interfaces.* I. C. Sanchez. Boston, Butterworth-Heinemann.

Simister, E. A., Lee, E. M. *et al.* (1992). *Journal of Physical Chemistry* **96**, 1373.

Sokolov, J., Rafailovich, M. H. *et al.* (1989). *Applied Physics Letters* **54**, 590.

Tabor, D. and Winterton, R. H. S. (1969). *Proceedings of the Royal Society* A **312**, 435.

Taunton, H., Toprakcioglu, C. *et al.* (1988). *Nature (London)* **332**, 712.

Taunton, H. J., Toprakcioglu, C. *et al.* (1990). *Macromolecules* **23**, 571.

Thomas, R. K. (1995). Neutron reflection from polymer bearing surfaces. *Scattering Methods in Polymer Science.* R. W. Richards. London, Ellis Horwood.

Vargo, T. G., Gardella, J. A. *et al.* (1993). Low-energy ion scattering spectrometry of polymer surface composition and structure. *Surface Characterisation of Advanced Polymers.* L. Sabbatini and P. G. Zambonin. Weinheim, VCH.

Vickerman, J. C., Brown, A. *et al.* Eds. (1989). *Secondary Ion Mass Spectrometry.* Oxford, Oxford University Press.

Vilar, M. R., Schott, M. *et al.* (1987). *Surface Science* **189**, 927.

Zhao, X., Zhao, W. *et al.* (1991). *Europhysics Letters* **15**, 725.

4

Polymer/polymer interfaces

4.0 Introduction

In technology the use of pure materials is rare. The idea that, in order to get desired properties, one needs to use not pure materials but mixtures dates to prehistoric times, when it was discovered how much more useful copper was if tin were added to make bronze. So it is a natural idea to assume that one can broaden the range of materials that can be made from a few basic polymer types by mixing or blending these basic types together. However, it turns out not to be so easy; if you take a pair of common polymers, poly(methyl methacrylate) and polystyrene, say, and mechanically mix them, you get a very brittle material whose properties are generally much worse than those of either of the pure ingredients. The reason for this is fairly clear – examination of the resulting material reveals that the two polymers separate to form coarse domains of pure polymer; there is no intimate mixing between the components. Moreover, the interface between the domains is very weak. The difficulty of making a strong bond between two polymers is something most people are familiar with; joining two different plastics is usually difficult with most glues, which form weak joints with plastics. It is possible to make useful materials from polymer mixtures, even when they do not mix; examples are ABS and high-impact polystyrene. It turns out that control of the interface is the overriding factor in making such materials useful.

We begin this chapter by discussing the factors that determine whether or not a given pair of polymers will form a molecular mixture at equilibrium. Then we will discuss the structure at equilibrium of the interface between two immiscible polymers, introducing as we go two powerful theoretical approaches that we will use later in the book for a variety of more complex surface and interface problems. Finally, we will discuss the ways in which interfaces arise and attain their equilibrium structure; firstly considering processes such as welding in which we take two pure polymers and attempt to form a join between them, such that equilibrium is achieved by the diffusion of chain segments across an initially sharp interface. We then discuss what is in a sense the opposite process, the formation of two-phase polymer mixtures, whereby we start out with a homogeneous mixture that then phase separates to form distinct domains of the two polymers separated by equilibrium interfaces.

4.1 Thermodynamics of polymer mixtures

The quantity that determines whether two species will mix or not is the free energy of mixing. Consider the situation in figure 4.1: in (a) we have two different species A and B, with a total free energy $F_A + F_B$. In (b) the two species form a single phase – an intimate mixture that is spatially homogeneous, in the sense that, if we take a small volume within the mixture and take an average over time of the composition, the value of composition that we find is constant throughout the volume[†]. The free energy of this mixture is F_{A+B}; if we can neglect surface effects we can write

$$F_{mix} = F_{A+B} - (F_A + F_B), \qquad (4.1.1)$$

where F_{mix} is the free energy change on mixing. All we need to know about the phase behaviour of the species A or B is contained in the behaviour of F_{mix} as a function of composition and temperature[‡].

Suppose a single phase has a composition ϕ_0. At equilibrium the system finds a state of minimum free energy, so, if there existed a pair of compositions ϕ_1 and ϕ_2 such that, if the system separated into two phases at these compositions, the total energy were lowered, then the single phase at composition ϕ_0

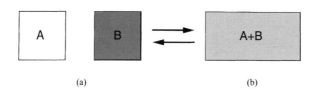

(a)   (b)

Figure 4.1. Free energies of mixing. In (a) two species A and B exist in an unmixed state, while in (b) they form a single, intimately mixed phase. The free energy of mixing is the change of free energy on going from state (a) to state (b).

[†] Note that instantaneously the composition is not the same throughout the volume – fluctuations of composition are inseparable from equilibrium.
[‡] All discussion is carried out under the assumption that we work at constant pressure, so it is the Gibbs free energy that is important.

would not be stable. The original total free energy of mixing is F_0. The total free energy of mixing of the separated system is given by F_0' (as can be seen by remembering that the total amounts of A and B are separately conserved). It is now obvious that, if the relationship of free energy with composition has a convex shape, then every single phase composition has a lower total free energy than any possible phase-separated composition – this is a single-phase mixture.

On the other hand, in figure 4.2b the free energy curve has a different shape. Between a certain pair of compositions the curve is convex; a system at a uniform composition ϕ_0 within this region will lower its free energy from F_0 to F_0' by separating into two phases with compositions ϕ_1 and ϕ_2. The lowest possible free energy is obtained when the phase-separated compositions are defined by the points at which a line is tangential to the free energy curve in two places; these two limits define the limits of composition within which a single phase is not stable.

However, even for compositions that fall inside the coexistence curve, another important distinction must be made. Figure 4.3 shows a magnification of the free energy curve just inside coexistence, showing that the curvature of the free energy function $\mathrm{d}^2 F/\mathrm{d}\phi^2$ may be either positive or negative. At a composition ϕ_b phase separation to two compositions close to ϕ_b results in a lowering of free energy from F_b to F_b'. At this composition the system is unstable with respect to small fluctuations in composition, and will immediately begin to undergo phase separation. However, at composition ϕ_a a similar small change in composition leads to an increase in free energy from F_a to F_a'; the system is locally stable with respect to such small composition fluctuations, even though it is still globally unstable with respect to separation into the two coexisting phases. There is an energy barrier that needs to be surmounted in

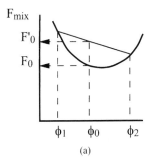
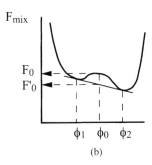

(a) (b)

Figure 4.2. Free energies of mixing as a function of the composition: (a) for conditions under which two species are miscible and (b) for conditions under which two species are immiscible.

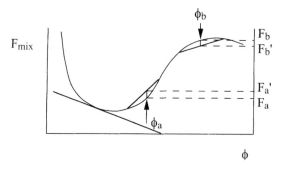

Figure 4.3. Enlargement of the free energy curve near one coexisting composition.

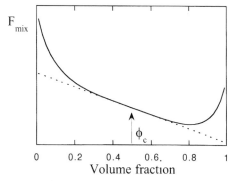

Figure 4.4. The free energy at the critical temperature, showing the vanishing of the curvature at the critical composition ϕ_c.

order to achieve the global energy minimum associated with phase separation and as a result this composition is metastable. Clearly the limit of local stability is defined by the condition that the curvature $d^2 F/d\phi^2 = 0$; the locus of these points is known as the *spinodal*.

Finally, we note that a *critical temperature* T_c separates the two types of situation shown in figures 4.2(a) and (b). In (a) the curvature of the free energy function $d^2 F/d\phi^2$ is always positive, whereas in (b) $d^2 F/d\phi^2$ is negative within a certain range of ϕ. The critical point is thus defined by the condition $(d^3 F/d\phi^3)_{\phi_c} = 0$, and it is the point at which the coexistence curve and spinodal line meet. Figure 4.4 shows the shape of a free energy curve at the critical point. The flatness of the free energy curve at the critical composition means that a system can make quite large fluctuations in composition at very little cost in free energy. The presence at equilibrium of very large composition fluctuations is a signature of the critical point and scattering of radiation from such fluctuations results in the phenomenon of *critical opalescence*.

The phase behaviour of binary mixtures is most conveniently summarised in a phase diagram, which can be constructed from the free energy as a function of temperature and composition as we have just discussed. Figure 4.5 shows such a diagram, in which the *coexistence curve* separates the one-phase region from the two-phase region. Within the coexistence curve the *spinodal* line separates compositions that are unstable from those that are metastable with respect to small composition fluctuations, with the coexistence curve and spinodal meeting at the critical point.

In order to predict phase diagrams it is sufficient to have a model for the free energy of mixing as a function of composition and temperature. A huge amount of effort has been put into building such models, but by far the most important historically has been the Flory–Huggins model. This is far from being the most sophisticated model available and its predictions are not terribly precise, but it is simple and tractable and, despite its quantitative shortcomings, it correctly embodies at least one qualitative truth about mixing in polymer systems.

The Flory–Huggins theory is in fact nothing more than a two-component polymer version of the simple lattice gas model introduced in section 2. We divide the free energy into an entropic part, which is assumed to take the simplest perfect gas form, while the enthalpic part is estimated using a typical mean-field assumption.

We saw in section 2.4 that the entropy of mixing of a polymer of degree of polymerization N and 'holes' was given by an expression of the form

$$\frac{S_{\text{mix}}}{k} = \frac{\phi}{N} \ln(\phi) + (1 - \phi) \ln(1 - \phi). \tag{4.1.2}$$

Exactly the same expression would hold for a polymer and a solvent. If instead

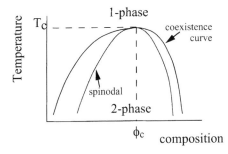

Figure 4.5. A schematic phase diagram, showing the coexistence curve separating the region in which compositions are stable as a single phase from the region in which phase separation into coexisting phases will lower the free energy and the spinodal line, which separates compositions that are unstable with respect to small composition fluctuations from compositions that are metastable.

of a polymer and a solvent we have two polymers with degrees of polymerisation N_A and N_B, the entropy of mixing *per monomer unit* is

$$\frac{S_{mix}}{k} = \frac{\phi}{N_A} \ln(\phi) + \frac{1-\phi}{N_B} \ln(1-\phi). \tag{4.1.3}$$

The translational entropy associated with each chain is reduced by a factor of N to reflect that monomers in the same chain are connected and cannot be positioned independently.

To complete the model of the free energy we need to consider the energetic part. Once again we simplify the problem by assuming a lattice model in which each site can only be occupied by a single segment. Interactions between neighbouring segments have an associated energy of ε_{AA} for two neighbouring A segments, ε_{BB} for neighbouring B segments and ε_{AB} for an A next to a B. Using once again a mean-field assumption the energy associated with a site becomes

$$U_{site} = z(\varepsilon_{AA}\phi_A^2 + \varepsilon_{BB}\phi_B^2 + 2\varepsilon_{AB}\phi_A\phi_B). \tag{4.1.4}$$

From this we need to subtract the value of the energy of the unmixed state to find the energy of mixing:

$$U_{mix} = z[\varepsilon_{AA}\phi_A(\phi_A - 1) + \varepsilon_{BB}\phi_B(\phi_B - 1) + 2\varepsilon_{AB}\phi_A\phi_B]. \tag{4.1.5}$$

If we now assume that the mixture is *incompressible*, every site must be occupied either by an A monomer or by a B monomer, so $\phi_A + \phi_B = 1$. We can express this as

$$U_{mix} = \phi_A\phi_B\chi kT, \tag{4.1.6}$$

where we have introduced the dimensionless *Flory–Huggins interaction parameter*

$$\chi = z(2\varepsilon_{AB} - \varepsilon_{AA} - \varepsilon_{BB})/(kT). \tag{4.1.7}$$

In words, χ is the energy change, in units of kT, when a segment of A is taken from an environment of pure A and swapped with a segment of B in an environment of pure B. If χ is positive, such a swap results in an increase in energy; mixing species with positive χ is energetically unfavourable (though as we shall see entropic factors may outweigh this). Conversely a negative χ means that mixing is energetically favourable.

Our free energy of mixing now has the form

$$\frac{F_{mix}}{kT} = \frac{\phi_A}{N_A} \ln(\phi_A) + \frac{\phi_B}{N_B} \ln(\phi_B) + \chi\phi_A\phi_B, \tag{4.1.8}$$

which is the *Flory–Huggins free energy of mixing*. It is now easy to compute the phase diagrams predicted by the Flory–Huggins model. The simplest case to consider occurs when the two polymers have identical degrees of polymerisation. Figure 4.6 shows free energy curves for various values of χ.

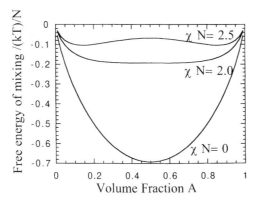

Figure 4.6. Free energy curves for various values of χN, calculated from the Flory–Huggins model.

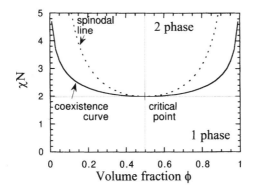

Figure 4.7. The phase diagram for the Flory–Huggins model with components of equal degrees of polymerisation.

For this symmetrical case the coexistence curve can be evaluated from the condition $dF/d\phi = 0$, which yields the inverse equation for ϕ

$$\chi_{\text{coex}} = \frac{1}{N} \frac{1}{2\phi - 1} \ln\left(\frac{\phi}{1 - \phi}\right). \qquad (4.1.9)$$

Solutions only exist for values of χ greater than the critical value $\chi_c = 2/N$. The spinodal line is easily found from the condition $d^2F/d\phi^2 = 0$, and the resulting phase diagram takes the form shown in figure 4.7. Phase separation occurs for all values of χ greater than $2/N$. The effect of the long-chain character of polymer molecules is to reduce the size of the unfavourable interactions needed to produce phase separation by a factor of N. This is a direct reflection of the reduction by this factor of the entropy of mixing. For small molecules even species with quite a large energetically unfavourable interaction will still mix because the gain in entropy on mixing outweighs the energetic penalty.

What range of values can χ take in practice? Firstly, note that, because it is a differential quantity, it can be either positive or negative – in this respect it is unlike the similar quantity ε which appears in the single-component lattice fluid model in chapter 2. Negative values of χ correspond to polymer pairs that are fully miscible; this is often due to some specific chemical interaction such as a hydrogen bond. In the absence of a specific interaction segments interact via the van der Waals force – this always leads to a positive χ value. Van der Waals interaction energies are essentially proportional to the product of the polarisabilities of the interacting species, so we would find thus that χ is always positive. So we can say that, in the absence of a specific interaction, for a high enough relative molecular mass polymers are always immiscible.

In our simple Flory–Huggins picture χ is entirely of energetic, rather than entropic, origin; from the way we have defined it it must depend on temperature as T^{-1}. This then implies that phase separation takes place as one lowers the temperature – the critical point is a so-called 'upper critical solution temperature'. In fact it is observed in many polymer mixtures that phases separate as the temperature is raised – they have a 'lower critical solution temperature' or LCST. This means that the interaction parameter must have a different temperature dependence and from this it follows that it is not purely energetic in nature – it must contain an entropic component, a so-called non-combinational entropy of mixing, which must arise from local packing constraints at the level of the polymer segments. Moreover, when χ is measured by scattering techniques it is often found that the value of χ depends quite strongly on composition (see for example Shibayama *et al.* 1985).

Thus we have strong experimental evidence that Flory–Huggins theory is inadequate as a quantitative description of mixing thermodynamics in polymer mixtures. From a theoretical point of view we can see four potential sources of error.

1. The assumption of unperturbed chain statistics. Implicit in Flory–Huggins theory is the assumption that the long-range chain statistics of polymer chains are ideal random walks. This is not to be expected; a polymer chain in a solvent collapses as conditions are changed to bring about phase separation between the polymer and the solvent (Grosberg and Khokhlov 1994). One would expect a polymer chain in a mixture to do the same as the conditions for phase separation were approached (Sariban and Binder 1987).
2. Neglect of fluctuations. Flory–Huggins theory is a mean-field theory and so neglects those large composition fluctuations that occur close to the critical point. This will result in the failure of the theory close to the critical point, but for

polymers one expects this region of failure to be rather small (Schwahn *et al.* 1987, Bates *et al.* 1990). Similarly, at very low concentrations chain connectivity imposes strong concentration fluctuations that are very important in polymer solution theory in the semi-dilute regime. For systems of high relative molecular mass screening effects reduce the significance of these fluctuations (de Gennes 1979).

3. Compressibility and 'equation of state' effects. In Flory–Huggins theory we implicitly assume that there is no change of volume on mixing the two species; we assume that no extra free space is created when the two polymers are mixed. This is known not to be the case experimentally and it is easy to understand the reason: if two species have a strongly unfavourable interaction, then, if we make a mixture it will be energetically favourable for the system to lower its density slightly, thereby reducing the number of unfavourable contacts and gaining some extra translational entropy associated with the 'vacancies'.

4. Local structure and packing. Finally, the local structure of monomers may lead to difficulties in packing, which will lead to a restriction of configurations available and thus to changes in the entropy of mixing. This is likely to be most important for pairs of polymers with very different local stiffnesses or with monomers with bulky side groups.

A huge amount of effort has gone into making a more refined theory of polymer mixtures. Although many of the resulting methods have had their successes, it is probably fair to say that no single improved method has achieved universal applicability. Most attempts deal with compressibility effects. Modifications of the equation of state methods discussed in section 2.5, such as the FOV model (Flory *et al.* 1964) and the Sanchez and Lacombe lattice gas model (Sanchez 1982) have had considerable empirical success. Local structure and packing effects can be dealt with using modified lattice models (Dudowicz and Freed 1991) and promising attempts to adapt methods successfully to deal with small-molecule liquids are now in progress (Schweizer and Curro 1994).

Despite all its shortcomings, Flory–Huggins theory provides the universal framework for considering polymer blend problems. This is partly because if one allows χ to vary with both composition and temperature in an empirical way one is using Flory–Huggins less as a theory and more as a convenient parametrisation of the free energy. Nonetheless, even in this debased form it does express one important truth about polymer mixing, which is the decreasing importance of translational entropy of mixing with increasing polymer molecular weight.

Polymer pairs fall into three broad classes.

1. Strongly miscible polymer pairs, with χ negative over some temperature range, e.g.

PS/PVME (Shibayama *et al.* 1985). Within the temperature range in which χ is negative these mixtures are miscible for all molecular weights. Because of the specific nature of the interactions both the temperature and the composition dependence of χ may be complex, and phase diagrams may have LCSTs, UCSTs or both.

2. Very similar polymer pairs. A very small, positive value of χ may be obtained with polymers that are nearly identical, the classical example being polymers that differ only by isotopic substitution. The carbon–deuterium bond is slightly shorter than the carbon–hydrogen bond and thus is slightly less polarizable, leading to values of χ of the order 10^{-4} (Bates and Wignall 1986). Although they are technologically irrrelevant, the simplicity of these polymer mixtures makes them very important for academic studies of miscibility; in particular the close structural similarities mean that χ can be expected to be largely energetic in origin, and that the Flory–Huggins theory can be expected to have more validity.

3. Almost all other polymer pairs. In the absence of close structural similarity or specific interactions χ will usually take a positive value in the range 0.001–0.1 (say); the polymer pair will be immiscible for all but the lowest relative molecular masses.

To these three categories we can add a fourth, more complex, situation – mixtures involving random copolymers. We can distinguish two situations.

(a) Homopolymer and copolymer with one similar monomer $(A_n + A_x B_y)$. An unfavourable interaction between A and B will be diluted by the presence of A units in the copolymer, leading to a value of χ for the homopolymer and the copolymer that is smaller than that for homopolymers A and B, but still positive.

(b) Homopolymer with copolymer containing two different monomers $(A_n + B_x C_y)$. Even if interactions among A, B and C are all unfavourable, we may still have a negative χ and overall miscibility if the interaction between B and C is more unfavourable than interactions between A and B and between A and C (Kambour *et al.* 1983).

4.2 Interfaces between weakly immiscible polymers: square gradient theories

We saw in the last chapter that, under a wide variety of circumstances, polymers will not mix at a molecular level at equilibrium. If we take such a pair of polymers and mix them mechanically, we will get domains of one polymer in the other; what will be the nature of the interface between the two coexisting phases and what determines the interfacial energy? The morphology of our mixture will be greatly influenced by the interfacial energy, which will control the domain size, while the microscopic structure of this interface will determine the degree of adhesion between the phases. Thus the mechanical properties of the whole mixture will be largely controlled by properties of these interfaces.

This interface will not be atomically sharp. Details of how its width may be calculated form the subject of this section and the next one, but we can get a feel for the physics involved with a very simple argument due to Helfand (1982). The width of the interface is determined by a balance of chain entropy, favouring a wider interface, and the unfavourable energy of interaction between the two different species, which favours a narrow interface. We can estimate the balance between these effects thus. Suppose a loop of the B polymer with N_{loop} units protrudes into the A side of the interface. Associated with the loop will be an unfavourable energy that we can estimate as

$$U_{loop} \sim N_{loop}\chi kT. \tag{4.2.1}$$

At equilibrium this unfavourable energy will be of order kT, so we find

$$N_{loop}\chi = 1. \tag{4.2.2}$$

The loop does a random walk, so the size of the loop and thus the width of the interface w has the form

$$w \sim a\sqrt{N_{loop}}, \tag{4.2.3}$$

where a is the statistical step length. Thus we find

$$w \sim \frac{a}{\sqrt{\chi}} \tag{4.2.4}$$

and the associated interfacial energy can be estimated by counting the number of unfavourable contacts per unit area, each of which has associated with it an energy χkT:

$$\frac{\gamma}{kT} \sim \rho a\sqrt{\chi}, \tag{4.2.5}$$

where ρ is the density.

As we shall see, crude though it is, this estimate gives the correct functional form for both w and γ for the limit of strongly immiscible polymers of high relative molecular mass.

To predict the width and interfacial tension of polymer/polymer interfaces in more detail we need to go beyond a description of the thermodynamics of spatially uniform mixtures to include the effect on the free energy of concentration gradients. This echoes the discussion of a single-component fluid in section 2; there we found that the expression for the free energy of a homogeneous system had to be augmented by a term that did not depend solely on the local composition; the simplest suitable form was a term proportional to the square of the concentration gradient. The same, essentially phenomenological, approach can be taken for polymer blends in the limit that the concentration gradients are relatively small. However, the question of the origin and magnitude of the coefficient of the gradient term for polymer mixtures remains. In the lattice fluid model introduced in chapter 2, we recall that the coefficient of the gradient term was given by $-z'b^2\varepsilon$ (equation (2.3.18)) – it arose from the non-locality of the energetic interactions between the molecules. We shall find for polymers that the gradient term has a different origin; it is dominated by entropic effects arising from the connectivity of the chain. Essentially, steep concentration gradients impose a constraint that limits the total number of configurations available to a chain and thus cause an increase in the free energy. We might expect that, over a whole chain, the coefficient of the gradient term, which has dimensions of length squared, would then be of order the squared radius of gyration of the chain, expressing the fact that serious distortions of the chain will start to arise if we impose gradients steep compared with its unperturbed dimensions. As we usually write the free energy on a per segment basis, this would lead us to expect that the coefficient of the gradient term is proportional to a^2, the square of the statistical segment length.

A detailed derivation of the square gradient term relies on the use of the random phase approximation, which is discussed in the appendix to this chapter. By this approach we find that we can write the free energy per segment of an inhomogeneous system as

$$\frac{F}{kT} = \int [f_{\text{FH}}(\phi) + \kappa(\phi)(\nabla\phi)^2]\,\mathrm{d}r, \tag{4.2.6}$$

where $f_{\text{FH}}(\phi)$ is the Flory–Huggins free energy for a homogeneous blend and the gradient term $\kappa(\phi)$ is given by

$$\kappa(\phi) = \frac{\chi r_0^2}{6} + \frac{a^2}{36\phi(1-\phi)}, \tag{4.2.7}$$

where r_0 gives the range of the energetic interactions (Binder 1983). The first term, then, is essentially that which we derived for a lattice gas model; however, in addition to this we now have a second, concentration-dependent, term. This is the term that expresses the entropic penalty arising from restriction of chain configurations in a steep concentration gradients and we see that it is proportional to a^2, as we expected.

The range of the intermolecular intractions, r_0, must be comparable in magnitude to the statistical segment length a. Thus, as long as χ is smaller than unity, which it will be for all but the most extreme cases of incompatibility, the entropic part of the gradient term dominates. This is particularly true close to the critical point, where χ approaches its critical value of $2/N$. Now we can use calculus of variations to find a concentration profile that minimises this free energy functional, exactly as we did in chapter 2 for the liquid/vapour interface of the lattice gas model. Restricting ourselves to one dimension z, the Euler–Lagrange equation for the volume fraction profile $\phi(z)$ is

$$2\kappa(\phi)\frac{d^2\phi}{dz^2} = \frac{d}{d\phi}f_{FH}(\phi); \qquad (4.2.8)$$

this needs to be solved subject to the boundary conditions

$$\phi(z \to -\infty) = \phi_1^b, \quad \phi(z \to \infty) = \phi_2^b, \quad \left.\frac{d\phi}{dz}\right|_{(z \to \pm\infty)} = 0,$$

where ϕ_1^b and ϕ_2^b are the two coexisting compositions.

We can always evaluate the profile numerically, but close to the critical point there is an analytical solution:

$$\phi(z) = \frac{1}{2}\left[\phi_1^b + \phi_2^b + (\phi_2^b - \phi_1^b)\tanh\left(\frac{z}{w_{crit}}\right)\right], \qquad (4.2.9)$$

where the critical interfacial width w_{crit} is given by

$$w_{crit} = \frac{a\sqrt{N}}{3}\left(\frac{\chi}{\chi_{crit}} - 1\right)^{-1/2}, \qquad (4.2.10)$$

where χ_{crit} is the value of χ at the critical point. Thus, near the critical point the interfacial width is larger than the polymer end-to-end distance, which justifies the use of the gradient expansion.

Similarly we can use the analogue of equation (2.3.9) to calculate the interfacial tension γ. This turns out to be given by

$$\gamma = \frac{9}{a^2\sqrt{N}}\left(1 - \frac{\chi_{crit}}{\chi}\right)^{3/2}. \qquad (4.2.11)$$

Thus, at the critical point the interfacial tension vanishes and the interfacial width becomes indefinitely wide as the two phases merge into one.

Of course, when the system approaches very close to the critical point, we must expect to have to reconsider our mean-field approach, which neglects the increasingly large concentration fluctuations, leading to qualitative and quantitative errors (Rowlinson and Widom 1982). The interfacial width still becomes very large and the interfacial tension does go to zero, but the exponents characterising these divergences take on different, non-classical values – for example the exponent of $3/2$ in equation (4.2.11) would be replaced by a value of approximately 1.26. An estimate of how close to the critical point one needs to go for such effects to become important is provided by the Ginzburg criterion:

$$\left| \frac{\chi}{\chi_c} - 1 \right| \leq \frac{1}{N}. \tag{4.2.12}$$

For a polymer pair that would undergo phase separation at high relative molecular masses, such as an isotopic polymer mixture like polystyrene/deuterated polystyrene, this criterion means that one would have to approach closer to the critical point than about 0.05 °C for these fluctuation effects to be important. Even for much more strongly immiscible pairs, that undergo phase separation at much lower relative molecular masses, the non-classical region will be within only a few degrees of the critical point (Schwahn *et al.* 1987, Bates *et al.* 1990). This is in sharp contrast to the situation for small-molecule liquids, for which fluctuations must always be taken into account in order to predict the phase behaviour. Mean-field theories – neglecting bulk concentration fluctuations – often can be expected to work quite well for polymers.

Experimental tests of equation (4.2.10) are rare. The most extensive study has been done on blends of polystyrene and deuterated polystyrene of high relative molecular mass (Chaturvedi *et al.* 1989). In this system the prefactor of equation (4.2.10) – essentially the radius of gyration of the polymer – is rather large, which makes it possible to measure the interfacial width using ^{3}He nuclear reaction analysis. Results are shown in figure 4.8, together with the square gradient prediction. The agreement with theory is satisfactory, though the uncertainty in the critical temperature makes rigorous comparison difficult.

Of course, the near-critical regime is a rather special case and we need to be able to make predictions about the interfacial widths for much more strongly immiscible mixtures. Still in the framework of the square gradient approximation, we can evaluate the concentration profiles by a numerical integration of equation (4.2.8) even when the analytical approximation of equation (4.2.9) is not valid. If the molecular weights of the two species are equal the profiles will always be symmetrical, and we can generalise our definition of the interfacial width, given by equation (4.2.10) for the near-critical case, in terms of the gradient of the profile at its centre:

$$w = \tfrac{1}{2}(\phi_2^b - \phi_1^b) \Big/ \frac{d\phi}{dz}\Big|_{z=0}, \tag{4.2.13}$$

where $z = 0$ at $\phi = 1/2$. In figure 4.9a interfacial widths calculated thus are plotted for a selection of different relative molecular masses as a function of the inverse of the square root of χ. This plot shows clearly two types of behaviour; for decreasing χ the interfacial widths diverge as one approaches the critical point. This behaviour is strongly dependent on the relative molecular mass. However, for much larger values of χ the curves for the different

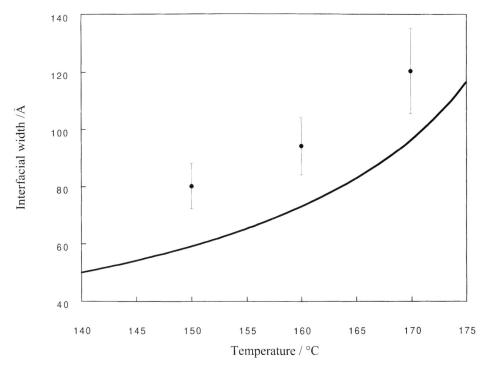

Figure 4.8. Equilibrium interfacial widths between high-relative-molecular-mass deuterated and normal polystyrene as a function of the temperature in a region of the phase diagram of partial miscibility. The interfacial profiles were analysed by nuclear reaction analysis and the solid line is the prediction of equation (4.2.10). After Chaturvedi *et al.* (1989).

relative molecular masses coincide and the interfacial width becomes simply proportional to the inverse of the square root of χ. In fact in this limit one can show that

$$w = \frac{a}{3\sqrt{\chi}} \qquad (4.2.14)$$

quite independently of relative molecular mass, exactly as suggested by the scaling argument given at the end of the last section.

The universal character of the divergence near the critical point is better seen in figure 4.9b, in which the interfacial width w, normalised by the overall chain dimension $a\sqrt{N}$, is plotted against $1/(\chi N)$. χN is sometimes referred to as the degree of incompatibility, because it is a measure of proximity to the critical point, at which it has the value 2 for a symmetrical polymer mixture. Equation (4.2.10) is a good approximation for values of χN close to criticality; interestingly, it also predicts the correct functional form for the interfacial widths far away from criticality, but the numerical coefficient is not correct.

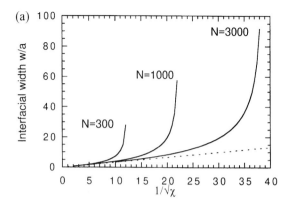

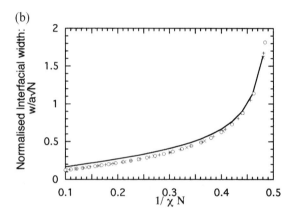

Figure 4.9. (a) Interfacial widths as a function of the interaction parameter χ calculated by square gradient theory for polymer mixtures with equal chain lengths N. The dashed line is the incompatible limit of equation (4.2.10). (b) Interfacial widths normalised by the chain dimensions $a\sqrt{N}$, plotted against the inverse degree of incompatibility $1/(\chi N)$. These are the same results as those shown in (a) (+, $N = 3000$; ○, $N = 1000$; and ×, $N = 300$); the solid line is the prediction of equation (4.2.10).

The shapes of the profiles are illustrated in figure 4.10 for two values of the degree of incompatibility, one quite close to criticality and one highly immiscible case. It can be seen that the tanh profile is in fact quite close to the directly computed profile in both cases. In fact, a more detailed look would reveal discrepancies at the edges of the profile, which would be most prominent for intermediate values of the degree of incompatibility. This is because a more detailed analysis shows that, in this intermediate regime, the profile is not governed by a single length scale (Binder and Frisch 1984). The interfacial

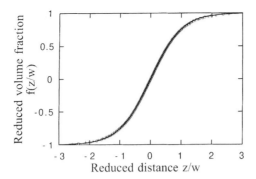

Figure 4.10. Interfacial profiles as calculated by integration of equation (4.2.8). The profiles are plotted as a reduced volume fraction $f(z) = (2\phi(z) - \phi_1^b - \phi_2^b)/(\phi_2^b - \phi_1^b)$ and the distance is plotted normalised with respect to the interfacial width w (+, $N = 10^3$ and $\chi N = 2.5$; and $\circ$, $N = 10^4$ and $\chi N = 25$). The solid line is the tanh profile.

width, measured in terms of the gradient at the centre of the profile, is independent of relative molecular mass, as expected on the basis of equation (4.2.10). However, at the edge of the profile, the way in which the composition approaches the bulk coexisting values is characterised by a length proportional to the radius of gyration of the polymer. The practical significance of this complication is not yet known.

However, we have more immediate cause to worry about the validity of this treatment. Figure 4.10 shows that the non-critical regime is characterised by interfacial widths that are already smaller than the overall size of the polymer, casting into doubt the validity of the square gradient approach, which of course was derived on the assumption that concentration gradients were small compared with the overall size of the chain. On this basis, then, we might worry that this entire treatment of the non-critical regime is invalid. More sophisticated analysis does in fact reveal that the simple square gradient treatment is quantitatively in error in this regime, but luckily its qualitative conclusions seem to be more robust.

The more correct analysis of the interface problem will require a much more sophisticated treatment of chain connectivity than our square gradient theory provides; one needs to use the self-consistent field methods outlined in the next section. However, in view of the fact that square gradient theory is so tractable and so convenient to use, it is worth asking whether there is any way in which the theory can be patched up, to extend its validity to length scales smaller than the radius of gyration. There is such a method (Roe 1986, Fredrickson 1992), though its reliability is somewhat questionable. If we use the random phase approximation (RPA) approach to deriving the square gradient term and instead of using the expansion of the RPA $S(q)$ for small q we make an expansion in the so-called intermediate scattering vector regime, valid for length scales intermediate between the statistical segment length and the radius of gyration, we find a gradient term identical in form to the long-wavelength expression, but with a different coefficient:

$$\kappa(\phi) = \frac{a^2}{24\phi(1-\phi)}. \tag{4.2.15}$$

Equation (4.2.6) with $\kappa(\phi)$ given by equation (4.2.15) is sometimes called the Roe functional. It follows that we can re-use all the results we have obtained so far by simply changing the numerical coefficients. In particular, we find for the interfacial width for the case in which χN is large

$$w = \frac{a}{2\sqrt{\chi}}, \tag{4.2.16}$$

which we simply obtain by multiplying equation (4.2.10) by $3/2$ (i.e. $36/24$). In fact, we shall see in the next section that the coefficient obtained from the much more rigorous self-consistent field theory is $1/\sqrt{6}$. The Roe functional thus slightly overestimates the free energy cost of sharp composition gradients. There is not much that can be done to improve the situation further in a simple way, because, as one can see from figure 4.10 and our discussion of it, a single profile will have regions where the composition profile has different gradients, so no single gradient coefficient can be valid across the whole profile. We should accept that square gradient theory, although it is a convenient way of getting qualitative results, cannot be relied on to be quantitatively accurate. The difference in results obtained with the two bounding cases will give some feel for the size of the errors thus introduced.

4.3 Self-consistent field theory

The physics underlying the square gradient treatments of inhomogeneous systems is that the imposition of composition gradients reduces the number of configurations available to the affected polymer chains and thus affects the free energy. This basic idea is undoubtedly correct, but the square gradient method treats it only at the lowest level of approximation. To obtain more accurate results, particularly for situations in which composition gradients are steep on the scale of the polymer radius of gyration, we need to keep track of the polymer configurations in more detail. The self-consistent field method provides a way of doing this.

In the mid-1960s Edwards pointed out a remarkable analogy between the problem of interacting polymers and the classical problem of interacting electrons (Edwards 1965). Methods that had been used to find approximate solutions to the problem of finding the wavefunction of an electron in the presence of many other interacting electrons, such as the Hartree method for solving the electronic structure of multi-electron atoms, could thus be adapted to deal with the problem of the conformation of a polymer chain that was surrounded by and interacting with many other polymer chains. These methods were particularly successful when applied to interfaces, first by Helfand (Helfand and Tagami 1971) and Edwards and Dolan (Dolan and Edwards 1974) and then by many others. In the following section we introduce the basic ideas of self-consistent field theory without any pretence at rigour.

We know that an ideal polymer chain is a random walk. Consider a distribution function $q(r, r', t)$, which tells us the probability that a chain that is t steps long has started at a position r and finished at a position r'. We know that a random walker in free space has a distribution function that obeys a diffusion equation, so for an ideal, isolated chain (with no excluded volume interaction) we can write

$$\frac{\partial q}{\partial t} = \frac{a^2}{6} \nabla^2 q(r, r', t). \tag{4.3.1}$$

The 'diffusion coefficient' is simply related to the square of the statistical step length a.

Now suppose the random walk is not in free space, but is affected by a spatially varying potential $U(r)$. This now makes the statistical weights of various possible steps in the random walk unequal via a Boltzmann factor and the effect of this is to modify the diffusion equation thus:

$$\frac{\partial q}{\partial t} = \frac{a^2}{6} \nabla^2 q(r, r', t) - \frac{U(r)}{kT} q(r, r', t). \tag{4.3.2}$$

This is the basic equation of the self-consistent field theory; the reader may notice that it has the same form as Schrödinger's equation. This is the basis of the analogy between polymer chain conformations in interacting systems and the properties of interacting electrons.

In an interacting system, each polymer chain will have a distribution function obeying equation (4.3.2), each with its own function $U_i(r)$ whose form can be obtained if the positions of all the other polymer chains are known. The resulting system of coupled differential equations, one for each chain in the system, is obviously completely intractable as it stands. However, we can make progress if we make a mean-field assumption – we assume that every polymer chain of the same chemical type experiences an average, mean-field potential $U(r)$. This potential has two parts. The first part arises simply from the hard core potential which prevents two segments occupying the same volume; in effect this part of the potential simply ensures that the polymer melt is incompressible. The second part arises from the net chemical attraction or repulsion between monomers that is expressed in the Flory–Huggins inter-action parameter χ. This second part of the potential, then, we can express at a mean-field level as being simply proportional to the average volume fraction of the chemically different species.

Let us see how this works for the problem of the interface between two immiscible polymers, A and B. For an interface with planar symmetry we need to consider only one spatial dimension, z, so the potentials felt by each species are

$$\frac{U_A(z)}{kT} = \chi\phi_B(z) + w(z), \tag{4.3.3a}$$

$$\frac{U_B(z)}{kT} = \chi\phi_A(z) + w(z), \tag{4.3.3b}$$

where the function $w(z)$ has to be chosen in order to ensure that the incompressibility condition is satisfied for all values of z:

$$\phi_A(z) + \phi_B(z) = 1. \tag{4.3.4}$$

The corresponding equations for the distribution functions are

$$\frac{\partial q_A(z, t)}{\partial t} = \frac{a^2}{6}\frac{\partial^2 q_A(z, t)}{\partial z^2} - \frac{U_A(z)}{kT}q_A(z, t), \tag{4.3.5a}$$

$$\frac{\partial q_B(z, t)}{\partial t} = \frac{a^2}{6}\frac{\partial^2 q_B(z, t)}{\partial z^2} - \frac{U_B(z)}{kT}q_B(z, t). \tag{4.3.5b}$$

Of course, the volume fraction profiles $\phi_A(z)$ and $\phi_B(z)$ that appear in equations (4.3.5) via the definitions of the potentials $U_A(z)$ and $U_B(z)$ in equations (4.3.3) are the very quantities we are trying to determine. But $\phi_A(z)$ and $\phi_B(z)$ are also related to the distribution functions $q_A(z, t)$ and $q_B(z, t)$ as follows:

$$\phi_A(z) = \frac{1}{N}\int_0^N dt q_A(z, N - t)q_A(z, t), \tag{4.3.6a}$$

$$\phi_B(z) = \frac{1}{N}\int_0^N dt q_B(z, N - t)q_B(z, t), \tag{4.3.6b}$$

where the integration is over all values of the number of steps along the chain t up to the total chain length N. The origins of this expression are illustrated in figure 4.11.

Equations (4.3.3), (4.3.5) and (4.3.6) now form a closed set; our aim is to find functions q_A and q_B that lead to self-consistent $\phi_A(z)$ and $\phi_B(z)$. In general we can use the following approach; guess trial functions for $\phi_A(z)$ and $\phi_B(z)$, solve (4.3.5) for $q_A(z)$ and $q_B(z)$ using (4.3.3) with the trial functions and then

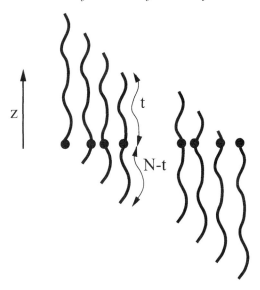

Figure 4.11. All A chains with a segment at position z situated anywhere along the length of the chain contribute to the volume fraction of A chains at z via equation (4.3.6b).

use (4.3.6) to calculate refined estimates of $\phi_A(z)$ and $\phi_B(z)$. The whole process can then be repeated until self-consistency has been achieved, at which point one has solved for the profiles and one can obtain the interfacial tension by an integration of the excess potential over the profile.

Usually this process needs the use of numerical methods to solve the partial differential equations (4.3.5). However, there is one important case in which an analytical solution is possible; this is for an interface between two highly immiscible polymers in the limit of infinite relative molecular mass, and was derived by Helfand (Helfand and Tagami 1971).

In the limit of infinite relative molecular mass, all chain segments are equivalent, insofar as none are influenced by the proximity of chain ends. Thus equations (4.3.6) are replaced by the much simpler relations

$$\phi_A(z) = q_A(z)^2, \tag{4.3.7a}$$

$$\phi_B(z) = q_B(z)^2 \tag{4.3.7b}$$

and, in the absence of any dependence on t, equations (4.3.5) become coupled ordinary differential equations

$$\frac{a^2}{6}\frac{d^2 q_A}{dz^2} = \chi q_B^2 q_A + w(z)q_A, \tag{4.3.8a}$$

$$\frac{a^2}{6}\frac{d^2 q_B}{dz^2} = \chi q_A^2 q_B + w(z)q_B, \tag{4.3.8b}$$

where we have already used equations (4.3.3) and (4.3.7).

In the limit of infinite relative molecular mass, the coexisting phases are the pure components, so we need a solution to (4.3.8) satisfying the boundary conditions

$$\phi_A(z \to -\infty) = 0, \quad \phi_A(z \to +\infty) = 1. \tag{4.3.9}$$

The reader may readily verify that the solution is

$$\phi_A(z) = \frac{1}{2}\left[1 + \tanh\left(\frac{z}{w_I}\right)\right], \tag{4.3.10}$$

where the interfacial width w_I is given by

$$w_I = \frac{a}{(6\chi)^{1/2}}. \tag{4.3.11}$$

The form of this volume fraction profile is illustrated in figure 4.12.

The interfacial tension can also be evaluated and is

$$\frac{\gamma}{kT} = \rho a \left(\frac{\chi}{6}\right)^{1/2}. \tag{4.3.12}$$

Notice that these expressions have the same functional form as the estimates that arise from the simple scaling argument presented in section 4.3.1.

These results are obtained under the assumption that the statistical segment lengths of the two polymers are identical, which is unrealistic in general. This approximation can be lifted (Helfand and Sapse 1975) to yield a more complex, but still tractable, expression for the interfacial tension:

$$\frac{\gamma}{kT} = [\chi(\rho_A\rho_B)^{1/2}]^{1/2}\left(\frac{\beta_A + \beta_B}{2} + \frac{1}{6}\frac{(\beta_A - \beta_B)^2}{\beta_A + \beta_B}\right), \tag{4.3.13}$$

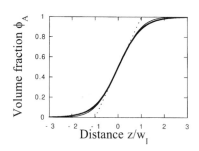

Figure 4.12. The interfacial profile predicted by self-consistent field theory for an interface between highly incompatible polymers of high relative molecular masses. The bold line is the self-consistent field solution of equation (4.3.10), whereas the finer line is the function $\phi_A(z) = 0.5\{1 + \mathrm{Erf}\,[z\sqrt{\pi}/(2w_I)]\}$, which has the same gradient at $z = 0$ as that of the tanh profile.

where ρ_i is the density of the polymer i and the parameter β_i is defined in terms of the statistical segment length of polymer a_i by

$$\beta_i = (\tfrac{1}{6}\rho_i a_i^2)^{1/2}. \tag{4.3.14}$$

The interfacial profile is given by a much more complex expression than equation (4.3.10), but qualitatively it retains the same form and is characterised by an interfacial width w_I given by

$$w_I = \left(\frac{\beta_A^2 + \beta_B^2}{2\chi(\rho_A\rho_B)^{1/2}}\right)^{1/2}, \tag{4.3.15}$$

which reduces to equation (4.3.11) when the densities and statistical segment lengths of the two polymers are equal.

The self-consistent field result is obtained in the approximation of an infinite relative molecular mass, but of course it is important to know how this result is changed for a real system with a finite relative molecular mass. Unfortunately this calculation is rather hard to do. The square gradient method predicts that the first correction to the limit of infinite relative molecular mass is inversely proportional to χN (Broseta *et al.* 1990), but as we have seen the coefficient one obtains by this method is unreliable. Tang and Freed (1991) have given interpolation formulae that have been compared with the results of numerical self-consistent field calculations and show good agreement over a wide range of parameters. This gives for the interfacial width

$$w_I(N) = \frac{a}{(6\chi)^{1/2}}\left[\frac{3}{4}\left(1 - \frac{2}{\chi N}\right) + \frac{1}{4}\left(1 - \frac{2}{\chi N}\right)^2\right]^{-1/2} \tag{4.3.16}$$

and for the interfacial tension

$$\frac{\gamma(N)}{kT} = \rho a \left(\frac{\chi}{6}\right)^{1/2}\left(1 - \frac{1.8}{\chi N} - \frac{0.4}{(\chi N)^2}\right)^{3/2}. \tag{4.3.17}$$

If we take typical values of χ and substitute into equations (4.3.11) and (4.3.12) we will typically find that we would expect the interfacial width to be a few tens of ångström units wide and the interfacial tension a few mJ m^{-2}. Experimental results are relatively sparse; these rather low values of interfacial tension are very difficult to measure in polymer melts with their very high viscosities even for relatively modest molecular weights. Interfacial widths of these thicknesses are only practicably measurable by neutron reflectivity.

Anastasiadis *et al.* (1988) performed an extensive experimental study of three polymer pairs, namely polystyrene and hydrogenated polybutadiene, polybutadiene and poly(dimethyl siloxane), and polystyrene and poly(methyl methacrylate). Concentrating on this last pair, they found that their data obeyed an empirical law of the form

$$\gamma = C_1 + C_2 M_n^{-z}, \tag{4.3.18}$$

where M_n is the number-average relative molecular mass, C_1 and C_2 are empirical constants and the exponent z was found to be 0.9 (see figure 4.13).

Experiments in which interfacial widths were directly measured have been summarised by Stamm and Schubert (1995). The most studied pair has been polystyrene and poly(methyl methacrylate), whose interfacial width has been measured directly using neutron reflectivity. The initial experiments were done by two groups, Anastasiadis, Russell and co-workers (Anastasiadis *et al.* 1990) and Fernandez, Higgins and co-workers (Fernandez *et al.* 1988); both obtained the virtually identical result that the interface was described by a tanh profile with a width of 25 Å[§]. This result is consistent with the interfacial tension measurement, but the prediction of equation (4.3.11) using an independently measured value of χ (Russell *et al.* 1990) is for a width of 14.9 Å.

One plausible explanation for this discrepancy is that the bare interface whose structure is predicted by self-consistent field theory is broadened by thermally excited capillary waves (Shull *et al.* 1993, Semenov 1994). The

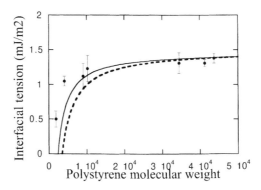

Figure 4.13. The interfacial tension between polystyrene and poly(methyl methacrylate). The relative molecular mass of the PMMA was 10 000 and the temperature of measurement was 199 °C. The solid line is the prediction of the Broseta equation (Broseta *et al.* 1990) with $\chi = 0.045$. The dashed line is the prediction of the Broseta equation with the coefficient changed from $\pi^2/6$ to 1.35, to bring it into line with the Freed interpolation formula (Tang and Freed 1991), equation (4.3.17), with $\chi = 0.055$. Data from Anastasiadis *et al.* (1988).

[§] Note that Fernandez *et al.* fitted their results to an error function profile, so their reported interfacial width of 20 Å needs to be multiplied by $\sqrt{2\pi}$ to be compared with the width of the tanh function used by Russell's group.

argument is identical to that used to deduce the form of the capillary wave roughening of the polymer melt surface discussed in section 2.6. Such thermally excited capillary waves will be present at a polymer/polymer interface, even far away from the critical point, and such capillary waves may make a significant contribution to the measured interfacial width. This may well be important for polymers, for their interfacial energy is rather small. Moreover, the apparent width of a capillary-wave-roughened interface will depend on the length scale over which the interface will be averaged by the measurement (see figure 4.14), which will differ according to the technique used. For example, an ion-beam or SIMS experiment will produce a lateral average over a macroscopic distance of order millimetres, whereas for a reflection technique, such as neutron reflection or ellipsometry, the distance over which the averaging is done is usually set by the coherence length of the radiation in the plane of the interface. For a neutron reflectivity experiment, for example, this is typically around 20 μm.

Using this kind of estimate to supply the value of the lower cut-off wavevector in equation (2.6.4), together with the interfacial energy predicted by equation (4.3.12), allows one to account very plausibly for the observed discrepancy between experimental and observed interfacial thicknesses. Strong evidence for the validity of this approach is provided by experiments in which the interfacial width between a thin polymer film and a polymer substrate is measured (Sferrazza *et al.* 1997). In these circumstances, dispersion forces acting across the thin films are important. They lead to a long-wavelength cut-off to the capillary waves that depends on the film thickness. We can define a dispersive capillary length l_{dis} in analogy to the gravitational capillary length of equation (2.6.5);

$$l_{\mathrm{dis}} = \left(\frac{4\pi\gamma d^4}{A}\right)^{1/2},\qquad(4.3.19)$$

where A is the Hamaker constant, γ is the interfacial tension and d is the film thickness. In combination with equation (2.6.4), this leads to the prediction

Figure 4.14. (a) An interface without capillary waves, with some intrinsic interface width w_{I}. In (b) the interface is roughened by thermally excited capillary waves. A technique that averages over a distance L will measure a larger apparent interfacial width $w(L)$.

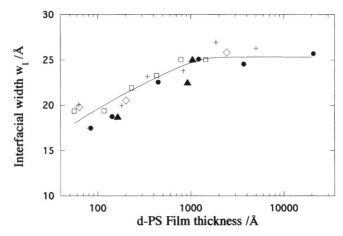

Figure 4.15. The interfacial width between a thin deuterated polystyrene film and poly(methyl methacrylate), measured by neutron reflectometry. The d-PS relative molecular masses were 244 000 (+), 387 000 (□), 641 000 (●) and 400 000 (◇ and ▲); and the PMMA relative molecular masses were 365 000 (+, □, ● and ◇) and 333 000 (▲). After Sferrazza *et al.* (1997).

that the interfacial width should vary logarithmically with the film thickness. Figure 4.15 shows that this is exactly what is found for film thicknesses below about 2000 Å; above this value the dispersive capillary length exceeds the coherence length of the neutrons. These measurements are consistent with a bare interfacial width of 11.7 ± 1.8 Å, compared with the prediction of equation (4.3.11) of 14.9 Å. Thus we can conclude that combining Helfand's mean-field theory of polymer interfaces with a capillary wave correction yields quite good predictions of interfacial widths between immiscible polymers.

4.4 Kinetics of formation of polymer/polymer interfaces

Up to now we have only discussed the structure of interfaces that have reached equilibrium. However, in many processes only a finite amount of time is available for an interface to be formed. For example, when a pair of polymers is coextruded, the time available for the interface to develop is limited to the time at the process temperature. Alternatively, if one polymer is applied as a solution to coat a second polymer, then molecular motion of the polymers near the interface will often only be possible before all the solvent has evaporated. This question of kinetics becomes particularly important when we come to consider interfaces between miscible polymers; here there is no equilibrium interface width at all and the width of the interface that is achieved in practice is solely determined by the kinetics of the interface formation and the time available.

When two polymers that are thermodynamically miscible are brought together the initially sharp interface between them broadens due to the mutual diffusion of the two types of polymer chain (figure 4.16a). But the process is very much more complicated than the simple prototypical diffusion experiment sketched in figure 4.16b. The chains themselves are complicated objects that strongly interact with many neighbours by being tangled up with them. Thus their motion is much more complicated than the simple Brownian motion of the dye molecules and requires theories such as reptation to describe it. In the dye diffusion case the environment through which the dye molecules move is

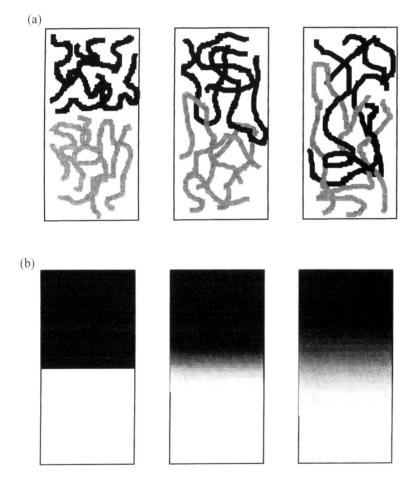

Figure 4.16. (a) Mutual diffusion between two polymers. (b) Diffusion of a dye molecule in dilute solution, initially confined to one side of a container.

essentially homogeneous, both in terms of its dynamic properties (i.e. the viscosity of the solvent) and in terms of its local thermodynamic properties – if the dye is dilute enough we do not need to worry about interactions of dye molecules with each other. In contrast, the polymer interface is highly inhomogeneous; the local mobility in the melt may be a strong function of the local composition and the thermodynamic interactions between neighbouring segments will certainly have to be taken into account. Finally, for our dye molecules, there is only one important length scale, the size of the molecules. As soon as the interface has become broad compared with molecular sizes, we can be confident about applying continuum approximations such as Fick's law. In the polymer case there is a hierarchy of length scales that need to be considered; from the segment size, to the tube diameter, to the radius of gyration. Different treatments need to be applied to each regime, with continuum concepts such as Fick's law being valid only above the largest length scale. By the time the interface has broadened to this degree many interesting phenomena, such as the development of strength, may have already come to fruition. We are in a position to say something useful about all aspects of the problem of polymer interface formation, but it is probably fair to say that no satisfactory unified treatment of the problem yet exists.

The motion of a single labelled polymer chain in a matrix of identical chains has been extensively studied both experimentally and theoretically and is well understood. For diffusion distances large compared with the radius of gyration Fick's law, with a constant diffusion coefficient, holds (see figure 4.17). Experimentally one can thus establish a sharp concentration gradient of a suitably labelled chain and then use a suitable depth-profiling technique to follow the broadening of the interface with time. This broadening can then be fitted with an appropriate solution of Fick's law to extract the diffusion coefficient. Figure 4.18 illustrates this for a forward recoil experiment. In this case deuterium has been used as the label; care is required whenever such a label is used to ensure that it does not perturb the measurement. As we saw in section 4.1, there is an

Figure 4.17. Self-diffusion of a single labelled chain in a chemically identical matrix. In time t the centre of mass of the chain has moved a distance x; when x is large compared with the chain dimensions the mean squared distance x is related to the time t via the self-diffusion coefficient D: $D = \langle x^2 \rangle / t$.

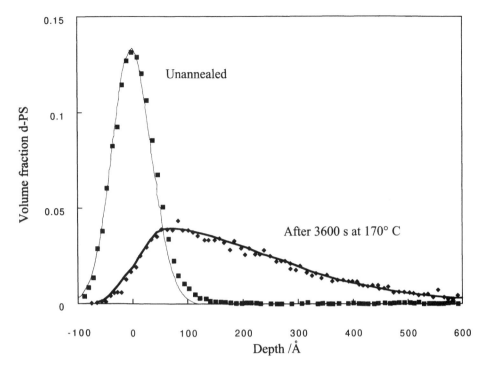

Figure 4.18. Diffusion of deuterated polystyrene (d-PS) in normal polystyrene (h-PS), measured by forward recoil spectrometry (FReS). A 10–20 nm film of d-PS of relative molecular mass 225 000 was floated onto a 2 μm film of h-PS; analysis of the deuterium depth profile shows that the d-PS layer was localised, to within the resolution of the technique, at the surface. After annealing at 170 °C for 3600 s FReS revealed that substantial diffusion into the film had occurred; the solid line is a fit to the solution of Fick's equation assuming that $D = 8 \times 10^{-14}$ cm^2 s^{-1}. After Mills *et al.* (1984).

unfavourable thermodynamic interaction between polystyrene and its deuterated analogue, which, as we shall shortly see, would cause a reduction in the measured diffusion coefficient compared with the true self-diffusion coefficient at high relative molecular masses.

Self-diffusion coefficients have now been measured for a variety of polymers, by a number of different techniques, as a function both of temperature and of relative molecular masses. For linear polymers the following picture emerges (Tirrell 1984).

As a function of the degree of polymerisation N there are two regimes; at low relative molecular mass the diffusion coefficient approaches an inverse dependence on N, whereas above a critical degree of polymerisation N_c (at which degree of polymerisation the dependence of the zero-shear viscosity η

changes from $\eta \propto N$ to $\eta \propto N^{3.4}$), the diffusion coefficient obeys a law of the form $D_{self} \propto N^{-2}$. At a given degree of polymerisation the temperature dependence of the diffusion coefficient does not follow a simple activated form, except at temperatures well above the glass transition temperature (in practice this means more than about 100 °C above the glass transition). Instead it follows a law of the so-called Vogel–Fulcher form (or equivalently a law described by the Williams–Landel–Ferry equation);

$$D_{self} = D_0 \exp\left(-\frac{B}{T - T_0}\right). \qquad (4.4.1)$$

B and T_0 are constants for a given polymer, which are independent of the degree of polymerisation for all but the shortest chains. T_0 is related to the glass transition temperature T_g, typically being about 50 °C below T_g. The independence of these parameters from the relative molecular mass means that we can factorise the effects of temperature and of the degree of polymerisation on the diffusion coefficient, writing, for high relative molecular mass,

$$D_{self} = D_{mono} \frac{N_e}{N^2}, \qquad (4.4.2)$$

where D_{mono} is an effective segment diffusion coefficient whose temperature dependence is given by

$$D_{mono} = D^0_{mono} \exp\left(-\frac{B}{T - T_0}\right). \qquad (4.4.3)$$

The dependence of the self-diffusion coefficient on the relative molecular mass is accounted for by the theory of reptation (Doi and Edwards 1986). Briefly, in this the constraining effect of the surrounding chains on the diffusion chain is to confine the motion of the latter to a 'tube' (figure 4.19); lateral excursions of the chain are only possible up to a certain distance that is set by the tube diameter, which is a material constant for a given polymer. To achieve motion of its centre of mass the chain has to wriggle along the tube; this motion is called reptation. The theory permits the description not only of the self-diffusion coefficient but also of the viscoelastic properties of a polymer melt in terms of two parameters; one is a microscopic friction factor, or alternatively an effective segment diffusion coefficient, and the other is the tube diameter, given by $a\sqrt{N_e}$. The parameter N_e is usually referred to as the number of monomer units between entanglement points, in reference to earlier theories of polymer viscoelasticity. It is a material constant, conveniently extracted from the plateau modulus of a polymer melt, which is related by a simple numerical factor to the critical degree of polymerisation N_c which marks the onset of entangled behaviour.

Chains shorter than the critical degree of polymerisation N_c do not move by

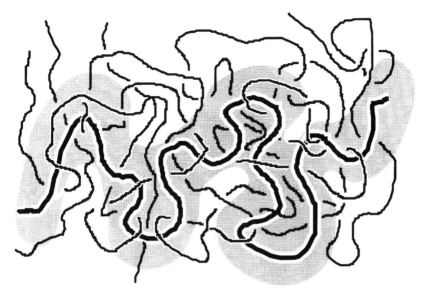

Figure 4.19. The tube model. The effect of the surrounding chains on any test chain (shown bold) is to confine its motion to a 'tube' (shown shaded). The test chain can move only by wriggling along the tube – this motion is called 'reptation'.

reptation; their motion is better described by the Rouse theory, which is based on an analysis of the overdamped normal modes of a Gaussian chain in a viscous medium. Reptation is also not possible for branched polymers; for such a molecule motion of the centre of mass is only possible by mechanisms that involve the withdrawal of an arm completely from its tube or the relaxation of the tube constraints. Space does not permit a detailed discussion of these interesting mechanisms; suffice it to say that such motion will be very slow compared with the reptative motion of a linear chain of similar relative molecular mass. Introducing branching into a molecule is an extremely effective way of suppressing the motion of its centre of mass (Klein 1986, Shull *et al.* 1990).

Self-diffusion coefficients describe the long-range motion of a single molecule moving in a matrix of identical molecules that is spatially homogeneous. When we create an interface between two different polymers then necessarily the region across the interface is inhomogeneous, because a gradient of polymer concentration is developed. Diffusion in the presence of gradients of chemical composition should be clearly distinguished from self-diffusion; to make this distinction we refer to *mutual diffusion* – see figure 4.20. We can clarify the distinction between self-diffusion and mutual diffusion by considering a very idealised case. Imagine a pair of polymers with identical degrees of

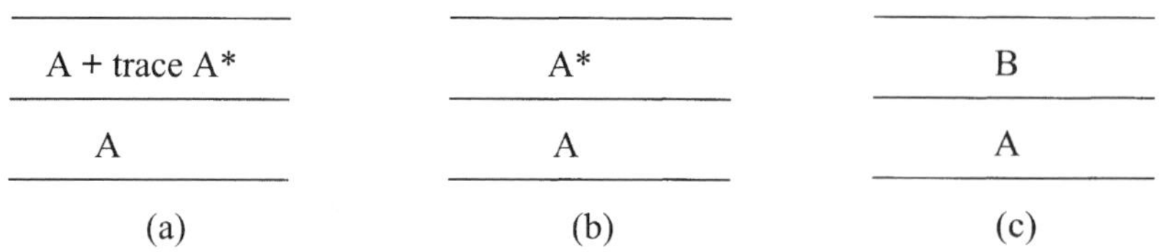

A + trace A*	A*	B
A	A	A
(a)	(b)	(c)

Figure 4.20. Experimental configurations for measuring self-diffusion and mutual diffusion coefficients. In (a) we make an interface between polymer A containing a trace amount of labelled, but chemically identical, A* and pure polymer A. Broadening of this interface is controlled by the self-diffusion coefficient. The self-diffusion coefficient is not dependent on the concentration, so, if the label is truly non-perturbing, situation (b), in which we have an interface between pure labelled A* and pure A, also measures the self-diffusion coefficient. However, if the two polymers making the interface are chemically different (c), then what controls broadening of the interface is a mutual diffusion coefficient.

polymerisation and identical mobility properties, between which we can turn on a chemical interaction. Let us further assume that the thermodynamics is well described by the Flory–Huggins model with a non-zero value of χ.

We need to recall that the fundamental quantity driving diffusion is not the concentration gradient, as assumed in Fick's empirical law, but rather the chemical potential gradient. Thus in linear irreversible thermodynamics the flux of material is taken to be proportional to the gradient of the chemical potential:

$$J = L\frac{d\mu}{dx}, \tag{4.4.4}$$

where L is the Onsager coefficient, which contains all the transport information about our diffusant (Tyrell and Harris 1984). If we assume that the blend is incompressible then μ is the exchange chemical potential; because the chains have identical mobility properties we can think of the broadening of the interface as an exchange of chains from one side to the other.

On evaluating the exchange chemical potential per chain from Flory–Huggins theory we find that

$$J = L\left(\frac{1}{\phi(1-\phi)} - 2\chi N\right)\frac{d\phi}{dx}. \tag{4.4.5}$$

If we imagine turning off the chemical interaction by setting $\chi = 0$ we recover the situation in which we have the broadening of an interface between chemically identical polymers; this is described by Fick's law with the self-diffusion coefficient. Thus we find the relationship between the Onsager coefficient L and the self-diffusion coefficient D_{self}:

$$L = D_{self}\phi(1-\phi). \tag{4.4.6}$$

Now turning our chemical interaction back on, we find the value of the mutual diffusion coefficient which controls the formation of the interface between the two species:

$$D_{\mathrm{mutual}} = D_{\mathrm{self}}[1 - 2\chi N\phi(1 - \phi)]. \qquad (4.4.7)$$

This relation has a number of interesting features. Firstly, we notice that, in addition to the information about polymer dynamics contained in the self-diffusion coefficient, it has a factor dependent on thermodynamics. This is a fundamental point that should be stressed – the kinetics of formation of interfaces between chemically different polymers is controlled by thermodynamic as well as kinetic factors.

Secondly, we can note the direction of the effect; if the interaction is unfavourable and the polymers are tending towards demixing, with a positive value of χ, the diffusion coefficient is reduced – this effect is known as thermodynamic slowing down. Conversely, if the polymers find it energetically favourable to mix, with a negative value of χ, the diffusion coefficient is increased by the interaction by a factor which, because of the presence of a factor of N, may be quite significant.

Thirdly, we note that the diffusion coefficient has a composition dependence, which has practical implications when we predict composition profiles across interfaces; we now need to use the non-linear version of Fick's second law to predict the shape of interfaces:

$$\frac{\partial \phi}{\partial t} = \frac{\partial}{\partial x}\left(D(\phi)\frac{\partial \phi}{\partial x}\right); \qquad (4.4.8)$$

the familiar solutions and methods of solution for the linear version of Fick's equation are no longer applicable and rather strange-looking interfacial composition profiles may result.

Although the idea of dynamically similar but thermodynamically different polymers seems at first to be a rather contrived and artifical model, there is one class of systems that are actually quite close realisations of it. These are pairs of isotopically substituted polymers of high relative molecular mass, approaching those relative molecular masses for which phase separation would take place. The composition dependence of the mutual diffusion coefficient for such diffusion couples has been measured directly for interfaces between polystyrene and deuterated polystyrene (Green and Doyle 1987) and the results are in excellent agreement with theory (figure 4.21). This also means that we can use equation (4.4.7) to check whether isotopic labelling is likely to perturb an attempted self-diffusion measurement.

Moving beyond isotopic blend systems, we find that most polymer pairs have markedly different dynamic properties, so we cannot hope to describe these

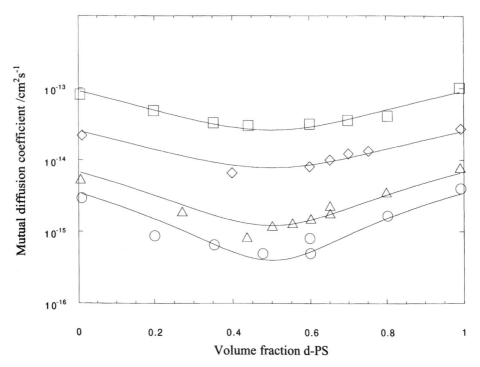

Figure 4.21. Mutual diffusion coefficients as functions of the concentration for blends of deuterated polystyrene ($N = 9.8 \times 10^3$) and normal polystyrene ($N = 8.7 \times 10^3$). The diffusion temperatures were 166 °C (○), 174 °C (△), 190 °C (◇) and 205 °C (□). The solid lines are the predictions of equation (4.4.11). The decrease in diffusion coefficient for volume fractions around a half is a direct result of the unfavourable thermodynamics of mixing – 'thermodynamic slowing down'. After Green and Doyle (1987).

properties in terms of a single self-diffusion coefficient. Not only will each chain have different diffusional properties, but also in each case the diffusional properties will depend on the environment. To make this clearer, we introduce another type of diffusion coefficient – the 'tracer' diffusion coefficient. This is defined by figure 4.22; imagine a diffusion couple in which the concentration of the second component (here labelled B) is constant, whereas a gradient between A and its labelled version A^* is established. Now one has diffusion of A without any chemical gradients – the diffusion coefficient thus defined will have no dependence on thermodynamics. A tracer diffusion coefficient for B can be similarly defined; each of these coefficients will be a function of composition.

The origin of this composition dependence is not, as yet, completely understood. Within the framework of the reptation theory we can expect that three

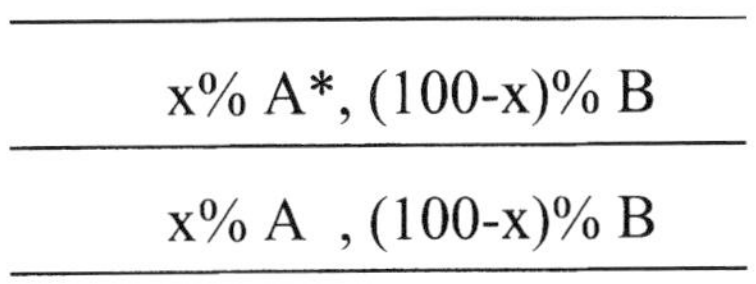

x% A*, (100-x)% B
x% A , (100-x)% B

Figure 4.22. A diffusion couple for measuring a tracer diffusion coefficient.

effects might be at work: the effective tube diameter or relative molecular mass for entanglement might be a function of composition; additional, non-reptative modes of diffusion may become important; and, possibly most importantly, the segmental friction factor may depend strongly on composition.

The composition dependence of the relative molecular mass for entanglement has been studied in the context of viscoelasticity; for example polystyrene (PS) and poly(xylenyl ether) (PXE) are fully miscible but have very different values of M_e; 19 100 for PS and 4300 for PXE. Measurements of the plateau modulus of mixtures suggested that a simple harmonic mean blending law was followed (Prest and Porter 1972).

Reptation assumes that the mobility of the matrix polymer plays no role in the relaxation of the tube constraint felt by a test molecule. However, if the matrix chains were very much more mobile than the test chain, additional lateral motion of the chain might be permitted by virtue of the constraining chains themselves moving away. This type of motion is called 'constraint release' or 'tube renewal', and may be operative if some of the matrix chains are significantly shorter or intrinsically more mobile than the test chain (Green 1991, Composto *et al.* 1992).

However, these two effects are likely to be relatively small compared with the third effect, the variation of the segment friction factor with composition. An appealing (but inadequate) *Ansatz* relates changes in the local friction factor with blend composition to changes in the blend glass transition temperature. If we use the Williams–Landel–Ferry form for the temperature dependence of the friction factor

$$\mu(T) = a_T \mu(T_0), \quad \log_{10} a_T = \frac{c_1^0 (T - T_0)}{c_2^0 + (T - T_0)} \tag{4.4.9}$$

and we take as the reference temperature T_0 the glass transition temperature, it turns out that the constants c_1^0 and c_2^0 are rather similar for a variety of different polymers (Ferry 1980). For a mixture of two miscible polymers, the glass transition temperature smoothly interpolates between the glass transition temperatures of the pure components, which suggests that we could estimate the composition dependence of the friction factor by using equation (4.4.9) with c_1^0 and c_2^0 constant, but with the reference temperature T_0 the composition-dependent glass transition temperature of the blend, varying with composition according to one of the many semi-empirical mixing laws that have been proposed, of which the simplest is (Fox 1956)

$$\frac{1}{T_g(\phi)} = \frac{\phi}{T_g^1} + \frac{1 - \phi}{T_g^2},\qquad(4.4.10)$$

where T_g^1 and T_g^2 are the glass transition temperatures of the pure materials.

This type of approach predicts order-of-magnitude variations of the segment friction factor with composition in blends whose components vary in glass transition temperature by 100 °C or so, a not unusual situation. A corollary of this approach would be the prediction that segment friction factors should be very weakly dependent on composition if measurements are made at a constant temperature increment above the blend's glass transition temperature. Experiments have, however, revealed that this is far from being the case (Composto *et al.* 1988, Kim *et al.* 1995). This confirms the suspicion that is aroused by other types of experiment, such as rheology (Zawada *et al.* 1992) and NMR (Menestrel *et al.* 1992), that the problem of segment-level dynamics in polymer blends is indeed rather complex and at present not fully understood.

Finally, if we have measured or estimated both tracer diffusion coefficients as a function of composition and we know the thermodynamic factor for our mixture, we can now ask whether we are in a position to calculate the kinetics of interfacial broadening that occurs when two pure polymers are placed in contact and annealed. To do this it is sufficient to know the mutual diffusion coefficient as a function of composition and use this to solve the non-linear version of Fick's second law, equation (4.4.8).

Our aim, then, is to take equation (4.4.7), in which the mutual diffusion coefficient is expressed as a product of thermodynamic and kinetic factors, but to replace the self-diffusion coefficient by some suitable weighted average of the concentration-dependent tracer diffusion coefficients of the two species. In fact it is by no means clear on the most fundamental grounds that we can expect to be able to do this at all; in order to produce such an expression some additional physical assumptions have to be made, either explicitly or implicitly, and the choice of these assumptions has in the past generated some controversy.

The root of the problem is illustrated in figure 4.23; imagine mutual diffusion taking place between two polymers, one of which is very much more mobile than the other (shown in figure 4.23 as the shorter chain). The tracer diffusion coefficient of the short chain is much larger than that of the long chains, so initially one might expect a build-up of total density on the left-hand side of the interface. One can now imagine two possibilities; this density build-up may

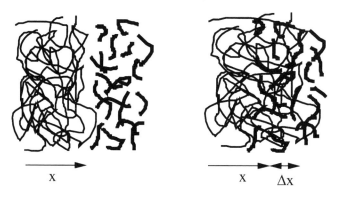

Figure 4.23. Mutual diffusion between two polymers of very differing mobilities. The less mobile polymer relaxes by bulk flow, causing the interface position to move Δx towards the side rich in the more mobile polymer.

prevent any further diffusion of the mobile chains until the slower chains had had a chance themselves to diffuse across the boundary. In this case we would expect the mutual diffusion coefficient to be dominated by the tracer diffusion coefficient of the less mobile species. On the other hand, it may be possible for the less mobile chains to relax in a non-diffusive way, resulting in a bulk flow of polymer that would move the whole interface towards the more mobile chains. If such bulk flow is possible then the mutual diffusion coefficient will be controlled by the tracer diffusion coefficient of the more mobile chains; indeed even if the less mobile chains were crosslinked so their tracer diffusion coefficient was actually zero it would still be possible for interfacial broadening to take place. To distinguish between these possibilities it is necessary to be able to say something about the rates of these non-diffusive modes of relaxation; this is the reason why it is not in principle possible to predict the mutual diffusion coefficient purely from tracer diffusion coefficients.

However, if we can assume that the non-diffusive relaxation of the less mobile polymers is fast compared with the diffusion of the more mobile chains, we can derive a simple equation relating the mutual diffusion coefficient to the tracer diffusion coefficient and a thermodynamic factor. This equation has been applied to liquid systems, in which context it is known as the Hartley–Crank equation (Tyrell and Harris 1984), and to metals, in which context it is known as the Darken equation (Haasen 1984). The polymer version was first stated by Kramer (Kramer *et al.* 1984) and is often known as the 'fast theory'. It is most compactly written in the form

$$D_{\text{mutual}} = (x_A D_B^* + x_B D_A^*)\phi(1 - \phi)\frac{\mathrm{d}^2 F}{\mathrm{d}\phi^2}, \qquad (4.4.11)$$

where D_A^* and D_B^* are the (ϕ-dependent) tracer diffusion coefficients of A and B, respectively, x_A and x_B are their mole fractions and F is the free energy of mixing.

This equation does have some convincing experimental support for the case of polymers of high relative molecular mass. Figure 4.24 shows the results of an experiment in which both tracer diffusion coefficients were measured as a function of composition (Composto *et al.* 1988). Equation (4.4.11) was then used to calculate the mutual diffusion coefficient, given a value of χ that had been measured independently. These calculated mutual diffusion coefficients were compared with the directly measured values and good agreement was found.

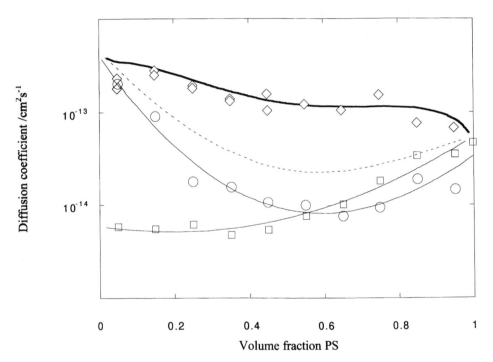

Figure 4.24. Diffusion coefficients as functions of the composition in the miscible blend polystyrene–poly(xylenyl ether) (PS–PXE) at a temperature 66 °C above the (concentration-dependent) glass transition temperature of the blend, measured by forward recoil spectrometry. Squares represent tracer diffusion coefficients of PXE ($N_{PXE} = 292$), circles the tracer diffusion coefficients of PS and diamonds the mutual diffusion coefficient. The upper solid line is the prediction of equation (4.4.11) using the smoothed curves through the experimental points for the tracer diffusion coefficients and an experimentally measured value of the Flory–Huggins interaction parameter. The dashed line is the prediction of equation (4.4.11), neglecting the effect of non-ideality of mixing, illustrating the substantial thermodynamic enhancement of the mutual diffusion coefficient in this miscible system. After Composto *et al.* (1988).

Largely because of the strong composition dependence of the tracer diffusion coefficients in cases in which the two polymers have different glass transition temperatures, the composition dependence of the mutual diffusion coefficient calculated from equation (4.4.11) can be very strong indeed. This, when used in Fick's equation to calculate the interfacial composition profile, can lead to profile shapes that depart very far from the classical error-function solution of Fick's equation for a constant diffusion coefficient. This is illustrated in figure 4.25, showing measured mutual diffusion coefficients for the miscible blend PVC/PCL, together with the measured interfacial profile compared with the solution of Fick's equation for the corresponding $D(\phi)$.

In the absence of bulk flow one is led by arguments using a dynamic version of the random phase approximation to an alternative expression to equation (4.4.11), in which it is the slow chains that dominate the kinetics (Brochard *et al.* 1983). This is sometimes referred to as the 'slow theory'. There have been suggestions (Akcasu *et al.* 1992) that mutual diffusion should approach the slow-theory limit as the temperature is reduced towards the glass transition temperature of one of the components. A definitive solution to this problem awaits more work, both theoretical and experimental.

A more extreme situation occurs when solvent diffuses into a glassy polymer. Here the tracer diffusion coefficient of the polymer is negligible compared with the tracer diffusion coefficient of the solvent. In addition the rate of non-diffusive relaxation of the polymer is also slow compared with the solvent diffusion, as well as being highly composition dependent, and it is this relaxation that is the rate-limiting step. The diffusion equation must now be modified by a term that expresses the mechanical response of the glassy polymer to the build-up of osmotic pressure caused by the ingress of the solvent (Thomas and Windle 1982).

The solution of the resulting modified equation takes the form of a sharp front that moves into the polymer, not as the square root of time, but at a constant velocity. Ahead of this front is a small Fickian precursor, which may be detected by ion beam analysis techniques. This type of diffusion – known as *case II diffusion* – is very important and widespread for the case of penetration of glassy polymers by solvent. Analogous phenomena will take place when two miscible polymers are brought together, one of which is below its glass transition temperature and one above. Such situations are not difficult to realise in practice (Jabbari and Peppas 1995).

The final complication in the problem of polymer interface formation is to consider the very early stages of interdiffusion, when the interface is broadened on a length scale comparable to or less than the radius of gyration of the polymers involved. At such small distances diffusion ideas, which deal solely

Polymer/polymer interfaces

(a)

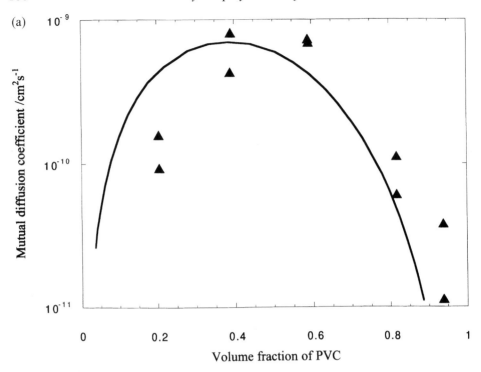

(b)

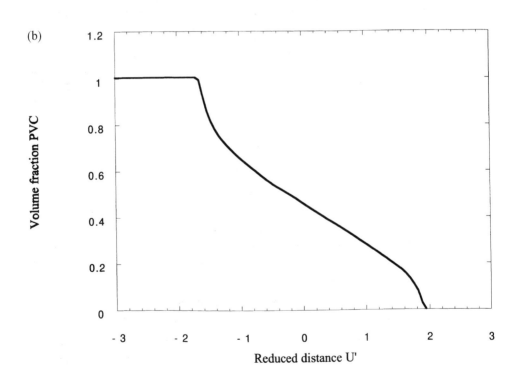

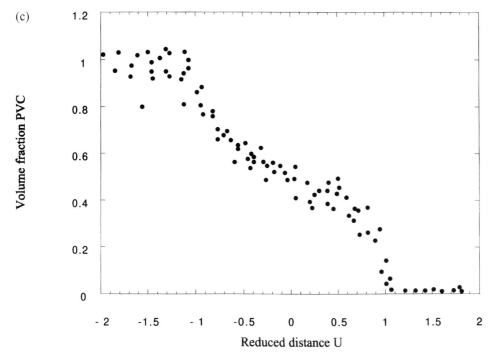

Figure 4.25. (a) The mutual diffusion coefficient in the miscible polymer blend poly(vinyl chloride)–polycaprolactone (PVC–PCL) at 91 °C, as measured by x-ray microanalysis in the scanning electron microscope (Jones *et al.* 1986). The solid line is a fit assuming that the mutual diffusion coefficient is given by equation (4.4.11), with the composition dependence of the tracer diffusion coefficient of the PCL given by a combination of equations (4.4.9) and (4.4.10). The tracer diffusion coefficient of the PVC is assumed to be small in comparison. (b) The calculated profile of diffusion between pure PVC and pure PCL, on the basis of the concentration dependence of the mutual diffusion coefficient shown in (a). The reduced length $u = \alpha x t^{1/2}$, where the time t is measured in seconds, the distance x is in centimetres and $\alpha = 5 \times 10^{-4}$ cm s$^{1/2}$. (c) The concentration profile as measured at a PVC/PCL interface after interdiffusion at 78 °C (S. Rocca and J. Klein, unpublished data). After Klein (1990).

with the centre-of-mass motion of polymer chains, are not sufficient. This is likely to be an important regime partly because the majority of strength development will take place during these early stages. Moreover, as we have seen, the equilibrium interface width between immiscible polymers is always less than the radius of gyration of a single chain (except close to a critical point) so centre-of-mass motion of chains can never be important in the kinetics of formation of immiscible interfaces (Steiner *et al.* 1990).

There will be two important differences between interfacial broadening on these small length scales and more conventional centre-of-mass diffusion. Firstly, because polymer chain conformations will be perturbed in the vicinity of an interface, extra thermodynamic driving forces will be present during the initial stages of diffusion. Secondly, at these small length scales internal motions of polymer chains can give rise to interfacial broadening, without the need for a chain to reptate out of its confining tube.

Interpretations of the experiments that have been performed in this area have concentrated on the second aspect. Karim and co-workers (Karim *et al.* 1990) and Stamm and co-workers (Stamm *et al.* 1991) both used neutron reflectivity to look at the very early stages of the interdiffusion of polystyrene and deuterated polystyrene. Figure 4.26 shows the results of Karim *et al.* The graph depicts the extent of interfacial broadening as a function of time and temperature, whereby a WLF shift factor (equation (4.4.9)) has been used to reduce the data for a number of different temperatures on to one curve. At very early times the interface very rapidly broadens by about 20 Å – in fact this initial broadening is so fast that it was impossible to catch it in progress. After this the interface broadens rather slowly and in a non-diffusive way – that is to say the interfacial width depends on time with a power law with an exponent less than

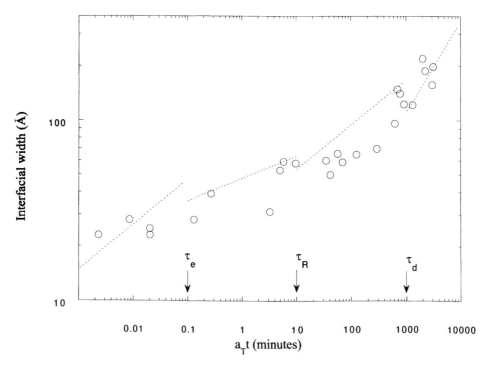

Figure 4.26. Interfacial broadening as a function of time for polystyrene and deuterated polystyrene, both of relative molecular mass 200 000, as measured by neutron reflectivity. Data taken at various temperatures have been reduced to a reference temperature of 120 °C by using a WLF shift factor. After Karim *et al.* (1990).

one half. This exponent appears to increase slowly, until, after a reptation time – the time it takes a polymer chain to move completely out of its restraining tube – we recover normal Fickian diffusion with the expected centre-of-mass diffusion coefficient.

A lot of effort has been made to attempt to understand this kind of data in terms of the reptation model and a rather complex sequence of power laws has been proposed to explain the features of curves such as figure 4.25. At very short times, a fast relaxation on the segment length scale leads to the fast initial broadening on essentially liquid-like time scales. For times up to the Rouse relaxation time for a chain length corresponding to the relative molecular mass for entanglement τ_e, one has effectively free Rouse motion, during which the chain feels the effect of its own connectivity, but does not yet feel the effect of entanglement. For times less than the Rouse relaxation of the whole chain, τ_R, the chain can undergo free Rouse relaxation parallel to the tube, but its motion is constrained in directions perpendicular to the tube. For times between the Rouse relaxation time for the whole chain and the reptation time τ_d the constraining effect of the tube is dominant, whereas after the reptation time we finally reach the limit of normal, centre-of-mass diffusion. The amount each of these different types of motion contributes to the observed interfacial profile at different times is not obvious and early straightforward interpretations of the kind of data shown in figure 4.26 perhaps should be treated as oversimplifications.

One striking prediction of the reptation models has been the question of the existence or otherwise of a discontinuity in the concentration profile at times earlier than the reptation time (de Gennes 1992). Figure 4.27 shows how the reptation model might lead to the prediction of such a gap. Direct observation of such a discontinuity has proved surprisingly difficult, because the contributions of other types of motion of the kind just discussed will have the effect of smearing out this 'gap'. However, Reiter and Steiner (1991) have analysed neutron reflectivity data by assuming a concentration profile consisting of the sum of two error functions, one relatively broad whose width increases with time and one rather narrower, whose width varies rather slowly with time, with the contribution of the narrow component decreasing to zero as the time increases past a reptation time.

A very elegant experiment to show a similar effect was conceived and executed by Russell, Wool and co-workers (Russell *et al.* 1993). They studied interdiffusion at an interface between two isotopic triblock copolymers of identical length. On one side of the initial interface the diblock had hydrogenated ends and a deuterated centre, whereas the polymer on the other side of the interface had the labelling scheme reversed, with deuterated end blocks and

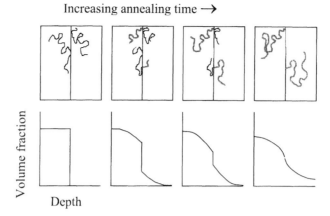

Figure 4.27. The interfacial 'gap' predicted by the reptation model at early times of diffusion. If polymer chains can move only by reptation, then, until a time such that the whole chain has been able to wriggle out of its initial tube, there will be a discontinuity in the concentration profile at the interface.

a hydrogenated centre. If the mode of diffusional broadening of the interface were reptation[§], then a transient ripple should appear in the concentration profile across the interface for times less than a reptation time (figure 4.28). When the experiment was carried out and the concentration profile measured using dynamic secondary ion mass spectrometry, such a ripple was in fact observed (see figure 4.29).

These experiments, which highlight the importance of chain ends during the very early stages of polymer diffusion, suggest one final question – would the kinetics of interfacial broadening at these very early stages be significantly modified if the distribution of chain ends near the interface were not uniform? As we discussed in section 2.6, it has been argued that this is the case; for entropic reasons it should be more favourable for the chains near a free surface to locate their ends close to that surface and there is some support for this view from computer simulations. De Gennes argues (de Gennes 1992) that the results of Reiter and Steiner discussed above support the hypothesis that chain ends are segregated at the surface. As we have discussed, however, direct evidence for chain-end segregation is frustratingly elusive and other groups have reported no evidence for such an effect. The situation is complicated by the fact that, as we shall see in chapter 6, rather small end groups on polymer chains can be very surface active. In real situations one often does not have much knowledge of or control over the chemistry of one's end groups, so it

[§] Or more generally any transport mechanism in which the ends of the chains have more mobility than the centre.

Increasing annealing time →

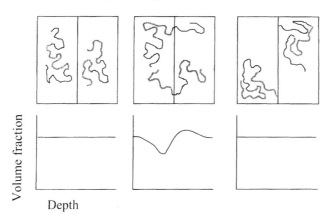

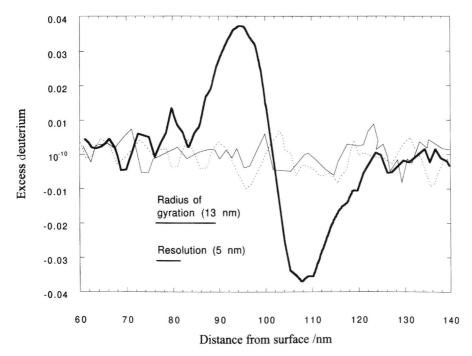

Figure 4.28. Annealing at an interface between two complementarily labelled triblock copolymers. On the left-hand side, the triblock is labelled in the sequence H–D–H, while on the right-hand side the sequence is D–H–D. Shortly after annealing has begun, because only the ends of chains can cross the interface, on the left-hand side there is an excess of D while on the right-hand side there is an excess of H. When the annealing time is long enough for the chains to have entirely escaped from their initial tubes, the concentration profile is once more uniform.

Figure 4.29. Depth profiles obtained by dynamic SIMS for broadening of the interface between complementarily labelled triblock copolymers (dotted line, 2880 min; bold line, 1080 min; and thin line, 30 min). After Russell *et al.* (1993).

may well be that a chemical group present adventitiously on some proportion of the chains in a sample may have a significant degree of surface activity. Until more information becomes available one should probably conclude only that the occurrence or otherwise of chain-end segregation will depend on fine details of the sample chemistry, the method of preparation of the surface and perhaps other as yet unknown factors.

4.5 The morphology of immiscible polymer blends

We saw in section 4.1 that phase separation in polymer mixtures is very common. The phase-separated structures that result will be characterised by the presence of interfaces between the coexisting phases; however, in polymer mixtures it will be very rare indeed to achieve the situation in which phase separation is complete and we have two macroscopic domains of the two phases separated by a single interface. Rather, the slow dynamics of polymers will mean that this equilibrium structure will not be reached and there will be a more complicated morphology consisting of much smaller domains of the two phases, which, although not at full equilibrium, may well be stable for practical purposes. Thus, to understand the origin of these structures, we need to understand the way in which phase separation is initiated and how the driving force provided by the interfacial tension leads to the coarsening of the partially phase-separated structure (Hashimoto 1993).

There are many ways of starting with a single phase and proceeding to a phase-separated structure; figure 4.30 illustrates three of them. In figure 4.30a we presuppose the existence of a phase diagram that has both one-phase and

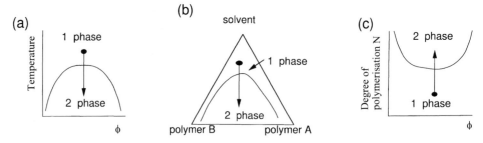

Figure 4.30. Schematic phase diagrams illustrating the different paths which may lead to phase separation. In (a), a mixture of two polymers undergoes phase separation on cooling. In (b), two polymers are dissolved in a common solvent and the solvent is then removed, leading to phase separation. In (c), one or both of the components polymerises, causing phase separation to occur when a certain critical degree of polymerisation has been reached.

two-phase regions of the phase diagram at an experimentally accessible temperature[§]. The sample is prepared in the one-phase region of the phase diagram as a homogeneous mixture and then the temperature is suddenly changed (quenched) to bring the sample into the two-phase region. Phase separation is initiated by the mechanism of spinodal decomposition or nucleation and growth, leading to the formation of domains of one phase in a matrix of the other. In figure 4.30b we make a ternary solution of the two polymers in a solvent that is a good solvent for both. At low total polymer concentrations the unfavourable polymer–polymer interactions are diluted by the solvent, and the system forms a single phase. However, as the solvent is removed – perhaps by evaporation if the solution has been applied as a coating – we pass into a two-phase region and the system phase separates. In figure 4.30c one or both components are originally present in low relative molecular mass form: for example one might have a polymer of high relative molecular mass dissolved in the monomer of a second polymer. As the second polymer is polymerised the relative molecular mass reaches a value such that the degree of incompatibility N exceeds its critical value and phase separation is initiated.

To these three possibilities we must add a fourth (figure 4.31) in which the two polymers are mechanically mixed. Work is done by the mixing device, part of which is converted into heat, but some goes into creating domains of one species dispersed in the other. Associated with the extra area of interface created is an interfacial energy that is supplied by the mixing device.

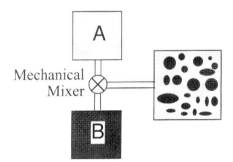

Figure 4.31. A two-phase mixture may be made by mechanically mixing two immiscible polymers. In this case the extra interfacial energy is supplied by the work done by the mixer.

[§] This is a demanding requirement; all polymers have a glass transition temperature below which all motion is frozen out, defining a lower limit on accessible temperatures; at higher temperatures chemical degradation of the polymers becomes important.

Of these methods of reaching a phase-separated state, the first method, via a temperature quench, is probably the least important in practice[§]. Nonetheless, it is the simplest to deal with theoretically and consequently it is the route to phase separation that has been most intensively studied. It is worth taking a more detailed look at the mechanisms by which phase separation occurs, for this emphasises the importance of the interfacial energy in determing the phase morphology.

We recall that the quality of central importance in constructing the phase diagram is the curvature of the free energy of mixing (see figure 4.3). The sign of the curvature determines whether a mixture that could lower its overall free energy by phase separating, and is thus in the two-phase region of the phase diagram, is locally unstable or locally metastable, with the two situations being separated by the spinodal line. Corresponding to this thermodynamic distinction, we expect to find two different mechanisms by which phase separation takes place.

Within the spinodal line, any small fluctuation in composition will lead to a lowering of the free energy; under these conditions phase separation will proceed immediately via a mechanism of amplification of random composition fluctuations called spinodal decomposition (Binder 1991). In the metastable part of the phase diagram a small composition fluctuation actually raises the free energy and, in order to begin the phase-separation process, a droplet of the minority phase, of a size greater than a critical size, has to be nucleated. Thus this mechanism of phase separation is known as nucleation and growth.

The mechanism operative in the unstable part of the phase diagram can be understood if we revisit equation (4.4.11) for the mutual diffusion coefficient of a polymer mixture, which we can rewrite in a simplified form as

$$D_{\text{mutual}} = D_0 \phi (1 - \phi) \frac{\mathrm{d}^2 F}{\mathrm{d}\phi^2}. \qquad (4.5.1)$$

All the information about the dynamics of the two polymers is contained in the prefactor D_0, which is always positive. The mutual diffusion coefficient is proportional to $\mathrm{d}^2 F / \mathrm{d}\phi^2$; within the spinodal this is negative, so as we cross the spinodal line the diffusion coefficient must change sign. The physical meaning of a negative diffusion coefficient is clear from Fick's first law; material diffuses not from regions of high concentration to regions of low concentration, but the other way, from regions of low concentration to regions of high concentration. Any random, thermally generated, concentration fluctuation will not die away, as it would in the one-phase part of the phase diagram;

[§] In contrast to the worlds of metallurgy and ceramics, in which it is virtually the only route!

rather, it will grow to form domains of the two coexisting phases. However, concentration fluctuations on different length scales do not grow at equal rates. Consider a fluctuation with a very long wavelength (figure 4.32b). The composition fluctuation can only grow in amplitude by transport of material from the troughs to the peaks; this happens by diffusion, so, as the wavelength becomes longer, the fluctuation will grow more slowly. On the other hand, for a fluctuation with a very short wavelength there will be additional contributions to the free energy beyond the simple Flory–Huggins terms; we know from expressions like equations (4.2.1) and (4.2.2) that concentration gradients which are steep compared with the length scale of the radius of gyration of a chain carry a free energy penalty. This has the effect of suppressing fluctuations

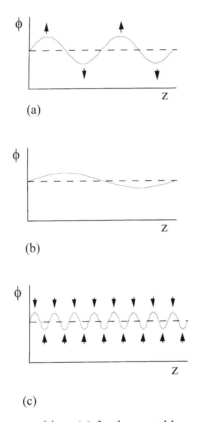

Figure 4.32. Spinodal decomposition. (a) In the unstable part of the phase diagram, random concentration fluctuations are unstable and grow in amplitude. Long-wavelength fluctuations (b) grow slowly because of the large distances through which material needs to be transported, while short fluctuations (c) are suppressed, because of the high free energy penalty associated with sharp concentration gradients. Thus fluctuations on an intermediate length scale (a) grow the fastest and dominate the resulting morphology.

of very short wavelength (figure 4.32c). Thus there is some intermediate length scale of fluctuations that will grow the fastest and the morphology of the phase-separating stucture will, at early times after phase separation has commenced, be dominated by this length (Cahn 1965).

It is possible to use the square gradient free energy expression to calculate what the fastest growing wavelength should be (de Gennes 1980, Binder 1983). For two polymers of equal degree of polymerisation quenched to a temperature not too far away from the spinodal we find

$$\lambda_{max} = (6\pi^2)^{1/2} R_g \left(1 - \frac{\chi_s}{\chi}\right)^{-1/2}, \tag{4.5.2}$$

where χ_s is the value of the Flory–Huggins interaction parameter on the spinodal line, which is, as a function of composition and the degree of polymerisation,

$$2\chi_s = \frac{1}{\phi N_A} + \frac{1}{(1 - \phi)N_B}. \tag{4.5.3}$$

Thus the characteristic size of the domains is predicted to be proportional to the overall chain dimension, multiplied by a factor expressing how deep into the two-phase region the quench is made. Typically one might expect this wavelength to range from a few hundred to a few thousand ångström units.

Because the dominant pattern arises from the selective amplification of random compositional fluctuations, the pattern itself is random – it can be thought of as a superposition of sinusoidal composition waves, with random directions and phases but with a fairly definite wavelength. If the starting composition is close to 50:50, one is likely to have a pattern in which the two phases are interconnected; as the starting composition becomes more and more different from the 50:50 one, one is likely to cross over to a situation in which one has disconnected domains of the minority phase in a matrix of the majority phase. However, observations of the phase-separation process in progress are often made not by direct observation in real space but by scattering techniques. Here what one sees is a relatively broad interference peak with a well-defined maximum at a scattering vector $q_{max} = 2\pi/\lambda_{max}$. According to the simplest version of the theory one would expect this peak to grow exponentially without its position changing; in practice with time the peak moves to smaller wave-vectors, corresponding to a coarsening of the domain structure. Figure 4.33 shows this effect very clearly for an isotopic polymer blend (Bates and Wiltzius 1989).

As one crosses the spinodal line – defined by the condition that $d^2 F/d\phi^2 = 0$ – the system is stable with respect to small compositional fluctuations and spinodal decomposition no longer takes place. Instead, for a pure system one

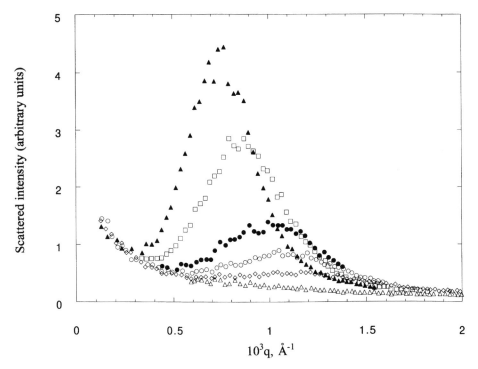

Figure 4.33. Light scattering patterns obtained after quenching an initially homo-geneous mixture of deuterated and normal poly-1,4-butadiene into the two-phase region of the phase diagram. Times after quenching to 322 K are $\triangle$, 200 s; $\diamondsuit$, 5300 s; $\circ$, 9800 s; $\bullet$, 17 600 s; $\square$, 46 300 s; and $\blacktriangle$, 84 700 s. After Bates and Wiltzius (1989).

needs to nucleate droplets of the minority phase that must be larger than a certain size in order to grow. We can estimate the likelihood of this process using classical homogeneous nucleation theory. In this region, if one forms a droplet of the minority phase of composition one lowers the free energy per monomer in the drop by ΔF_v, which is given in terms of the Flory–Huggins free energy $F_{FH}(\phi, \chi, N)$ as

$$\Delta F_v = F_{FH}(\phi, \chi, N) - F_{FH}(\phi_0, \chi, N), \qquad (4.5.4)$$

where the volume fraction on the binodal, ϕ_b, is given by the solution of equation (4.1.9) for a mixture with equal relative molecular masses. However, by making a droplet one also creates a certain amount of interface, with associated energy γ per unit area; thus the net energy change associated with a droplet of radius r is

$$\Delta F(r) = \tfrac{4}{3}\pi r^3 \, \Delta F_v + 4\pi r^2 \gamma. \qquad (4.5.5)$$

The droplet needs to be of at least a critical radius r^* before the system's total free energy is lowered by the growth of the droplet. This critical radius size is

$$r^* = \frac{2\gamma}{\Delta F_{\mathrm{v}}} \qquad (4.5.6)$$

and to nucleate such a critical droplet the system must first *increase* its free energy by

$$\Delta F^* = \frac{16\pi\gamma^2}{\Delta F_{\mathrm{v}}}. \qquad (4.5.7)$$

Formation of such droplets must then be an activated process whose rate is proportional to $\exp[-\Delta F^*/(kT)]$. We can estimate this rate using equation (4.2.6) for the interfacial energy γ, and the result is that the rate of homogeneous nucleation we should expect for polymer systems is vanishingly small. In practice nucleation is usually aided by the presence of other interfaces, for example impurity particles such as dust or the container walls may well be able to nucleate critical droplets with much lower activation energies (heterogeneous nucleation) or indeed with no activation energy at all. We will return to this subject in section 5.3 when we discuss the effects of surfaces on phase separation.

Whether the initial mechanism of phase separation is spinodal decomposition or nucleation (and in passing it is worth mentioning that the clear distinction between these mechanisms that is maintained in elementary discussions like the forgoing cannot be sustained in more rigorous treatments (Binder 1991)), the final morphology one achieves in practice is a long way from that predicted by the early stage models. Typically one sees domains that are much bigger than the size scale that emerges as the fastest growing wavelength of the linear theory of spinodal decomposition and than the critical droplet size of nucleation theory – what are the processes that control the growth of the domains? This is an extremely complex area, but it seems, at least for the late stages of growth that are often practically relevant, that some simplicity re-emerges.

The picture of spinodal decomposition assumed in figure 4.32, in which one imagines composition fluctuations growing in amplitude but remaining with constant wavelength, cannot be sustained for long. As the peak compositions approach the coexisting compositions, the amplitude of the composition wave cannot go on increasing, yet the system is still very far from equilibrium. The only possibility is for the size of the domains to start to grow. Thus we are led to the picture of the stages of spinodal decomposition shown in figure 4.34.

Two length scales are important in this problem; R, which characterises the size of the domains and their average separation, and w, which characterises

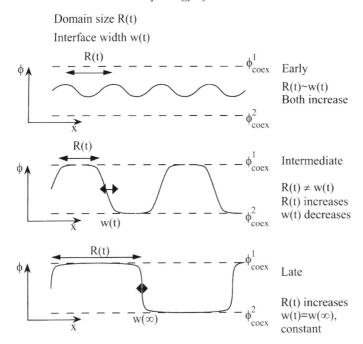

Figure 4.34. The behaviour of the interface width w and the average domain size R at various stages of spinodal decomposition: (top) the early stage – well described by the linear (Cahn) theory of spinodal decomposition, (centre) the intermediate stage and (bottom) the late stage – dynamic scaling should hold.

the width of the interface between the domains. During the early stage, both R and w are essentially constant and are related to the fastest growing wavelength λ_{max}. When the peak compositions approach the coexisting compositions we enter a complicated, intermediate stage of phase separation, during which R, characterising the size and separation of the domains, increases, while the interfacial width w decreases towards its equilibrium value. During the late stage of phase separation we have a simpler situation, in which discrete domains with compositions close to the coexisting compositions are separated by interfaces essentially at the equilibrium width w_I. The only length scale characterising the morphology is R, which characterises both the average size and the separation of the domains. An identical situation would be reached during the late stages of phase separation that has occurred by a nucleation mechanism.

The fact that the morphology at this late stage is characterised by only one length scale leads to the important idea that the domain pattern is self-similar in time. That is to say, statistically the domain pattern at a later time is simply a blow-up of the pattern at an earlier time. This so-called dynamical scaling hypothesis allows one to construct an elegant argument for the rate at which the domain coarsens, due to Huse (1986).

Near one of the domains, the local composition is slightly increased above the coexisting composition due to the Gibbs pressure associated with a curved interface. The magnitude of these composition differences is of order

$$\Delta\phi \sim \frac{\gamma}{R(t)} \qquad (4.5.8)$$

so the system contains concentration gradients that will drive diffusion. The typical concentration gradient will be of order

$$\frac{\Delta\phi}{R(t)} \sim \frac{\gamma}{R(t)^2}. \qquad (4.5.9)$$

These concentration gradients lead to fluxes of material J whose typical magnitudes are given by Fick's first law:

$$J \sim \frac{D\gamma}{R(t)^2}. \qquad (4.5.10)$$

In this expression D is a mutual diffusion coefficient; even though we are within the two-phase part of the phase diagram this is still positive because we are in the metastable region. These diffusional fluxes cause the growth of the domains and, on dimensional grounds, it is clear that the rate of growth of the average domain size is simply proportional to the flux. Thus we find

$$\frac{\mathrm{d}R(t)}{\mathrm{d}t} = \frac{D\gamma}{R(t)^2}, \qquad (4.5.11)$$

which we can integrate to yield

$$R(t) \sim (D\gamma)^{1/3} t^{1/3}. \qquad (4.5.12)$$

This characteristic growth of the domain size with the cube root of time has often been observed and is known as the Lifshitz–Slyozov law (Lifshitz and Slyozov 1961). An alternative, much faster, mechanism of coarsening exists if it is possible for material to be transported by bulk flow; there it can be shown that domain growth is linearly proportional to time (Siggia 1979).

This argument shows us that the driving force for the coarsening of the domains is the interfacial energy associated with them. In practice the final steady-state domain size one finds in a polymer mixture is a result of the fact that the mixture has a finite time for conditions under which molecules can move about. This might be a finite time of processing at high temperature for a system which undergoes phase separation as the temperature is reduced, a finite time before all solvent evaporates and the system becomes glassy or otherwise immobile, for a system cast from a solvent, or a finite time before such a high relative molecular mass is reached that the system becomes immobile in a reactive system in which phase separation is driven by polymerisation of one or both of the components. In all these cases the value of the interfacial energy plays a crucial role in determining how big the domains can grow in the time available.

4.6 Appendix to chapter 4. The random phase approximation

In order to derive expressions for the free energy cost of sustaining composition gradients in polymer mixtures, it is helpful to use a mean-field treatment of correlations in polymer mixtures known as the random phase approximation (RPA) (de Gennes 1979, Binder 1983, Doi and Edwards 1986). We start by asking how the local composition changes in response to a change in the local chemical potential. For an ideal polymer mixture, for which the Flory–Huggins interaction parameter $\chi = 0$, we can write the chemical potential of a monomer of type A as

$$\mu_A = \frac{kT}{N_A} \ln \phi_A + \text{constant} \tag{4.6.1}$$

and similarly for monomer B. So the necessary response functions are

$$\frac{\partial \phi_A}{\partial \mu_A} = \phi_A \frac{N_A}{kT}, \tag{4.6.2a}$$

$$\frac{\partial \phi_B}{\partial \mu_B} = \phi_B \frac{N_B}{kT}. \tag{4.6.2b}$$

Now assume that our polymer blend is *incompressible* – the two volume fractions ϕ_A and ϕ_B cannot change independently and we know that

$$\phi_A + \phi_B = 1. \tag{4.6.3}$$

Now it is convenient to define an *exchange chemical potential* $\Delta\mu$, the free energy change on replacing an A monomer by a B monomer. Using (4.6.2) we can write the response of the system to a change in exchange chemical potential as

$$\delta(\Delta\mu) = \delta\mu_A - \delta\mu_B = kT \left(\frac{\delta\phi_A}{\phi_A N_A} - \frac{\delta\phi_B}{\phi_B N_B} \right), \tag{4.6.4}$$

which we can rearrange using (4.6.3) to yield the response function for the exchange chemical potential as

$$\frac{\partial \phi}{\partial(\Delta\mu)} = \frac{1}{kT} \left(\frac{1}{\phi N_A} + \frac{1}{(1-\phi)N_B} \right)^{-1}. \tag{4.6.5}$$

We know from thermodynamics that there is a general relation between this type of response function and equilibrium fluctuations; for example the mean square size of composition fluctuations is given by

$$\langle(\Delta\phi)^2\rangle = \frac{kT}{V}\left(\frac{\partial\phi}{\partial(\Delta\mu)}\right). \qquad (4.6.6)$$

However, we can only use this formula for volumes large compared with the chain size, because at distance scales smaller than a chain size there are concentration fluctuations arising simply from the connectivity of the chain that are not taken into account in fluctuation formulae derived from the response functions of (4.6.2). For a single chain the quantity in which we are interested is the pair correlation function: if we choose one monomer in a chain as the origin and ask what the density of monomers belonging to that chain at distance r is, the result, when averaged over all starting positions, is the single-chain correlation function $g(r)$. It is convenient to work in Fourier space, and the Fourier transform of $g(r)$ is the single-chain structure factor $S(q)$. We need to generalise equations (4.6.2) to arbitrary wave-vectors q using this quantity;

$$\frac{\partial\phi_A(q)}{\partial\mu_A(q)} = \frac{\phi_A S_A(q)}{kT}, \qquad (4.6.7a)$$

$$\frac{\partial\phi_B(q)}{\partial\mu_B(q)} = \frac{\phi_B S_B(q)}{kT}. \qquad (4.6.7b)$$

For an ideal random walk we know the explicit forms of $S_A(q)$ and $S_B(q)$; they are given in terms of the Debye function $f_D(x)$ as

$$S_A(q) = N_A f_D(x),$$

where

$$f_D(x) = \frac{2}{x}\left\{1 - \left(\frac{1 - e^{-x}}{x}\right)\right\}$$

with

$$x_A = \frac{1}{6}N_A a^2 q^2. \qquad (4.6.8)$$

Now, by the same reasoning that led to equation (4.6.5), we can write for the q-dependent response function of the ideal mixture

$$\frac{\partial\phi(q)}{\partial(\Delta\mu(q))} = \frac{1}{kT}\left(\frac{1}{\phi S_A(q)} + \frac{1}{(1 - \phi)S_B(q)}\right)^{-1}, \qquad (4.6.9)$$

and thus for the structure factor for the ideal blend system $S_{ni}(q)$

$$S_{ni}(q) = \left(\frac{1}{\phi S_A(q)} + \frac{1}{(1 - \phi)S_B(q)}\right)^{-1}. \qquad (4.6.10)$$

This function, then, describes both the response to an external perturbation and the equilibrium concentration fluctuations of a mixture of polymers that have no thermodynamic interaction between them. However, even in the absence of a thermodynamic interaction there is a very strong interaction that acts on all segments equally; these are the intermolecular forces of cohesion which act to maintain a constant density in the mixture. To extend our treatment to systems that do have a thermodynamic interaction expressed by a non-zero Flory–Huggins interaction parameter χ, we assume that this interaction is only a small perturbation on the strong cohesive forces which maintain constant density. We note that the analogue of equation (4.6.5) for a non-zero χ is

$$\frac{\partial \phi}{\partial (\Delta \mu)} = \frac{1}{kT} \left(\frac{1}{\phi N_A} + \frac{1}{(1 - \phi) N_B} - 2\chi \right)^{-1}, \qquad (4.6.11)$$

as can simply be derived from the Flory–Huggins free energy expression. This suggests that equation (4.6.10) should be modified to account for a non-zero χ as

$$\frac{1}{S(q)} = \frac{1}{S_{ni}(q)} - V(q), \qquad (4.6.12)$$

where $V(q)$ is the Fourier transform of the nett thermodynamic interaction between chemically different monomers, whose value at zero q is 2χ. Its detailed form must depend on all the details of the intersegment potentials that give rise to the interaction, but for small q we must be able to expand it as

$$V(q) = 2\chi(1 - \tfrac{1}{6} q^2 r_0^2), \qquad (4.6.13)$$

where r_0 is a measure of the range of the intersegment forces, which must be of order the segment size.

In equation (4.6.12), combined with equations (4.6.10) and (4.6.8), we have a prescription giving the average fluctuations in composition in a polymer mixture at any wave-vector q. This formula is of central importance in the application of scattering methods to polymer blends and, in particular, it forms the basis for using neutron scattering to determine the Flory–Huggins interaction parameter χ. It is worth stressing the two major assumptions that underlie it; firstly that chain conformations in a polymer blend are ideal and secondly that polymer blends are incompressible. Of course neither assumption is in general correct and failures of the RPA will occur when these assumptions break down.

For our purposes we need to use these equations to find the effect of concentration gradients on the free energy. To do this, note that we can expand the Debye function at small q as

$$f(x) \cong 1 - \frac{x}{3} \qquad (4.6.14)$$

so the inverse structure function can itself be expanded for small q as

$$\frac{1}{S(q)} = \frac{1}{\phi N_A} + \frac{1}{(1-\phi)N_B} - 2\chi + \frac{a^2 q^2}{18\phi(1-\phi)} + \frac{q^2 r_0^2 \chi}{3}. \qquad (4.6.15)$$

This expression can be derived from a free energy that is written as a gradient expansion; if

$$\frac{F}{kT} = \int [f_{FH}(\phi) + \kappa(\phi)(\nabla\phi)^2]\,\mathrm{d}r \qquad (4.6.16)$$

then we can identify $\kappa(\phi)$ as

$$\kappa(\phi) = \frac{\chi r_0^2}{6} + \frac{a^2}{36\phi(1-\phi)}. \qquad (4.6.17)$$

4.7 References

Akcasu, A. Z., Bahar, I. *et al.* (1992). *Journal of Chemical Physics* **97**, 5782.

Anastasiadis, S. H., Gancarz, I. *et al.* (1988). *Macromolecules* **21**, 2980.

Anastasiadis, S. P., Russell, T. P. *et al.* (1990). *Journal of Chemical Physics* **92**, 5677.

Bates, F. S., Rosedale, J. H. *et al.* (1990). *Physical Review Letters* **65**, 1893.

Bates, F. S. and Wignall, G. D. (1986). *Physical Review Letters* **57**, 1429.

Bates, F. S. and Wiltzius, P. (1989). *Journal of Chemical Physics* **91**, 3258.

Binder, K. (1983). *Journal of Chemical Physics* **79**, 6387.

Binder, K. (1991). Spinodal decomposition. *Phase Transformations in Materials.* P. Haasen. Weinheim, VCH.

Binder, K. and Frisch, H. L. (1984). *Macromolecules* **17**, 2929.

Brochard, F., Jouffroy, J. *et al.* (1983). *Macromolecules* **16**, 1638.

Broseta, D., Fredrickson, G. H. *et al.* (1990). *Macromolecules* **23**, 132.

Cahn, J. W. (1965). *Journal of Chemical Physics* **42**, 93.

Chaturvedi, U. K., Steiner, U. *et al.* (1989). *Physical Review Letters* **63**, 616.

Composto, R. J., Kramer, E. J. *et al.* (1988). *Macromolecules* **21**, 2580.

Composto, R. J., Kramer, E. J. *et al.* (1992). *Macromolecules* **25**, 4167.

Doi, M. and Edwards, S. F. (1986). *The Theory of Polymer Dynamics.* Oxford, Oxford University Press.

Dolan, A. K. and Edwards, S. F. (1974). *Proceedings of the Royal Society, London,* A **337**, 50.

Dudowicz, J. and Freed, K. F. (1991). *Macromolecules* **24**, 5076.

Edwards, S. F. (1965). *Proceedings of the Physical Society, London* **85**, 613.

Fernandez, M. L., Higgins, J. S. *et al.* (1988). *Polymer* **29**, 1923.

Ferry, W. D. (1980). *Viscoelastic Properties of Polymers.* New York, John Wiley.

Flory, P. J., Orwoll, R. A. *et al.* (1964). *Journal of the American Chemical Society* **86**, 3515.

Fox, T. G. (1956). *Bulletin of the American Physical Society [2]* **2**, 123.

Fredrickson, G. H. (1992). Theoretical methods for polymer surfaces and interfaces. *Physics of Polymer Surfaces and Interfaces.* I. C. Sanchez. Boston, Butterworth-Heinemann.

de Gennes, P. G. (1979). *Scaling Concepts in Polymer Physics.* Ithaca, NY, Cornell University Press.

de Gennes, P. G. (1980). *Journal of Chemical Physics* **72**, 4756.

de Gennes, P. G. (1992). in *Physics of Polymer Surfaces and Interfaces*. I. C. Sanchez. Boston, Butterworth-Heinemann.

Green, P. F. (1991). *Macromolecules* **24**, 3373.

Green, P. F. and Doyle, B. L. (1987). *Macromolecules* **20**, 2471.

Grosberg, A. Y. and Khokhlov, A. R. (1994). *Statistical Physics of Macromolecules*. New York, AIP Press.

Haasen, P. (1984). *Physical Metallurgy*. Cambridge, Cambridge University Press.

Hashimoto, T. (1993). Structure of polymer blends. *Structure and Properties of Polymers*. E. L. Thomas. Weinheim, VCH.

Helfand, E. (1982). Polymer interfaces. *Polymer Compatibility and Incompatibility*. K. Solc. Chur, Harwood.

Helfand, E. and Sapse, A. M. (1975). *Journal of Chemical Physics* **62**, 1327.

Helfand, E. and Tagami, Y. (1971). *Journal of Chemical Physics* **56**, 3592.

Huse, D. A. (1986). *Physical Review* B **34**, 7845.

Jabbari, E. and Peppas, N. A. (1995). *Macromolecules* **28**, 6229.

Jones, R. A. L., Klein, J. *et al.* (1986). *Nature (London)* **321**, 161.

Kambour, R. P., Bendler, J. T. *et al.* (1983). *Macromolecules* **16**, 753.

Karim, A., Mansour, A. *et al.* (1990). *Physical Review* B **42**, 6846.

Kim, E., Kramer, E. J. *et al.* (1995). *Macromolecules* **28**, 1979.

Klein, J. (1986). *Macromolecules* **19**, 105.

Klein, J. (1990). *Science* **250**, 640.

Kramer, E. J., Green, P. *et al.* (1984). *Polymer* **25**, 473.

Lifshitz, I. M. and Slyozov, V. V. (1961). *Journal of Physics and Chemistry of Solids* **19**, 35.

Menestrel, C. L., Kenwright, A. M. *et al.* (1992). *Macromolecules* **25**, 3020.

Mills, P. J., Green, P. F. *et al.* (1984). *Applied Physics Letters* **45**, 957.

Prest, W. M. and Porter, R. S. (1972). *Journal of Polymer Science* A-2 **10**, 1639.

Reiter, G. and Steiner, U. (1991). *Journal de Physique II* **1**, 659.

Roe, R. J. (1986). *Macromolecules* **19**, 728.

Rowlinson, J. S. and Widom, B. (1982). *Molecular Theory of Capillarity*. Oxford, Clarendon Press.

Russell, T. P., Deline, V. R. *et al.* (1993). *Nature (London)* **365**, 235.

Russell, T. P., Hjelm, R. P. *et al.* (1990). *Macromolecules* **23**, 890.

Sanchez, I. C. (1982). Equation of state theories of polymer blends. *Polymer Compatibility and Incompatibility: Principles and Practice*. K. Solc. Chur, Harwood.

Sariban, A. and Binder, K. (1987). *Journal of Chemical Physics* **86**, 5859.

Schwahn, D., Mortensen, K. *et al.* (1987). *Physical Review Letters* **58**, 1544.

Schweizer, K. S. and Curro, J. G. (1994). *Advances in Polymer Science* **116**, 321.

Semenov, A. N. (1994). *Macromolecules* **27**, 2732.

Sferrazza, M., Xiao, C. *et al.* (1997). *Physical Review Letters* **78**, 3693.

Shibayama, M., Yang, H. *et al.* (1985). *Macromolecules* **18**, 2179.

Shull, K. R., Kramer, E. J. *et al.* (1990). *Nature (London)* **345**, 790.

Shull, K. R., Mayes, A. M. *et al.* (1993). *Macromolecules* **26**, 3929.

Siggia, E. (1979). *Physical Review* A **20**, 595.

Stamm, M., Huttenbach, S. *et al.* (1991). *Europhysics Letters* **14**, 451.

Stamm, M. and Schubert, D. W. (1995). *Annual Review of Materials Science* **25**, 325.

Steiner, U., Krausch, G. *et al.* (1990). *Physical Review Letters* **64**, 1119.

Tang, H. and Freed, K. F. (1991). *Journal of Chemical Physics* **94**, 6307.

Thomas, N. L. and Windle, A. H. (1982). *Polymer* **23**, 529.
Tirrell, M. (1984). *Rubber Chemistry and Technology* **57**, 523.
Tyrell, H. J. V. and Harris, K. R. (1984). *Diffusion in Liquids*. London, Butterworths.
Zawada, J. A., Ylitalo, C. M. *et al.* (1992). *Macromolecules* **25**, 2896.

5

Adsorption and surface segregation from polymer solutions and mixtures

5.0 Introduction

If we study the composition of the surface of a polymer solution or a polymer mixture we should expect to find that the composition of the region near the surface is not the same as that of the bulk. The same applies to an interface between such a system and a solid; for example a colloidal particle suspended in a polymer solution will often be coated with a layer of polymer, that may well be able to stabilise the colloid, preventing its aggregation. In polymer solutions this phenomena is known as *adsorption*, if the polymer accumulates at the surface, or *depletion*, if it is the solvent that is favoured there; in polymer blends one often refers to the *surface segregation* or *surface enrichment* of one component of the mixture. However, these different names describe what is essentially the same phenomenon, the subject of this chapter.

Segregation phenomena are universal in mixed systems; the surface of a mixture of small-molecule liquids will usually be enriched in the species of low surface energy and the surface composition of a metallic alloy will rarely be the same as the bulk composition. However, mixtures containing polymers are special in two ways. Rather large degrees of segregation or adsorption can be obtained even when the interaction driving the segregation is small; offset against any gain in energy obtained on adsorption is a loss of translational entropy of the molecule in the bulk, but for a large molecule this entropy is rather small. The length scale characterising the distance from the surface over which the composition is perturbed is generally much larger than that for small molecules, either liquid or solid, because this length scale is itself set by the size of the molecules themselves. The importance of adsorption for polymers owes much to these special factors.

In this chapter we will first discuss surface segregation phenomena in miscible polymer blends. We will find that the theoretical tools needed to understand these phenomena are largely those introduced in the last chapter, square gradient theory and self-consistent field theory. We will then be able to introduce the much richer and more important topic of polymer adsorption from solution. Here many of the same physical principles as those discussed for surface segregation continue to apply, but the problem is much more complex because the thermodynamics of polymer solutions in the bulk are more difficult than the thermodynamics of polymer blends and in particular the mean-field ideas on which our approach for blends is based can no longer be relied on. Finally, we shall consider what happens when the solution or blend is not miscible. Here we have to move beyond considering equilibrium effects to consider the role the surface or interface plays in the kinetics of phase separation.

5.1 Surface segregation in polymer mixtures

Some of the earliest quantitative work on the surface composition of polymer mixtures was carried out by Prest, Pan and Koberstein (Pan and Prest 1985, Bhatia *et al.* 1988), who used XPS to study the surface of blends of polystyrene (PS) and poly(vinyl methyl ether) (PVME). This pair of polymers is, at room temperature, completely miscible; yet it was found that, for a blend of 50% PS with 50% PVME, the concentration of PVME at the topmost surface was nearly 95%. The explanation of this effect lies in the values of the surface energies of the two polymers; depending on relative molecular mass the surface energy of PVME may be as much as $7 \, \mathrm{mJ \, m^{-2}}$ lower than that of PS. The driving force for the segregation of PVME to the surface of the mixture is clearly related to the lowering of the system's free energy that is possible by having a larger concentration of the material of lower surface energy at the surface.

However, surface energy differences are by no means the only factor that determines the extent of surface segregation. There must also be a free energy cost associated with maintaining a certain region of the system with a different composition from that of the bulk, by definition, because the system is thermodynamically miscible. One can think of the near-surface region, with its different composition from that of the bulk, as being a region of incipient phase separation; because the system is miscible, creating such a volume of phase-separated material must increase the overall free energy of the system and the bigger the volume at a composition different from that of the bulk the bigger the free energy cost.

One may wonder why then the system does not simply have a very thin layer of the material of lower surface energy at the surface, to get the free energy benefit of the low-energy surface without paying the price of having a large volume of material not at the bulk composition. The answer lies in the gradient energy. As we saw in section 4.2, the presence of steep concentration gradients imposes its own free energy penalty; such a concentration gradient restricts the number of possible conformations available to the chain, thus reducing its entropy. As a result of this effect the distance over which the concentration returns to its bulk value cannot be very much shorter than the size of the polymer chain.

Figure 5.1 is a schematic diagram of the kind of volume fraction profile one might expect near the surface of a miscible polymer blend. The component of lower surface energy is enriched from its bulk value of ϕ_∞ to a surface value of ϕ_1; the volume fraction will relax to its bulk value over a length scale λ that we might expect to be comparable to the overall dimensions of the polymer, if gradient energy penalties are not to be too large, and the total excess amount of the polymer of lower surface energy is z^*, defined in terms of the volume fraction/depth profile $\phi(z)$ as

$$z^* = \int [\phi(z) - \phi(\infty)]\,\mathrm{d}z. \tag{5.1.1}$$

Experimental techniques might be sensitive to ϕ_1 or z^*, whereas our aim in a theory would be to be able to calculate the whole profile $\phi(z)$.

The simplest such theory, due originally to Cahn (1977) and adapted for polymer blends by Schmidt and Binder (1985), uses the square gradient approach of section 4.2; one writes down the free energy as a functional of the volume fraction profile, including a term proportional to the square of the

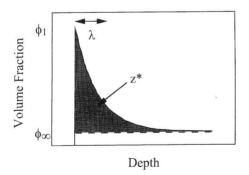

Depth

Figure 5.1. A schematic diagram of the volume fraction/depth profile near the surface of a polymer mixture in which one component has segregated to the surface. The volume fraction relaxes from its surface value ϕ_1 to its bulk value ϕ_∞ over a characteristic distance λ, resulting in a total excess z^* of the lower surface energy component (shown shaded).

gradient of the profile; the equilibrium volume fraction profile is the one that minimises the free energy. The simplest way to account for the saving of free energy obtained by having the material of lower surface energy at the surface is to include a term that is a function only of the surface volume fraction ϕ_1, the so-called bare surface energy $f_s^{(b)}(\phi_1)$. This can be thought of as the surface energy that the mixture would have if there were no surface segregation and the surface composition were the same as the bulk composition. A zeroth approximation would be to assume that the bare surface energy can be interpolated linearly between the surface energies of the two pure components of the mixture; a phenomenological extension adds a correction term quadratic in the volume fraction to give

$$f_s^{(b)}(\phi) = -\mu_1\phi - \tfrac{1}{2}g\phi^2, \tag{5.1.2}$$

where we have assumed that the pure component at $\phi = 1$ has the lower surface energy. The difference in surface energies $\Delta\gamma$ between the two components is now related to the coefficients μ_1 and g by

$$\Delta\gamma = \frac{kT}{b^3}\left(\mu_1 + \frac{g}{2}\right). \tag{5.1.3}$$

Here b^3 is the volume of the Flory–Huggins lattice cell (usually taken as the monomer volume – note that b should not be confused with the statistical step length a); for convenience we have absorbed a factor of kT into the definitions of μ_1 and g.

Some insight into the physical origin of these two coefficients may be obtained by a rather crude argument (Jones *et al.* 1989a, b), which combines the spirit of the discussion of the origin of the surface energy in terms of bond-counting in section 2.1 with the definition in terms of a lattice model of the Flory–Huggins interaction parameter χ given in section 4.1. Suppose that the energy of contact between two monomers of A is ε_{AA}, that between two monomers of B is ε_{BB} and that for a contact between A and B is ε_{AB}. If the coordination number of the lattice is z and z' bonds are cut when a new surface is made, then the surface energy of a blend whose volume fraction of A at the surface is ϕ is

$$f_s^{(b)}(\phi) = \frac{z'b}{2}[\phi^2\varepsilon_{AA} + (1 - \phi^2)\varepsilon_{BB} - 2\phi(1 - \phi)\varepsilon_{AB}]. \tag{5.1.4}$$

Recalling that the surface energies of the pure materials are given by

$$\gamma_A = \frac{z'}{2}\frac{\varepsilon_{AA}}{b^2} \tag{5.1.5}$$

and similarly for γ_B and rewriting ε_{AB} in terms of the Flory–Huggins interaction parameter χ using equation (4.1.7), we can identify the coefficients μ_1 and g as follows:

$$\mu_1 = \frac{b^3 \Delta\gamma}{kT} + \frac{z'}{2z} b\chi, \tag{5.1.6}$$

$$g = -\frac{z'}{z} b\chi. \tag{5.1.7}$$

Of course this argument neglects any entropic factors arising from changes of correlations near the surface and from local packing effects and cannot be expected to be more than a very crude approximation.

Because the volume of material associated with the surface is usually small compared with the bulk volume, it is convenient to work in the grand canonical ensemble, considering the bulk to be an infinite reservoir at a fixed chemical potential. Writing down the surface excess grand potential $\mathcal{J}$ for an area of surface A we have

$$\frac{\mathcal{J}}{AkT} = \int_0^\infty dz \left[f_{FH}(\phi) - \Delta\mu\,\phi(z) + \kappa(\phi)\left(\frac{d\phi(z)}{dz}\right)^2 \right] + f_s^{(b)}(\phi), \tag{5.1.8}$$

where $f_{FH}(\phi)$ is the Flory–Huggins free energy of mixing defined in equation (4.1.8), $\Delta\mu$ is the exchange chemical potential $df_{FH}(\phi)/d\phi$ evaluated at the bulk volume fraction ϕ_∞ and the gradient coefficient $\kappa(\phi)$ is given by equation (4.2.7) as

$$\kappa(\phi) = \frac{a^2}{36\phi(1-\phi)}, \tag{5.1.9}$$

where we have made the (good) approximation that χ is much less than unity and the energetic part of the gradient term can be neglected compared with the entropic part.

Finding $\phi(z)$ to minimise equation (5.1.8) presents a problem in the calculus of variations; the solution for the equilibrium surface volume fraction ϕ_1 is

$$\frac{df_s^{(b)}}{d\phi_1} = \pm\frac{a}{3}\left(\frac{f_{FH}(\phi_1) - f_{FH}(\phi_\infty) - (\phi_1 - \phi_\infty)\Delta\mu}{\phi_1(1-\phi_1)}\right)^{1/2}. \tag{5.1.10}$$

The surface excess free energy per unit area F_s corresponding to this solution is given by

$$\frac{F_s}{kT} = \int_{\phi_1}^{\phi_\infty} d\phi \left[\frac{a}{3}\left(\frac{f_{FH}(\phi_1) - f_{FH}(\phi_\infty) - (\phi_1 - \phi_\infty)\Delta\mu}{\phi_1(1-\phi_1)}\right)^{1/2} - \frac{\partial f_s^{(b)}(\phi)}{\partial\phi}\right], \tag{5.1.11}$$

which allows us to make a simple physical interpretation of equation (5.1.10) in terms of its graphical representation, that is often called the *phase portrait*:

this is shown in figure 5.2. The left-hand side is the derivative of the surface free energy that would be saved by virtue of having a given value of the surface volume fraction; the right-hand side is the derivative of the free energy cost of maintaining a surface-segregated layer with this surface volume fraction. Surface segregation thus leads to the maximum free energy saving when the two lines cross, corresponding to the equality of equation (5.1.10) and the shaded area corresponds to the nett amount by which the system's free energy is lowered as a result of the surface segregation.

The actual concentration–depth profile near the surface is given implicitly by an integration:

$$z = \frac{a}{6} \int_{\phi_1}^{\phi(z)} d\phi \, \{\phi(1 - \phi)[f_{FH}(\phi) - f_{FH}(\phi_\infty) - (\phi - \phi_\infty)\Delta\mu]\}^{1/2}; \quad (5.1.12)$$

this will usually have to be done numerically, but in many situations it will be reasonably well approximated by an exponential profile

$$\phi(z) = \phi_\infty + (\phi_1 - \phi_\infty)\exp\left(\frac{z}{-\xi}\right), \quad (5.1.13)$$

where the decay length of the profile ξ is the correlation length for bulk concentration fluctuations

$$\xi = \frac{a}{6} \Big/ \left(\frac{1 - \phi_\infty}{2N_A} + \frac{\phi_\infty}{2N_B} - \chi\phi_\infty(1 - \phi_\infty)\right)^{1/2}. \quad (5.1.14)$$

From this it can be seen that the basic length setting the scale of the decay of the concentration profile is related to the overall chain dimensions of the polymers involved, as we anticipated.

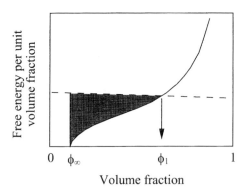

Figure 5.2. The 'phase portrait' for a miscible polymer blend; a graphical representation of equation (5.1.10). The dashed line represents the left-hand side of equation (5.1.10) and the solid line represents the right-hand side. The shaded area represents the amount by which the system's free energy is lowered when the surface volume fraction is ϕ_1.

A final experimental quantity of interest is the total surface excess z^* defined in figure 5.1. Of course one can always calculate this by solving for the concentration profile $\phi(z)$ and integrating this, but it can also be obtained more directly as

$$z^* = \frac{a}{6} \int \frac{\mathrm{d}\phi(1-\phi)}{\{\phi(1-\phi)[f_{\mathrm{FH}}(\phi) - f_{\mathrm{FH}}(\phi_\infty) - (\phi - \phi_\infty)\Delta\mu]\}^{1/2}}. \qquad (5.1.15)$$

One interesting feature both of this equation and of equation (5.1.12) for the volume fraction profile is that the bare surface energy, that expresses the difference in surface energy between the two components, enters only via the surface volume fraction ϕ_1, that itself enters only as a limit on an integral. The shape of the segregation profile depends only on the bulk thermodynamics of the polymer mixture.

Some of these results have been tested for blends of normal and isotopic polystyrene (Jones *et al.* 1989a, b). We have seen that the small difference in polarisability between C—H and C—D bonds leads to a very small, positive value of χ; it should not surprise us that there is a small difference in surface energy arising from the same origin. Near-surface depth profiles were measured by forward recoil spectrometry for a blend of normal and deuterated polystyrene of high relative molecular mass (see section 3.2.5). As the sample is prepared one sees a uniform depth profile, but after annealing above the glass transition temperature to allow the sample to approach equilibrium one sees that a substantial accumulation of the deuterated component has occurred at the surface. With this technique the resolution is too poor to discriminate the form of the concentration profile, but nonetheless the surface excess z^* may be accurately determined. If one has an independent measure of the value of the Flory–Huggins interaction parameter χ, perhaps from neutron scattering, one can use the values of z^* as a function of bulk volume fraction ϕ_∞ to deduce the surface volume fractions ϕ_1 via equation (5.1.15) and then deduce the form of the derivative of the bare surface energy via equation (5.1.10). In this way, for this system it was possible to estimate the difference in surface energies between deuterated and normal polystyrene as $0.08\ \mathrm{mJ\,m}^{-2}$. It is not possible to measure the surface energies of polymers of such high relative molecular masses with sufficient precision to check this value directly, but it is comparable with the difference in surface energies between analogous small-molecule hydrocarbons and their deuterated analogues. What is striking is the size of the segregation effect that is driven by such a small surface energy difference; this reinforces the point made above about the special importance of surface-segregation phenomena for polymers.

In order to make a more demanding test of theories of surface segregation one needs to use techniques that can resolve finer details of the depth profile, rather than simply measuring the surface excess. Possible higher resolution techniques include neutron reflectivity and dynamic secondary ion mass spectrometry (DSIMS). In figure 5.3, neutron reflectivity data are shown for the same blends of deuterated and normal polystyrene of high relative molecular mass (Jones *et al.* 1990). Two fits are shown; one corresponds to the best fit profile found from solving equation (5.1.12) with χ and ϕ_1 allowed to vary, while the other is an empirically found function that produces a slightly better fit. We will return below to the physical significance of this small discrepancy, but for the moment the most important point illustrated is the sensitivity of the neutron reflection technique to quite subtle changes in profile shape.

The observed profile, then, is quite close to the exponential form of equation (5.1.13) and is characterised by a single length λ. In figure 5.4 the variation of this length is shown as a function of the bulk volume fraction for this system, measured both by neutron reflectivity and by dynamic secondary ion mass spectrometry. Reasonable agreement with theory is indicated, as well as a reassuring consistency of the results of the two techniques. The decay length is comparable to the overall chain dimensions of the polymers involved and it increases slightly as the bulk volume fraction increases, reflecting the fact that an increase in ϕ_∞ corresponds to a move closer to criticality.

These experiments seem to show that the simple square gradient theory works reasonably well. However, as hinted above there seem to be problems in detail; one finds, as we saw in figure 5.3, systematic deviations of the profile shape from that predicted by the square gradient theory. This can be seen more clearly in the experiments of Norton *et al.* (1995), who looked at surface segregation in another isotopic blend, poly(ethylene propylene) and its deuterated analogue. Figure 5.5 shows a concentration profile from this system, which shows a pronounced flattening near the surface. At the time of writing, no definitive explanation for these shortcomings is available. Nonetheless, there are a number of areas in which it is clear that the simple square gradient theory might fail. There are two main areas of difficulty; firstly the possibility that the square gradient theory leads to qualitatively incorrect profile shapes due to shortcomings in the way chain connectivity is dealt with and secondly

Figure 5.3. (a) Neutron reflectivity from a 15% blend of d-PS (relative molecular mass 1.03×10^6) with h-PS (relative molecular mass 1.8×10^6), with (b) deduced volume fraction profiles. The dashed curves are the profile and fit corresponding to the best fit of equation (5.1.11), whereas the solid line is an empirical function that produces a closer fit to the neutron data. After Jones *et al.* (1990).

(a)

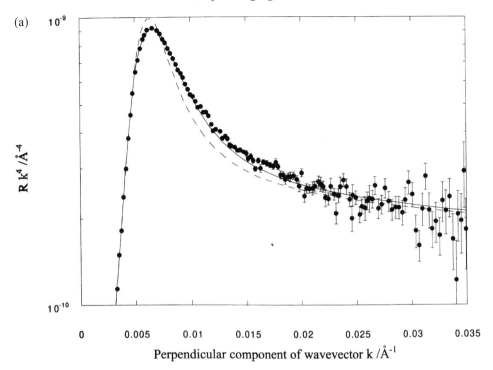

(b)

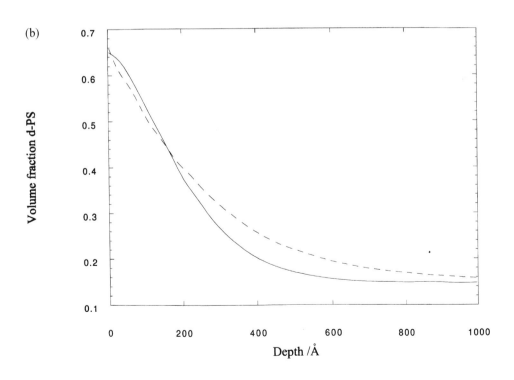

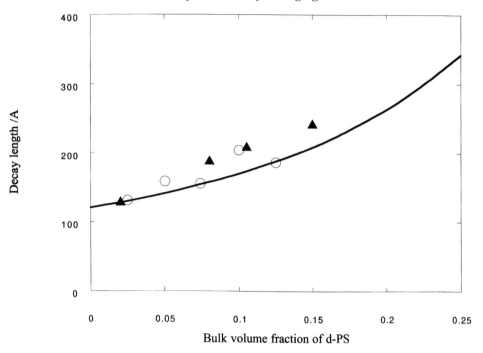

Figure 5.4. Decay lengths versus bulk volume fraction for blends of d-PS (relative molecular mass 1.03×10^6) with h-PS (relative molecular mass 1.8×10^6), as measured by neutron reflectivity (circles) and dynamic SIMS (triangles). After Jones *et al.*(1990).

difficulties that arise when one tries to deal with the surface boundary condition at any more detailed a level than that of the crudest arguments given above.

In our discussion of the polymer/polymer interface problem in section 4.2, we found that one difficulty in the use of square gradient theory for all cases except those very close to criticality was that a profile contained concentration gradients too steep for the gradient expansion to be strictly valid. We face the same potential problem for surface segregation. We should expect a self-consistent field approach to provide a more satisfactory treatment of chain connectivity that overcomes this difficulty. The self-consistent field method outlined in section 4.3 can indeed be applied to the problem of surface segregation (Carmesin and Noolandi 1989, Genzer *et al.* 1994), but unlike the polymer/polymer interface problem no analytical results are available. However, it is relatively straightforward to implement numerical solutions of the self-consistent field equations. The results for surface segregation are different in detail from the predictions of the square gradient theory; for example square gradient theory seems to overestimate the value of the surface volume fraction resulting from a given derivative of the bare surface energy. However, the

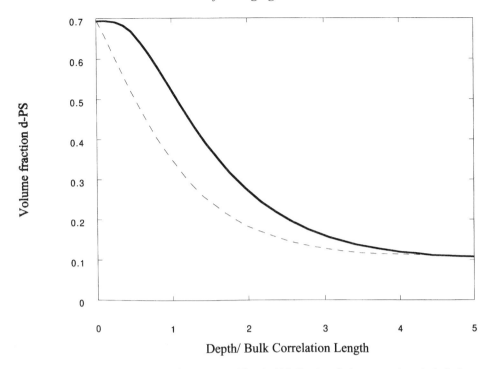

Figure 5.5. The surface enrichment profile (solid line) of deuterated poly(ethylene propylene) at the surface of a blend with normal poly(ethylene propylene), measured by neutron reflectivity. The degree of polymerisation of each component was 2250 and the sample was annealed at 70 °C. The depth is normalised by the bulk correlation length and the dashed line is the prediction of equation (5.1.11). After Norton *et al.* (1995).

shapes of the profiles that are calculated do not seem to be qualitatively different from those calculated from square gradient theory, so it does not seem possible to account for profiles such as those in figure 5.5 in this way.

Another potential shortcoming arises from the whole principle of treating the various surface interactions via a bare surface interaction that is a function solely of the surface volume fraction. This approach may be criticised on two grounds; firstly the phenomenological form of the bare surface energy does not account for entropic factors, while secondly, even for differences in surface energy of a purely energetic origin, it fails to account for the fact that the forces from which the surface energy differences arise are of long range. For example, if the surface energy differences were to arise from van der Waals forces the bare surface energy should depend on the distance from the surface, z, as z^{-3} (Israelachvili 1991). Perturbation calculations show that inclusion of long-range surface forces should indeed lead to a flattening of the profile near the surface (Chen *et al.* 1991), but explicit numerical calculations (Jones 1993) suggest that the magnitude of the effect is rather small and cannot account for the size of the effect seen in figure 5.5.

Entropic factors in the bare surface energy should arise because polymer chains must have their conformations perturbed near a surface. It has been argued that this effect should lead to terms in the bare surface energy proportional to the concentration gradient at the surface. There is some experimental evidence for such terms derived from experiments on surface segregation in blends of deuterated polystyrene and a copolymer of styrene and 4-bromostyrene (Bruder and Brenn 1993). In one theory the size of this term is proportional to the difference in flexibility between the two polymer species; this provides a driving force for the more flexible component of a polymer blend to segregate to the surface even in the absence of any difference in surface energy (Fredrickson and Donley 1992). In contrast to this result, computer simulations have suggested that it is the stiff chains that are preferred at a surface (Yethiraj *et al.* 1994). This suggests that it might be risky to push an essentially continuum theory to the point where it is attempting to deal with essentially local, monomer structure; at the level of detail at which one is considering differences in entropy that arise from the way monomers can pack near a surface one should bear in mind that a full theory incorporating the details on a molecular level of the polymers may be required, probably in combination with computer simulations. Some progress has been made towards such a theory (Kumar *et al.* 1995); from this kind of work a complex picture emerges, in which the details of local packing of chain segments, loss of overall configurational entropy near a surface and local entropic effects interact with compressibility effects in a way that makes prediction of which component is favoured at a wall rather sensitive to the approximations used. However, it should be stressed that these purely entropic effects are unlikely to occur on their own, without additional enthalpic effects, in real systems. One system that has been studied experimentally comprises blends of polyolefins with various architectures (Steiner *et al.* 1992); here it was found that the more branched microstructure was favoured at the surface. However, even in this apparently simple case, in which the two components are identical chemically but differ in architecture, it has been argued that only by the inclusion of a bulk enthalpic thermodynamic interaction (i.e. a non-zero value of χ) can the results be rationalised (Yethiraj 1995). This emphasises the way in which surface-segregation bulk and surface thermodynamics are intimately linked.

Entropic factors are likely to be central to one very important problem: surface segregation in mixtures of chemically identical polymers of different relative molecular masss. A practically important aspect of this would be the question of whether, in a polydisperse polymer, the component of low relative molecular mass segregates preferentially to the surface. Intuitively one would expect this to be so, on two grounds. Empirically, we know that the surface energy of polymer melts increases with increasing relative molecular mass (recall the discussion in chapter 2); also it is easy to argue qualitatively that shorter chains would suffer less of an entropy penalty by being near a surface and thus having the number of possible conformations reduced than would a long chain. However, having made these qualitative arguments it seems at the moment surprisingly difficult to make a more quantitative statement. Experiments are difficult because, in order to be able to distinguish among chemically identical chains of different relative molecular masses, we need some form of labelling, but we have seen that polymer thermodynamics is so delicate that even deuterium labelling, which one might have thought of as being non-perturbative, itself can lead to substantial segregation phenomena. On the other hand, theoretical treatment is difficult because the origins of the surface energy difference between short and long chains are intrinsically entropic, as well as being tied up with the differences in compressibility between long and short chains (as discussed in section 2.5) so that they cannot possibly be well represented by a short-range bare surface energy of the type shown in equation (5.1.2). In one experiment that does convincingly demonstrate the existence of an effect (Hariharan *et al.* 1993) segregation in blends of deuterated and normal polystyrene was re-examined. As discussed above, when polymers of equal relative molecular masses are mixed it is the deuterated species that is favoured at the surface. However, when the normal polystyrene was much lower in relative molecular mass than was the deuterated polystyrene the effect of the relative molecular mass can outweigh the isotopic effect; for example for deuterated polystyrene of relative molecular mass 500 000 a normal polystyrene of relative molecular mass less than 52 000 will segregate to the surface.

Of course the limiting case of one polymer in a mixture being much shorter than the other is a polymer solution. However, segregation and depletion of polymers at the surface of polymer solutions is both more complex than segregation from polymer mixtures and arguably more important, so it forms the subject of the next section.

5.2 Adsorption from polymer solutions

5.2.1 Introduction

Polymers are frequently encountered in solutions or dispersions and are often applied to surfaces from these vehicles, for example paints and anti-static sprays. Therefore it is pertinent to consider the basis of the adsorption of a polymer from solution on to a surface or interface. The surface or interface may be an interface with a solid with which the polymer is in contact or the free surface of the solution. Adsorption and depletion of polymers from hard surfaces play important roles in the stabilisation and flocculation of colloidal dispersions; we do not explicitly discuss this aspect and refer the interested reader to the authoritative text by Napper (1983). In many cases when polymers are used to stabilise a colloid, they are grafted by one end. This situation – the so-called polymer brush – is discussed in chapter 6.

Depending on the nature of the polymer–interface interaction, we may have either a region where the polymer concentration exceeds that of the bulk solution, a surface excess, or a depletion region where the polymer concentration is lower than that of the solution. In this, adsorption from solution resembles the surface-segregation phenomenon discussed in section 5.1. Here we found that the length scale over which the composition of the solution was perturbed from its bulk value was comparable to the overall size of the molecule. In very rough terms the same is true for polymer solutions, but in defining more precisely what we mean by the molecular size we run into serious difficulties. In many cases adsorption takes place from a dilute solution. Here molecules are swollen and have an overall size that varies as roughly the 3/5th power of the relative molecular mass. However, close to the interface the local volume fraction may be very much bigger, approaching a concentrated solution. In such a concentrated solution the excluded volume interactions that lead to an expansion of the coil are screened out and the molecular size approaches that of an ideal random walk, varying with the square root of the relative molecular mass. So for polymers of high relative molecular mass the molecular size can vary a great deal between the dilute and concentrated solution cases and we are left with the question of which one characterises the size of the adsorbed layer.

Polymer adsorption from solution is a very large subject and we cannot provide an exhaustive treatment here. A much more complete treatment may be found in a recent monograph (Fleer *et al.* 1993). Here we describe the scaling and self-consistent field descriptions of homopolymer adsorption, together with experimental data selected to illustrate the important aspects. A brief outline of mean-field (Flory–Huggins) and scaling theories of polymer solutions is provided, to establish the parameters that recur when homopolymer adsorption is discussed. Much more complete discussions of mean-field theory of polymer solutions are available (Flory 1953, Yamakawa 1971, Kurata 1982) while the scaling theories of polymer solutions are dealt with in other books (de Gennes 1979, Freed 1987, des Cloizeaux and Jannink 1990).

5.2.2 Bulk thermodynamics of polymer solutions

A mean-field theory of polymer solution thermodynamics can be developed in an exactly analogous way to the lattice fluid model of polymer melts discussed in chapter 2 and the Flory–Huggins model of polymer mixtures discussed in chapter 4. In fact, the Flory–Huggins theory of polymer solutions may simply be regarded as the limit of the Flory–Huggins model of a polymer mixture where the degree of polymerisation of one of the components of the mixture is reduced to one. Here we reiterate the way in which an expression for the Gibbs free energy change on mixing of a polymer with a solvent is derived. This free energy expression can then be used to obtain the osmotic pressure and phase behaviour of the solution.

As discussed in chapters 2 and 4, the Gibbs free energy change on mixing a polymer of degree of polymerisation N and solvent is

$$\Delta G_{\mathrm{m}} = k_{\mathrm{B}} T \left(\phi_1 \ln \phi_1 + \frac{\phi_2}{N} \ln \phi_2 + \phi_1 \phi_2 \chi \right), \tag{5.2.1}$$

where ϕ_1 and ϕ_2 are the volume fractions of solvent and polymer, respectively, and χ is the Flory–Huggins interaction parameter. If χ is not too large the dominant term in equation (5.2.1) is the $\phi_1 \ln \phi_1$ contribution to the entropy, which is large and negative. This suggests that dissolution of the polymer in the solvent is due to solvent molecules being able to be dispersed throughout the spatial region occupied by the polymer, rather than to any increase in configurational entropy of the polymer molecules.

We now focus on the solvent in the polymer solution. Equation (5.2.1) is a specific form of the more general relation between thermodynamic state functions;

$$\Delta G_{\mathrm{m}} = \Delta H_{\mathrm{m}} - T \Delta S_{\mathrm{m}}. \tag{5.2.2}$$

An even more general relation for ΔG_{m} is in terms of the Gibbs free energy of the mixture, G_{12}, and those of the pure components

$$\Delta G_{\mathrm{m}} = G_{12} - G_1 - G_2. \tag{5.2.3}$$

In the mixture the chemical potential of the solvent, μ_1, is $(\partial G_{12} / \partial N_1)_{N_2}$ where N_1 is the number of *moles* of solvent. The partial differential of equation (5.2.3) with respect to N_1 yields

$$\left(\frac{\partial \Delta G_m}{\partial N_1} \right)_{N_2} = \mu_1 - \mu_1^0, \tag{5.2.4}$$

where μ_1^0 is the chemical potential of the solvent in its standard state. Using equation (5.2.1) to obtain $(\partial \Delta G_m / \partial N_1)_{N_2}$ we obtain

$$\mu_1 - \mu_1^0 = RT[\ln\phi_1 + (1 - 1/N)\phi_2 + \chi\phi_2^2]. \qquad (5.2.5)$$

Now the osmotic pressure π of a polymer solution is equivalent to the pressure in the gas laws and the osmotic pressure is related to $\mu_1 - \mu_1^0$ by the van 't Hoff relation

$$\pi = -(\mu_1 - \mu_1^0)/\overline{V}_1, \qquad (5.2.6)$$

where V_1 is the molar volume of the solvent and thus we find

$$\pi = -\frac{RT}{\overline{V}_1}\left[\ln(1 - \phi_2) + \left(1 - \frac{1}{N}\right)\phi_2 + \chi\phi_2^2\right] \qquad (5.2.7)$$

(note the correspondence to $pV = nRT$ in this equation). Expanding the logarithm up to terms in ϕ_2^2 and noting that $r \simeq \overline{V}_2/\overline{V}_1$ and $\phi_2 = c_2v_2$ (with v_2 the partial specific volume of the polymer whose solution concentration is c_2) we eventually obtain

$$\frac{\pi}{c_2} = \frac{RT}{M_2} + RT\left(\frac{\overline{v}_2^2}{\overline{V}_1}\right)(\tfrac{1}{2} - \chi)c_2 + \cdots . \qquad (5.2.8)$$

Although this expression is the basis for the interpretation of osmotic pressures of polymer solutions to obtain relative molecular masses (M_2) of polymers, this is not why it is reproduced here. The right-hand side of equation (5.2.8) has two components, the first term (RT/M_2) is that expected for an ideal solution at infinite dilution. The second component is an excess term arising from two-body interactions generally represented by the second virial coefficient;

$$A_2 = \left(\frac{\overline{v}_2^2}{\overline{V}_1}\right)(\tfrac{1}{2} - \chi). \qquad (5.2.9)$$

The Flory–Huggins theory assumes that each polymer molecule experiences a local chemical potential that is uniform throughout the solution. This is the mean-field assumption that underlies equation (5.2.1), with its perfect gas entropy of mixing and the assumption of no correlation between neighbouring sites that underlies the calculation of the enthalpy of mixing. This is a good approximation when the concentration of polymer in the solution is relatively high, but as the solution becomes more dilute the connectivity of the polymer chains enforces larger and larger fluctuations of concentration in space, which means that the mean-field approach becomes increasingly inaccurate. This is illustrated in figure 5.6. When the solution is so dilute that, on average, the distance between molecules is greater than the radius of gyration of a single molecule (a), then necessarily the concentration of segments is not spatially uniform. The gradient of concentration around the centre of each chain leads to an osmotic pressure that tends to swell the chain, so the dimensions of the

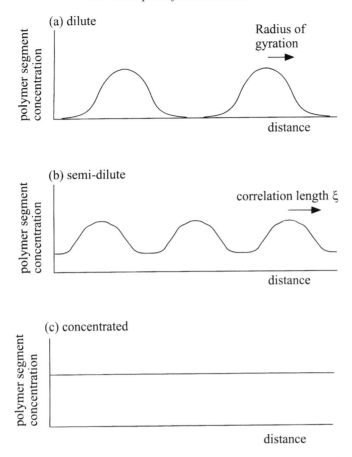

Figure 5.6. Schematic plots of the polymer segment concentration as a function of the distance in bulk polymer solutions. In (a) the concentration of chains is low enough that on average different polymer chains do not overlap. The segment concentration has large spatial fluctuations; this is a dilute solution. In (b) chains start to overlap, but there are still strong composition fluctuations imposed by the connectivity of the chains characterised by a correlation length ξ. This is the so-called semi-dilute concentration regime. In (c) the solution is concentrated and there are no concentration fluctuations on length scales larger than the monomer size.

polymer are no longer given by the random walk formula $R_g \sim aN^{1/2}$, but rather by the formula for a self-avoiding random walk, $R_g \sim aN^v$. Here the exponent v, known as the excluded volume exponent, is rather greater than the random walk exponent of $1/2$; a very simple argument due to Flory predicts an exponent $v = 3/5$, whereas more accurate theory reveals that its true value is closer to $v = 0.588$. In this dilute concentration regime the thermodynamics retains some simplicity because individual chains may be treated as essentially non-interacting. When the concentration is increased and the chains begin to overlap (b), the fluctuations of segment concentration become less marked and the chains are less swollen than they are in the dilute case. The thermodynamics in this so-called 'semi-dilute' regime are particularly difficult; the fluctuations in segment density are large enough to render mean-field approaches invalid, while, in contrast to the dilute regime, the chains interact with each other quite strongly. It is in this regime that one must abandon mean-field approaches and resort to the scaling methods described below. Finally, when the polymer concentration becomes large enough (c), the spatial concentration fluctuations become negligible; it is in this concentrated regime that mean-field approaches may be used with confidence.

Dilute solutions were treated in the mean-field framework by Flory (Krigbaum and Flory 1952, Flory 1953). The concept is that although the individual polymer molecules are widely separated from each other in dilute solution and exclude other polymer molecules from the space pervaded by the molecule (the excluded volume principle) within the space the average chain unit concentration is uniform and the previously determined relations can be applied with little modification. The major innovation for dilute solutions is the introduction of dilution parameters. The first term on the right-hand side of equation (5.2.8) is identical to the expression obtained for an *ideal* solution. Consequently the second term on the right-hand side is identified as an *excess* contribution. Using the relation of equation (5.2.7) we can write down the excess chemical potential as

$$(\mu_1 - \mu_1^0)^{\mathrm{EX}} = -\pi^{\mathrm{EX}}\overline{V}_1 = -RT\,\overline{v}_2^2 c_2^2 (\tfrac{1}{2} - \chi) = -RT\phi_2^2(\tfrac{1}{2} - \chi). \quad (5.2.10)$$

Since the excess chemical potential is a free energy then χ must also be a free energy, i.e. it has both an entropy and an enthalpy component. Flory defined enthalpy and entropy of dilution parameters, κ_1 and ψ_1, respectively, such that the partial molar enthalpy of dilution was

$$\Delta\overline{H}_1 = RT\kappa_1\phi_2^2 \quad (5.2.11)$$

and the corresponding partial molar entropy

$$\Delta\overline{S}_1 = R\psi_1\phi_2^2. \quad (5.2.12)$$

We have

$$\Delta\overline{H}_1 - T\,\Delta\overline{S}_1 = (\mu_1 - \mu_1^0)^{\mathrm{E}} \quad (5.2.13)$$

whence

$$\psi_1 - \kappa_1 = (\tfrac{1}{2} - \chi). \quad (5.2.14)$$

A temperature, θ, is defined as that at which the excess contribution to the chemical potential is zero:

$$\theta = \kappa_1 T / \psi_1, \tag{5.2.15}$$

therefore

$$\tfrac{1}{2} - \chi = \psi_1 (1 - \theta/T). \tag{5.2.16}$$

When the actual experimental temperature used is equal to θ, $\chi_1 = 1/2$, at which point all excess contributions to the solution thermodynamics disappear and the solution exhibits ideal behaviour since the second virial coefficient has a value of zero. At this point the excluded volume effects that cause an expansion of the polymer molecule are exactly balanced by the unfavourable polymer–solvent interactions and the molecule adopts unperturbed, random walk dimensions. The influence on polymer dimensions and the highly detailed theories of polymer configuration in relation to the excluded volume parameter are beyond the scope of this book but are extensively covered by Yamakawa (1971) and to some extent by des Cloizeaux and Jannink (1990).

Whereas concentrated solutions may be satisfactorily dealt with using mean-field theory and dilute solutions have some simplicity arising from the weak interactions between polymer chains, the behaviour of polymer solutions in the semi-dilute regime is much more difficult to treat theoretically. A key insight was that, insofar as the character of semi-dilute solutions is dominated by concentration fluctuations, their properties should be amenable to the same kind of renormalisation group methods as those that were developed for critical phenomena (Wilson and Kogut 1974). A rigorous (and demanding!) exposition of the application of these methods has been provided by des Cloizeaux and Jannink (1990). A more accessible discussion is available in the seminal book by de Gennes (1979), in which one particularly powerful and transparent aspect of these methods, the use of scaling laws, is emphasised.

We will illustrate this approach by considering the variation with concentration of the correlation length in semi-dilute solutions, assuming to begin with that the solvent is a thermodynamically good one. Let us return to figure 5.6. As we discussed above, in dilute solution (figure 5.6a) the molecules are distant from each other and behave like independent spheres, each with an overall dimension given by the formula for a self-avoiding random walk, $R_g \sim aN^n$. As the concentration increases above a certain value the polymer coils begin to overlap (figure 5.6b); the intramolecular interactions responsible for the expansion of the polymer molecule become screened from each other by intervening polymer segments. The excluded volume effect becomes attenuated and the polymer is less swollen than it would be in a dilute solution. The concept of screening in moderately concentrated solutions was introduced by Edwards (1966). The concentration at which this overlap between molecules begins is symbolised by c^* and may be defined in a number of ways. Conceptually, it is the concentration at which the 'hard sphere' molecules in solution just fill the available volume. Thus, for a polymer of relative molecular mass M with a radius of gyration, R_g,

$$c^* \approx \frac{3M}{4\pi R_g^3 N_A}. \tag{5.2.17}$$

It may also be defined using hydrodynamic properties such as the intrinsic viscosity, $[\eta]$; in that case

$$c^* \approx 1/[\eta]. \tag{5.2.18}$$

Note that the two definitions do not give identical values for c^* because the hydrodynamic volume is generally larger than the apparent occupied volume of a molecule.

Scaling arguments can be used to quantify the degree to which polymer coils shrink with increasing concentration. Figure 5.7 is a sketch of a region of the polymer solution with a concentration greater than c^*; the solution resembles a network with an average distance, ξ, between points where two different chains interact.

This network is of course not permanent, but a continually fluctuating entity wherein intersections are continually being formed and destroyed. If we select one particular molecule and decrease the magnification (figure 5.8), our selected molecule can be viewed as a sequence of 'blobs' each having (on average) the same number of chain units with each blob of diameter ξ. Within the blob excluded volume interactions exist, but there are no such interactions *between* blobs due to the screening of intervening molecules. Because of this the distance ξ is termed the screening or correlation length.

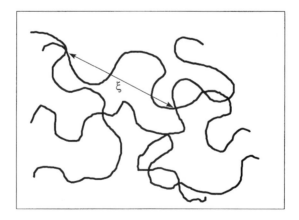

Figure 5.7. A schematic diagram of a semi-dilute solution, which may be thought of as a transient network characterised by a screening or correlation length ξ.

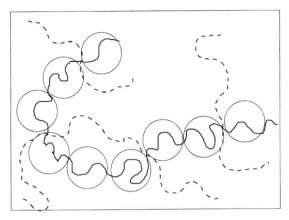

Figure 5.8. The 'blob' picture of a chain in a semi-dilute solution. The average blob diameter is given by the correlation length ξ; some of the interacting chains are shown as dashed lines.

We can write a scaling relation for the correlation length, that must take the form

$$\xi(c) \sim R_g \left(\frac{c}{c^*} \right)^m, \tag{5.2.19}$$

where R_g is the dilute solution radius of gyration ($\equiv \xi$ in dilute solution) that scales with relative molecular mass as $R_g \sim N^\nu$, where N is the degree of polymerisation and ν the excluded volume exponent. From equation (5.2.17) we obtain a relation between c^* and N as $c^* \sim N^{4/5}$ and thus

$$\xi(c) \sim N^{3/5} \left(\frac{c}{N^{-4/5}} \right)^m. \tag{5.2.20}$$

We can determine the value of the unknown exponent m by the following argument. At least in the limit of a high relative molecular mass, we would expect ξ to depend only on c and not on the relative molecular mass. Thus the scaling exponent m must have a value that removes any N dependence from (5.2.20) and therefore $m = -3/4$. We conclude that, in the semi-dilute region for polymers in thermodynamically good solvents,

$$\xi(c) \sim c^{-3/4}. \tag{5.2.21}$$

We can now use the 'blob' picture of a single chain to get the scaling relation for the radius of gyration. On average each blob contains g chain units of length a and because excluded volume conditions prevail *within* the blob, then

$$\xi \approx a g^{3/5}. \tag{5.2.22}$$

Assuming that the blobs have a Gaussian distribution and the Kuhn model of the blob chain applies then

$$R_g^2(c) \approx \frac{1}{6} \frac{N}{g} \xi^2$$

$$\sim \frac{N}{g} \xi^2. \tag{5.2.23}$$

By replacing for g from equation (5.2.22) and using $\xi(c)$ from (5.2.21) we obtain

$$R_g^2(c) \sim Na^2 c^{-1/4}. \tag{5.2.24}$$

We conclude that, in the semi-dilute region, both the correlation length and the radius of gyration decrease as the polymer concentration increases, the correlation length exhibiting the faster decrease. Daoud and Jannink (1975) developed the scaling arguments further to incorporate the influence of temperature (and thus the excluded volume parameter) and thereby defined a state diagram for the behaviour of a polymer solution in the temperature–concentration plane. Figure 5.9 reproduces the essential features.

Region I' is the theta or tricritical region where the polymer displays all the characteristics associated with theta solution behaviour. Region I is the dilute solution–good solvent region; region II corresponds to semi-dilute behaviour. Between region II and region III there is another crossover concentration, c^{**}; in region III the polymer attains its unperturbed dimensions and exhibits no dependence of R_g on c – this is the concentrated regime. Table 5.1 reproduces the scaling relations in each of these regions for the case of a thermodynamically favourable solvent and three dimensions.

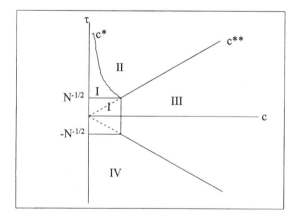

Figure 5.9. Scaling regimes in the concentration–reduced temperature ($\tau = 1 - \theta/T$) plane, as explained in the text. After Daoud and Jannink (1975).

Table 5.1. *Scaling relations*

Region	R_g^2	ξ
I'	N	
I	$N^{6/5}\tau^{2/5}$	
II	$Nc^{-1/4}\tau^{1/4}$	$c^{-3/4}\tau^{-1/4}$
III	N	c^{-1}

$\tau = (1 - \theta/T)$

There is ample experimental evidence in support of these different regions of solution behaviour (Daoud *et al.* 1975, Cotton *et al.* 1976). Subsequently an extrapolation formula was obtained (Edwards and Jeffers 1979) that spanned the dilute solution and bulk amorphous state behaviour by *gradual* changes of the parameters, which is intuitively more physically acceptable than the sharp divisions implied in figure 5.9 (in truth it was pointed out from the beginning that the divisions would always be diffuse). Some evidence in support of additional behaviour regimes proposed by the Edwards and Jeffers model has been obtained (Richards *et al.* 1978, 1981a, b, c, Schaefer 1984). Much of the experimental evidence supporting scaling laws has been summarised by des Cloizeaux and Jannink (1990).

5.2.3 Mean-field theories of polymer adsorption from solution

As we saw in the last section, the thermodynamics of polymer solutions is substantially more complicated than is that of polymer mixtures; rather than having a single theory that works well for all concentrations, different theoretical approaches are needed in different concentration regimes. This leads to a problem when we try to construct a theory of polymer adsorption, because here all the concentration regimes may be involved in a single experiment. Consider a situation in which a polymer is adsorbed at a surface from a dilute solution. We need to understand the thermodynamics of the dilute solution in order to understand the thermodynamic driving force for adsorption. On the other hand, the local concentration of polymer in the region immediately adjacent to the surface may correspond to a semi-dilute or even a concentrated regime. As we move away from the surface the concentration must decrease continuously. Thus it will not usually be possible to say that the adsorbed polymer is in any single concentration regime and it will not be possible to find a single theoretical approach that will be rigorously appropriate when applied to all parts of the adsorbed layer.

In the literature two approaches have been used. The first is to apply mean-field methods of the type developed in chapter 4, in the hope that many properties of the adsorbed layer will be dominated by that part of the concentration profile in which the polymer is in the concentrated solution regime and that those effects of concentration fluctuations that mean field theories neglect, that we know to be of dominant importance in the semi-dilute regime, will not prove too serious. One over-riding advantage of the mean-field methods is that computationally they are straightforward to apply to realistic situations. Thus, even though these methods are fundamentally theoretically flawed, one can at least use them to predict quantities that one will measure in real experiments. The other approach is to adapt the scaling methods that are so successful for bulk solution behaviour to interfacial problems. This approach has the advantage that it tackles head-on the known failures of mean-field theory and it has produced results of immense elegance that bring great clarity to the underlying physics. The difficulty with the scaling approach is that its results are rigorously valid only in the limit of very large relative molecular masses and very low solution concentrations. It is likely that these limits are infrequently or never approached in actual experiments.

Before going on to discuss the details of these theories, we should make some general points about polymer adsorption. What, for example, is the driving force for polymers to adsorb at a surface? Adsorption of a polymer on to an impenetrable surface results in a loss of conformational entropy for the polymer; the presence of the hard wall reduces the number of conformations the molecule may adopt. Additionally, localisation at the wall implies a loss of translational entropy. Thus, in the absence of any specific interaction between polymer segments and the surface, adsorption appears to be energetically unfavourable. On the other hand adsorption may be accompanied by the displacement of many small molecules from the surface and the nett entropy change for the system may be favourable. In fact there are many examples of polymers being adsorbed on to a surface in what appears to be a permanent manner. This arises from a favourable adsorption energy for a polymer segment over that for a solvent molecule which can be characterised in terms of a dimensionless energy parameter, χ_s, defined as

$$\chi_s = (U_{1a} - U_{2a})/k_B T, \qquad (5.2.25)$$

where U_{ia} is the adsorption energy of species i, 1 being solvent and 2 a polymer segment. A positive value of χ_s provides a driving force for polymer adsorption; the analogy of χ_s with the Flory–Huggins interaction parameter is evident[*].

Adsorption of a polymer necessarily implies a change in the conformation; the most common description is the loop–train–tail model (Jenkel and Rumbach 1951) shown schematically in figure 5.10. The trains are made up of segments in direct contact with the surface, whereas loops have no direct contact with the surface but are in close proximity. Tails are non-adsorbed chain ends. Although tail segments may constitute a small proportion of all segments, they determine the hydrodynamic layer thickness of the adsorbed polymer. Many other properties of adsorbed polymers are determined by the total segment concentration profile as a function of the distance from the surface.

One very important quantity that characterises an adsorption process is the surface excess (z^*) of polymer. This has the same definition as was introduced in section 5.1 (see figure 5.1):

$$z^* = \int_0^\infty (\phi(z) - \phi_\infty)\,dz. \qquad (5.2.26)$$

Here $\phi(z)$ is the volume fraction as a function of the distance z from the surface, with ϕ_∞ the bulk volume fraction. If the polymer solution is sufficiently dilute then z^* can be interpreted as the adsorbed amount of polymer. A plot of the equilibrium amount adsorbed as a function of bulk solution

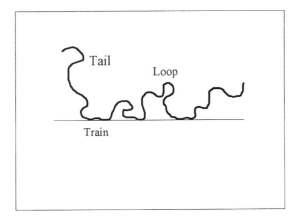

Figure 5.10. The loop–tail–train model of an adsorbed polymer.

---

[*] Although χ_s is referred to as an energy parameter and U_{ia} are adsorption energies it is clear that they are actually viewed as *enthalpies* by the original authors of the self-consistent field theory of polymer adsorption. This needs to be appreciated in reading the literature since great confusion can arise otherwise.

concentration is known as an adsorption isotherm. Adsorption isotherms of polymers are typically of the so-called high-affinity type, i.e. there is an increase to a plateau value in a very short range of ϕ_∞, as sketched in figure 5.11. (Volume fractions may be converted to mass per unit volume and z^* to mass per unit area by multiplication by the partial specific volume of the polymer in the solution.)

The shape of the adsorption isotherm is clearly affected by the magnitude of the segmental adsorption energy χ_s defined above. The thermodynamics of the bulk solution also influence the adsorption; for polymers in θ solutions the amount adsorbed is higher and generally this tends to increase continuously with increasing solution concentration. If the solvent quality is worse still, then one may have a very thick adsorbed layer indeed; effectively one has bulk phase separation initiated at the surface. This situation is discussed in section 5.3.

The simplest theories of polymer adsorption are mean-field theories. In fact, it is possible simply to apply the square gradient theory of surface segregation developed in section 5.1 to polymer solutions by setting the degree of poly-merisation of one of the components to unity. An analogous theory was developed by de Gennes (1981) and applied to adsorption from theta solvents (under which circumstances mean-field theory can be expected to be reason-ably accurate) by Klein and Pincus (1982).

It is, however, the self-consistent field method that has been most extensively applied to polymer adsorption and in particular the lattice-based discretisation of the Edwards modified diffusion equation associated with Scheutjens and Fleer (1979, 1980). In this model the solution up to the impenetrable adsorbing surface is modelled as a lattice of equal volume cells. All of the lattice layers

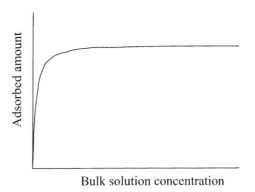

Figure 5.11. A sketch of a typical polymer adsorption isotherm for a polymer in a good solvent adsorbed onto a solid surface, showing the characteristics of a high-affinity isotherm.

parallel to the surface are planes of equal potential energy for the polymer segments (the potential energy of relevance here being the Gibbs free energy of each plane). At large distances from the surface this potential is set to zero; we are concerned with the excess Gibbs free energy over that of the bulk solution. As a result each polymer segment experiences a different potential depending on its distance from the surface and they distribute themselves over the lattice according to a Boltzmann distribution, resulting in a concentration profile normal to the surface. The energy of a polymer molecule is the sum of all the potential energies of the individual segments. Consequently a particular conformation defined by the lattice cells occupied may have a high or low probability of occurring in the total population of conformations. The statistical weight of any conformation is generated from $\prod \lambda_i G_i(z)$, where λ is the fraction of nearest neighbour lattice cells to that occupied by the polymer segment in layer i (the bond weighting factor) and $G_i(z)$ is the segment weighting factor in layer i a distance z from the surface, which is the usual Boltzmann factor

$$G_i(z) = \exp\left[-U_i(z)/(k_B T)\right]. \qquad (5.2.27)$$

In effect a step-weighted walk on the lattice is carried out; here lies the connection with the Edwards modified diffusion equation (equation (4.3.2)) discussed in section 4.3, which is the mathematical description of such a random walk in continuous space (this correspondence has been made explicit by Shull (1991)). The concentration profile of polymer normal to the impenetrable surface is obtained by calculating the average occupancy of each lattice layer by the segments and this is obtained from a recurrence relation. This recurrence relation expresses the average statistical weight of a particular segment being in a selected lattice plane as the product of the average occupancy of the layer by the preceding chain segment and the segment weighting factor, $G_i(z)$. By this means the distribution of each individual segment is obtained and the volume fraction as a function of the distance, z, from the absorbing surface is calculated by summing the recurrence relations for each segment and multiplying by a normalisation constant to conserve mass. The self-consistency is imported via the potential profile, $U(z)$, since only one profile will be compatible with the necessary condition that the sum of the volume fraction of all species over all lattice planes must be unity. Although $U(z)$ has contributions from many sources such as external fields and the internal energy of each segment, the most important are the pseudo-hard-core repulsions engendered in the excluded volume effects. In the absence of specific interactions between polymer and surface, calculations of this type show that, in polymer solutions, there is a depletion zone near the wall (figure 5.12). This is due to an

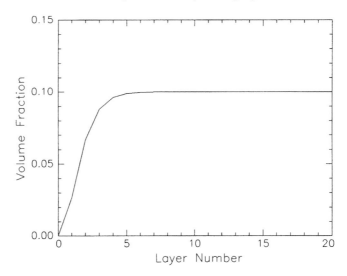

Figure 5.12. The volume fraction profile near a non-interacting wall for a polymer with 1000 monomer units per molecule, calculated by the Scheutjens–Fleer method,showing a depletion zone near the wall.

unfavourable loss of conformational entropy because polymer molecules become less like a random coil near the surface and have a more planar conformation.

Two factors contribute in influencing the adsorption from solution, both of which contribute specific interaction terms to the mean field, $U(z)$, used in calculating the statistical weight terms. The most immediately apparent interaction term is the existence of a favourable adsorption energy, per segment, U_2, i.e. U_{2a} is more negative than the adsorption energy for a solvent molecule, U_{1a} (note that $-(U_1^a - U_2^a) \simeq$ change in interfacial tension). Even where this difference is small there is a considerable effect on the adsorption of the polymer on to a surface. This is because the adsorption energy is *per segment* and adsorption of many chain units is accompanied by a large decrease in the potential energy because many chain units are adsorbed and as one unit becomes detached others become attached. This is the reason why removal of adsorbed polymer chains from a surface is often difficult.

The second contribution influencing polymer adsorption from solution is the Flory–Huggins interaction parameter between polymer and solvent. Such an enthalpy of mixing term adds a contribution of $\chi_1(\phi_2(z) - \phi_2^b)$ to the interaction energy contribution, where ϕ_2^b is the bulk solution concentration of the polymer (2) and, when χ_1 approaches thermodynamically poor values, then the adsorbed amount of polymer increases significantly. Figure 5.13 shows experi-

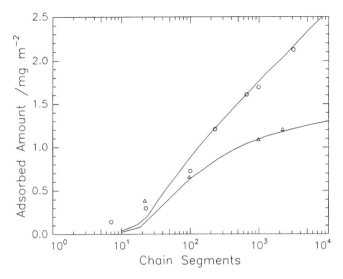

Figure 5.13. The amount of polystyrene adsorbed on to SiO_2 from cyclohexane at 35 °C (circles) and carbon tetrachloride (triangles), as a function of the degree of polymerisation. Lines are self-consistent mean-field calculations using $\chi_1 = 0.5$ for cyclohexane and $\chi_1 = 0.4$ for carbon tetrachloride solutions. Adapted from van der Linden and van Leemput (1978).

mental data compared with self-consistent field calculations for which only the adsorption energy has been used as an adjustable parameter. Note the weak dependence on the relative molecular mass of the adsorbed amount for the thermodynamically good condition ($\chi_1 = 0.4$) whereas, under the theta condition ($\chi_1 = 0.5$), a strong dependence on the polymer relative molecular mass is predicted and observed.

5.2.4 Scaling theory of polymer adsorption

As we have seen, mean-field theories can conveniently compute predicted volume fraction profiles for any set of experimental conditions. However, a fundamental question mark hangs over their validity for situations in which the polymer concentration falls into the semi-dilute regime over any part of the profile. In these situations, we can obtain some insight from scaling theory. The quantity we must focus on is the correlation length ξ, which, as we saw in section 5.2.3, is a strong function of the concentration or volume fraction, varying with the inverse $3/4$ power. Since the volume fraction of polymer decreases with the distance from the adsorbing surface, then the screening length, $\xi(z)$, also increases with the distance z. De Gennes (1981) argued that because z is the only length scale available, then $\xi(z)$ must scale in the same way as z itself ($\xi(z) \sim z$). This implies, in conjunction with equation (5.2.21) for the concentration dependence of ξ, namely $\xi(\phi) \sim \phi^{-3/4}$, that

$$z \sim \phi(z)^{-3/4} \tag{5.2.28}$$

and hence we can write

$$\phi(z) \sim z^{-4/3}. \tag{5.2.29}$$

An interesting feature of equation (5.2.29) is that it represents a *self-similar* solution for the adsorption profile. This is illustrated in figure 5.14.

Another way of thinking about this result draws on the notion we introduced in section 5.1 that the concentration perturbation introduced by the surface decays to the bulk over a distance characterised by the bulk correlation length. For a system such as a symmetrical polymer blend, for which the correlation length is a weak function of the concentration, this leads to an exponential adsorption profile

$$\phi(z) \approx \phi_1 \exp(-z/\xi). \tag{5.2.30}$$

If, however, the correlation length is a strong function of the concentration, then we must replace equation (5.2.30) by an implicit equation for $\phi(z)$:

$$\phi(z) \approx \phi_1 \exp(-z/\xi(\phi(z))), \tag{5.2.31}$$

where we assume that $\xi(\phi(z))$ takes the same form as the concentration dependence of ξ for the homogeneous case $\xi(\phi)$. This equation is approximately solved by equation (5.2.29) if we neglect logarithmic terms.

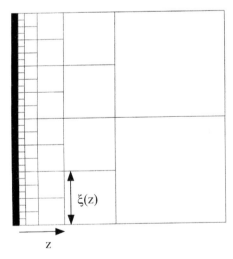

Figure 5.14. The self-similar concentration profile for an adsorbed polymer in the semi-dilute regime. At a distance z from the wall, the size of the correlation length is itself of order z. After de Gennes (1981).

For equation (5.2.29) to hold exactly, one would need the relative molecular mass of the polymer to be infinitely great, so that even vanishingly small concentrations remained in the semi-dilute regime, and the adsorption energy to be relatively weak, so that the solution remains semi-dilute even close to the wall. In reality neither of these limits applies, so the applicability of equation (5.2.29) must be restricted to an intermediate range of distances, $D < z < \xi(z = \infty)$. The short-range cut-off D is the distance at which the adsorption energy per segment has a dominant effect (this is called the proximal region), whereas the upper cut-off is defined by the condition that z approaches the correlation length of the bulk solution: $\xi(z = \infty)$. In this so-called distal region the adsorption profile is once again characterised by a single length scale, the correlation length of the bulk solution, and the volume fraction is given by

$$\phi(z) = \phi_b[1 + \exp(-z/\xi(z = \infty))]. \tag{5.2.32}$$

For proximal and central regions, de Gennes originally proposed a single form for $\phi(z)$:

$$\phi(z) \sim \phi_s\left(\frac{4D/3}{z + 4D/3}\right)^{4/3}, \tag{5.2.33}$$

with the surface volume fraction ϕ_s given by

$$\phi_s \sim \left(\frac{a}{4D/3}\right)^{4/3}, \tag{5.2.34}$$

where a is the statistical step length. Subsequently, alternative forms of these equations were proposed by de Gennes (1983):

$$\phi(z) \sim \phi_s \quad (0 < z \leqslant a), \tag{5.2.35}$$

$$\phi(z) \sim \phi_s(a/z)^{1/3} \quad (a < z \leqslant D), \tag{5.2.36}$$

$$\phi(z) \sim \phi_s\left(\frac{a}{z}\right)^{1/3}\frac{2D}{z + D}, \tag{5.2.37}$$

with $\phi_s \sim a/D$. An expression for the adsorbed amount can be obtained by integrating each equation over the specified region and adding the terms together. For good solvent conditions (and we remind ourselves that the expression $\xi \sim \phi^{-3/4}$ assumes such conditions) no dependence on the relative molecular mass of the adsorbed polymer is predicted. If the scaling relation for poor solvent conditions is used ($\xi \sim \phi^{-1}$), then the adsorbed amount has a natural logarithmic dependence on the degree of polymerisation. Figure 5.15 shows the form of the volume fraction profiles predicted by equations (5.2.33) and (5.2.35)–(5.2.37).

The scaling and self-consistent field approaches to polymer adsorption

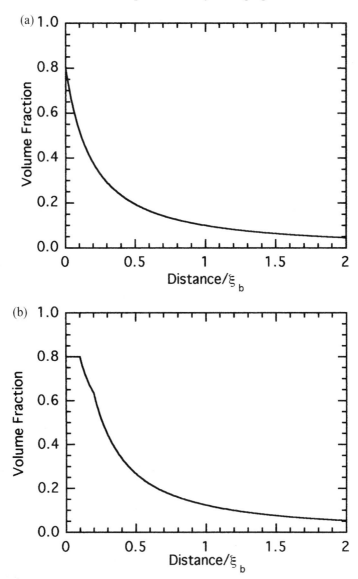

Figure 5.15. Volume fraction profiles for an adsorbed polymer predicted by two forms of the scaling theory, represented by equations (5.2.33) and (5.2.35)–(5.2.37).

theory are very different in philosophy, so it is interesting to try to find some points of contact. Although the scaling theory as presented above is specifically constructed to include non-mean-field effects, it is possible to construct a scaling version of a mean-field theory whose results can be directly compared with the outputs of self-consistent mean-field calculations. This can be done by exploiting a well-known result from the theory of phase transitions, that is that in general mean-field theories would be accurate if we lived in a world with four space dimensions. In such a world, even in a good solvent, polymers would follow ideal random walk statistics (the extra space provided by the fourth dimension disposes of the excluded volume problem) and the correlation length would depend on the inverse square root of concentration. Thus a mean-field concentration profile should scale like the inverse square of the distance:

$$\phi(z) \sim z^{-2}. \tag{5.2.38}$$

This scaling law was compared with the results of self-consistent field theory by van der Linden and Leermakers (1992); they found that the profiles did follow a power law over the central region. In the limit of vanishing bulk volume fraction and infinitely long chains the power law exponent did indeed tend towards 2 as predicted by equation (5.2.38), but the corrections for finite relative molecular mass and bulk volume fractions are considerable. For calculations on a cubic lattice they found that the power law exponent α could be represented by

$$\alpha = -2 + 1.87(\ln \phi_\infty)/r^{0.5}, \tag{5.2.39}$$

where r was the number of lattice units of the polymer, proportional to the degree of polymerisation. As yet it is not known whether a similar expression to modify the scaling results for finite relative molecular mass and concentration can be found, but this work certainly suggests that substantial corrections to the scaling results should be expected.

One strength of the numerical self-consistent field method over the scaling approach as originally formulated is that the self-consistent field method allows one to calculate explicitly and separately the contributions of loops and tails to the volume fraction profile. This is important for comparing theory with experiment, because several properties of adsorbed layers, particularly their hydrodynamic properties, are likely to be dominated by the effect of the tails. A version of the scaling theory that accounts for loops and tails separately has been constructed (Semenov and Joanny 1995); this showed that equation (5.2.2) was applicable only for distances closer to the wall than a characteristic length that scales as the square root of the degree of polymerisation (note that this is much smaller than the size of the polymer in a good solvent). Within this region, the profile is dominated by loops, whereas the outermost part of the profile is dominated by tails. This picture is qualitatively in agreement with the results of numerical self-consistent field theory (Johner *et al.* 1996); once again a direct comparison between the numerical results and scaling laws constructed for the mean field case can be made.

5.2.5 Polymer adsorption: comparison between theory and experiment

A huge volume of data on polymer adsorption has now been gathered, but comparison between theory and experiment is not usually straightforward. Few techniques give direct access to volume fractions and in any case extraction of volume fraction profiles from experimental data may be incomplete because the tails became so dilute that their contribution to the observed signal is minimal. However, it is the tails that dominate the hydrodynamic properties due to interactions between segments distant from the two absorbing surfaces of two particles. A summary discussion of the structure of adsorbed polymer layers from both theoretical and experimental viewpoints emphasises the role of the various interactions in the system (e.g. polymer–solvent, solvent–surface etc.) (Cosgrove 1990).

Much experimental work is reported for adsorption on to solid particles; this has the benefit that adsorbed amounts can easily be obtained, but the results may sometimes be surprising. An example of this are the results for the adsorption of polydimethyl siloxane on to silica from solutions in hexane and carbon tetrachloride (Patel *et al.* 1991), the latter being the poorer solvent. In these circumstances a greater adsorbed amount is expected from the solution in the poorer solvent. This was not observed, because carbon tetrachloride also has an affinity for the silica surface and was not displaced by the polymer. A further interesting comparison made in this paper, that bears directly on the discussion above about the relative importance of loops and tails, is the difference in adsorption from a good solvent between cyclic and linear polymers. Figure 5.16 summarises their findings. There is a crossover in adsorbed amount for linear and cyclic polymers at a certain relative molecular mass.

Cyclic polymers are unable to form tails and hence the conformational energy change on adsorption is less for the cyclic polymer at low relative molecular masses when the surface concentration is small. As the relative molecular mass increases, cyclics form larger loops that reduce the entropy change on adsorption whereas in the linear polymer the contribution of the tails to the entropy change becomes diluted at higher relative molecular masses. Figure 5.17 shows the volume fraction profiles calculated from self-consistent field theory, with the separate contributions from loops and tails.

Cosgrove *et al.* (1990) used small-angle neutron scattering on a polystyrene latex in water to obtain the adsorbed layer thickness of polyethylene oxide at three different temperatures. At higher temperatures a greater adsorbed amount was observed, which was consistent with water becoming a poorer solvent for the polyethylene oxide, but in all cases the layer thickness was about 20–30 Å, an order of magnitude less than the hydrodynamic layer thickness. Neutron

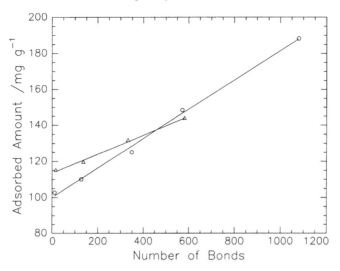

Figure 5.16. Maximum amounts of cyclic (△) and linear (○) poly(dimethyl siloxane) adsorbed onto silica from hexane solution. Adapted from Patel *et al.*(1991).

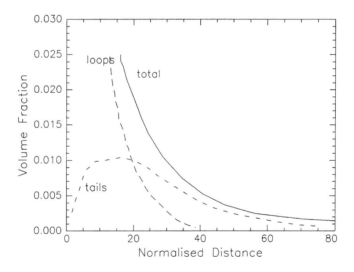

Figure 5.17. The volume fraction profile of a linear adsorbed polymer calculated from self-consistent field theory, with the separate contributions from loops and tails. Adapted from Fleer *et al.* (1993).

reflectivity in principle should provide a more definitive description of the volume fraction profile but questions about uniqueness enter in interpreting the data, as remarked by Richardson *et al.* in interpreting their data for the adsorption of polystyrene from cyclohexane on to mica (Cosgrove *et al.* 1991). Mica has the benefit of being molecularly smooth; other surfaces can be rough and the reflectivity data for the adsorption of polyethylene oxide on to quartz illustrate some problems that may occur (Lee *et al.* 1990). Figure 5.18 shows the volume fraction profiles obtained from the reflectivity data, with the polymer apparently penetrating the 'impenetrable' surface.

Although there has been continuing interest in the shape of the volume fraction profile of the adsorbed polymer, it is arguable that the adsorbed amount is the more important parameter. For small values of the adsorbed amount the molecules are relatively isolated on the surface and all segments become attached. As the adsorbed amount increases, loops and tails form and the concentration of these in the solution immediately adjacent to the surface determines whether dilute, semi-dilute or concentrated conditions prevail and thus the form of the concentration profile is also controlled by the adsorbed amount. When the adsorbed amount becomes very large the adsorbed layer will adopt a structure that is more brush-like than a loop, train, tail type of structure. Consequently, it is disappointing that, as Fleer *et al.* (1993) comment, examples of good experiments on well-defined systems investigating a wide range of parameters are scarce. Figure 5.19 shows the dependence of the adsorbed amount on relative molecular mass for polystyrene adsorbing on to silica. The adsorbed amounts are taken from the plateau regions of the adsorption isotherm for each relative molecular mass. The theoretical predictions (see figure 5.13) are that for good solvents the adsorbed amount becomes independent of relative molecular mass, whereas in theta solvents it apparently increases without limit. The results for polystyrene in carbon tetrachloride (a good solvent for polystyrene) illustrate the different results that can be obtained on different systems when different measuring techniques are used.

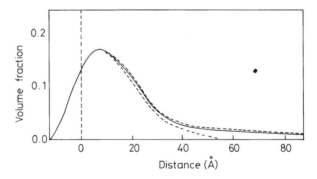

Figure 5.18. Volume fraction profiles obtained by neutron reflectivity for poly(ethylene oxide) adsorbed at the quartz/water interface. The vertical dashed line indicates the position of the quartz/water boundary. Dashed lines on the volume fraction curve show the limits of uncertainty at long distances from the boundary. After Lee *et al.* (1990).

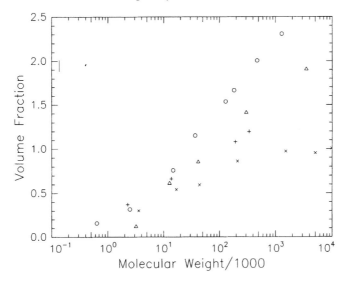

Figure 5.19. The dependence of the amount adsorbed on the relative molecular mass for polystyrene adsorbing onto silica from theta solvents ($\circ$, cyclohexane and Δ, decalin) and a good solvent ($+$ and $\times$, carbon tetrachloride). Adapted from Fleer *et al.*(1993).

A major experimental complication for adsorption at a solid/liquid interface is that such interfaces can be very variable due to differences in cleaning procedures. Additionally, because solid surfaces are rigid, the attainment of an equilibrium adsorbed conformation may be difficult. Furthermore, adsorption may be located in regions of high specificity and thus the experimental observation may correspond to a thick layer (i.e. the hydrodynamic radius) whereas the adsorbed amount suggests a much flatter conformation.

Some of these issues have been addressed in a series of experiments using infra-red dichroism on poly(methyl methacrylate) adsorbing on to oxidised silicon from carbon tetrachloride (Frantz and Granick 1995). An earlier paper (Frantz and Granick 1992) discussed the subject of preparing the silicon in a reproducible manner for adsorption of polymer. Using FTIR-ATR they were able to determine the surface excess, the bound fraction of carbonyl groups and the dichroic ratio of the adsorbed poly(methyl methacrylate). Figure 5.20 shows the change of bound fraction with adsorption time.

These results show that polymer molecules arriving first on the surface are adsorbed with a flat conformation. Later adsorbing chains have fewer surface sites available and thus the bound fraction decreases. Over a time period of 3 h no conformational rearrangements were observed, suggesting that rearrangement of chains adsorbed early to form loops and tails in response to higher

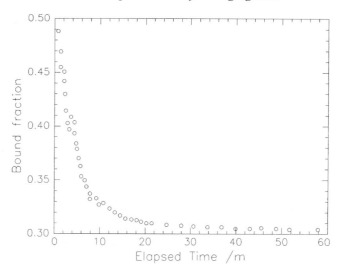

Figure 5.20. The evolution of the bound fraction with time for the adsorption of poly(methyl methacrylate) onto silica from a solution in carbon tetrachloride at 0.01 mg mol^{-1}. Adapted from Frantz and Granick (1995).

surface concentrations does not occur. The extended volume fraction profiles obtained for adsorbed polymers are due to the chains adsorbing later that are forced to occupy an extended spatial region because very few adsorbing sites are available.

The comments above regarding the possible problems with solid/liquid interfaces indicate the advantages of the liquid/air interface as an absorbing surface, notably that it is flat (but see chapters 3 and 8 regarding capillary waves) and reproducible. Only a few studies have been undertaken at such interfaces. Perhaps the most complete of these studies is a neutron reflectivity study of poly(dimethyl siloxane) in toluene (Guiselin *et al.* 1991). Solution concentrations were such that only the dilute range was studied but the range of relative molecular mass was from $19\,000-4\,200\,000$ and both deuterated and hydrogenous combinations of polymer and solvent were employed. Apart from the scaling law relation (equation (5.2.33)) for the volume fraction profile, a two-layer and an exponential profile were employed to fit to the reflectivity data. From the values of the fitting criterion (χ^2) it was concluded that the scaling law profile was the best description and from their data they obtained for the profile

$$\phi(z) = \phi_s \left(\frac{d}{(z-D)+d} \right)^{\mu}, \quad D < z < L$$

with $\mu = 1.3$, $d = 5.2$ Å, $D = 5.1$ Å and ϕ_s, the surface volume fraction i.e.

between $z = 0$ and $z = D$, was 0.96. The dependence of the central region length scale, L, on relative molecular mass is shown in figure 5.21. The parameter L is proportional to the relative molecular mass, in the lower range L equals the radius of gyration whereas in the higher range the value is somewhat smaller.

Neutron reflectometry has also been used to determine the structure of the surface excess layer in aqueous solutions of polyethylene oxide (Lu *et al.* 1996). Only two relative molecular masses were used ($\simeq 20\,000$ and $\simeq 90\,000$) but no differences in the reflectivities and the volume fraction profiles were observed for the 0.1 wt% solutions used. Volume fraction profiles that were either three layers or the sum of two half Gaussians were both found to represent the data well; figure 5.22 shows both of these models of the surface excess for the poly(ethylene oxide) with a relative molecular mass of about 20 000.

Lastly in this section we note that Kent *et al.* (1992) have used x-ray evanescent (XEWIF) wave-induced fluorescence to study the adsorption of poly(dimethyl siloxane) at the air/liquid interface, the liquids being bromoheptane (a good solvent) and bromocyclohexane (a theta solvent at 29 °C). The signal is rather too noisy to allow a definitive and quantitative comparison with theory. In the good solvent the surface region was inferred to have a polymer

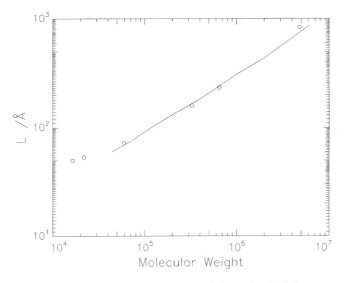

Figure 5.21. The dependence of the thickness of the adsorbed layer on the relative molecular mass of poly(dimethyl siloxane) adsorbed from toluene solution to the solution/air interface, as measured by neutron reflectivity. The solid line represents $L \sim 0.13 M^{0.57}$. Adapted from Guiselin *et al.*(1991).

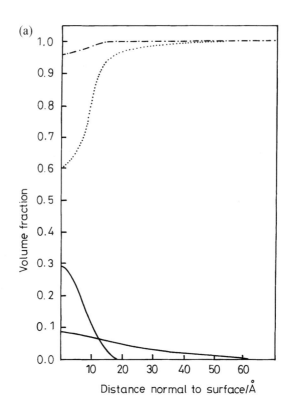

(a)

Volume fraction

Distance normal to surface/Å

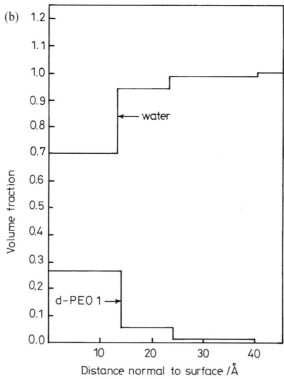

(b)

water

d-PEO 1

Volume fraction

Distance normal to surface /Å

volume fraction of about 0.9 that fell to about 0.2 within the first 20 Å from this surface region. For near-theta conditions the adsorbed amount was significantly larger as predicted both by scaling law and by self-consistent field theories.

5.3 Wetting and surface-driven phase separation in polymer solutions and mixtures

In section 5.1 we introduced the idea of surface segregation as the result of a free energy trade-off between surface terms, favouring segregation of the low-energy species, and a bulk term, that resists the creation of a large volume of material not at the bulk composition, the connectivity of the chain imposing a constraint that penalises concentration gradients too steep compared with the overall chain dimensions. The same principle survives for polymer adsorption from solution, but, as we have seen, it is much less easy to implement it as a basis for calculating adsorption profiles due to the intrinsic failure of mean-field theories for polymer solutions. Nonetheless, both for melts and for solutions we might expect that, as our mixture became closer and closer to the point of phase separation, the amount of material adsorbed or segregated would become greater, as the energetic cost of maintaining a layer at different composition to the bulk becomes smaller.

We can illustrate this by using the mean-field square gradient theory of section 5.1 to calculate surface segregation profiles in a polymer mixture approaching phase separation. Figure 5.23 shows how the profiles, calculated using equation (5.1.11), vary as the point of phase separation is approached in two ways. If we approach the critical point, the segregation profile is always monotonically decreasing, but the length over which the adsorption profile decays to the bulk value increases as the critical point is approached. We might expect this qualitative behaviour, because we know that the correlation length for bulk concentration fluctuations diverges at the critical point and it is this correlation length that characterises the distance over which the perturbation induced by the surface is healed. Of course, close to the critical point the kind of mean-field theory we have used is invalid; instead we expect this phenomenon of *critical adsorption* to be characterised by a power law decay with a non-classical exponent (Fisher and de Gennes 1978, Zhao *et al.* 1995).

Figure 5.22. The surface excess layer in a 0.1% wt/vol solution of poly(ethylene oxide) in water. (a) Segment distributions modelled as two half-Gaussians (solid lines); the dotted line is the volume fraction of water, whereas the dash–dot line is the total volume fraction. (b) The result of a three-uniform-layer fit to the same data. Adapted from Lu *et al.* (1996).

(a)

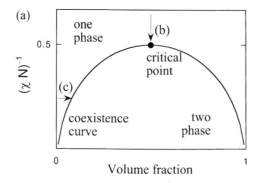

(b)

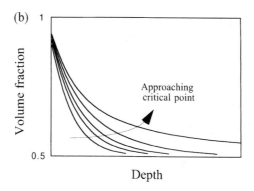

(c)

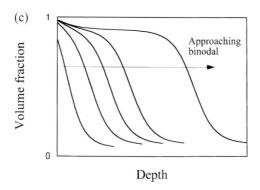

Figure 5.23. Segregation or adsorption profiles as the point of phase separation is approached, calculated for a symmetrical polymer blend whose phase diagram is shown in (a) (top), calculated using equation (5.1.11). In (b) we approach the critical point by changing the temperature or interaction parameter while keeping the concentration constant at the critical value, resulting in critical adsorption. In (c) the interaction parameter or temperature is kept constant and the concentration is increased towards its value on the binodal or coexistence curve, where we have a wetting layer at the surface.

In the much more common situation in which the two-phase region is approached not through the critical point, but across the binodal or coexistence curve, we still expect to find that the total amount adsorbed becomes very large, but the way that this happens is very different from the critical adsorption case. Instead of having a monotonic decay with an increasing decay length, we find that, as the binodal is approached, a thick layer of material at the other coexisting composition breaks away from the surface. These profiles are now characterised by two length scales. The correlation length, that of course is finite on the coexistence curve, characterises the width of the interface between the surface layer and the bulk. This interface is exactly the same as the interface between the corresponding bulk coexisting phases. Meanwhile another length, that does diverge as the coexistence curve is approached, describes the thickness of the layer. At equilibrium, then, in a phase-separated system, we expect the surface to be coated by a macroscopically thick layer of the phase of lower surface energy. In practice the thickness of this layer will be limited by gravity or by the amount of material actually present in a finite system. Such a layer is called a *'wetting layer'* (Cahn 1977, de Gennes 1985).

Because the layer is macroscopically thick and the interface between the surface layer and the bulk is identical to that between bulk coexisting phases, we can write down an inequality for the surface tensions of the two phases and the interfacial tension between the phases, on the basis of the idea that, to have perfect wetting, the Young–Dupré equation must not have a solution corresponding to a finite contact angle. Writing the surface energies of the two phases as σ_β and σ_γ and the interfacial tension between them as $\sigma_{\beta\gamma}$, we have

$$\sigma_\gamma > \sigma_\beta + \sigma_{\beta\gamma}. \tag{5.3.1}$$

Note that we can always expect this inequality to be satisfied close enough to the critical point, because, as we saw in section 4.2, as one approaches the critical point the interfacial tension goes to zero[§]. Further away from the critical point, this inequality will not be satisfied and the phases must meet at a finite contact angle, as shown in figure 5.24b. Instead of having a uniform wetting layer, we will have lenses of one phase at the surface, with a contact angle θ_c given by

$$\sigma_\beta \cos(\theta_c) + \sigma_{\beta\gamma} = \sigma_\gamma. \tag{5.3.2}$$

How can we reconcile this with our microscopic picture of the wetting layer derived from the square gradient theory? To do this we must go back to

---

[§] Of course the difference between the surface tensions of the two phases also goes to zero at the critical point, but it can be shown that the interfacial tension goes to zero more quickly (Cahn 1977).

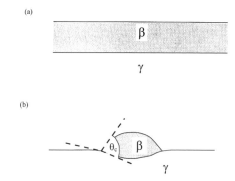

Figure 5.24. (a) Phase β completely wets phase γ: the inequality (5.3.2) holds. (b) Phase β only partially wets phase γ; the contact angle ϕ_c is given by the Young–Dupré equation (5.3.2).

equation (5.1.10) that defines the surface volume fraction and its graphical interpretation in figure 5.2.

Plots of these so-called phase portraits are shown in figure 5.25. If the bulk volume fraction is on the coexistence curve, then the bulk term of equation (5.1.10), that expresses the differential free energy cost of maintaining a layer at a different composition from that of the bulk, develops another zero at the other coexisting volume fraction. This merely follows from the definition of coexistence; the two phases have equal chemical potential so some of one phase can be converted to the other with no free energy cost. This means that equation (5.1.10) can now have two possible solutions instead of one. One solution corresponds to a surface volume fraction higher than the other coexisting volume fraction – this is the wetting phase. However, the other solution has a surface volume fraction less than the other coexisting phase; here we have a concentration profile with a small degree of surface segregation, which decays back to the bulk composition over a distance characterised by the polymer radius of gyration. This corresponds to the phase marked γ in figure 5.24b. One important point to remember is that, even when two polymers are very immiscible, at equilibrium there will be a thin layer of the polymer of low surface energy coating the surface of the incompletely wet phase; this means that the surface tension will be lower than that of the pure material, even though the solubility of the minority polymer may be vanishingly small.

To calculate under what conditions the transition between these two situations occurs, it is necessary to compute the surface excess free energy associated with each of the two solutions obtained from the phase portrait. The surface excess free energy per unit area F_s is given by equation (5.1.11); in figure 5.26 the resulting surface excess free energies are plotted for both

(a)

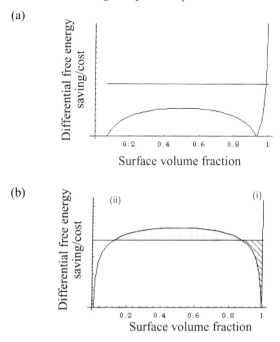

(b)

Figure 5.25. Phase portraits when the bulk concentration is at coexistence. In (a), not too far from the critical point, there is only one solution for the surface volume fraction, that corresponds to the case of complete wetting. In (b), for a more highly immiscible situation, there are two solutions, (i) and (ii). The other intersection corresponds to a maximum, not a minimum, of the free energy. Solution (i) corresponds to complete wetting, whereas solution (ii) corresponds to partial wetting; the two solutions have the same free energy when the two shaded areas are equal.

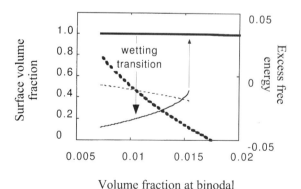

Figure 5.26. A wetting transition. The solid, bold line is the surface volume fraction as a function of the volume fraction at the binodal as the interaction parameter is changed, whereas the bold, dashed line is the corresponding free energy for the wet solutions. The thinner solid line and the thinner dashed line are the surface volume fraction and corresponding free energy for the partially wet solution. The point at which the free energies of each solution are equal marks the wetting transition.

solutions for the surface volume fraction. When the volume fraction on the binodal is small, that is when the value of χ is large, the solution with a lower surface volume fraction has a lower free energy. Thus, in this highly incompatible situation, the equilibrium situation is for the surface to be partially wet. However, if conditions such as temperature are changed to make the system more miscible, so that χ decreases and the volume fraction on the binodal is increased, we reach a point at which the completely wet solution has the lower excess free energy. The conditions under which the free energies of the partially wet and completely wet solutions are the same mark the wetting transition, that one can see in this example is a true thermodynamic first-order phase transition.

The wetting transition has attracted a huge amount of theoretical study (see for example Dietrich 1987); it is closely related to a wide range of other interesting phenomena in surface physics, such as surface magnetic phase transitions, transitions to multilayer adsorption in the study of clean surfaces in UHV and surface melting. Topics discussed include the situations in which the transition may be second order rather than first; in that case interesting fluctuation phenomena might be expected and one has the possibility of a precursor transition in the one-phase region – the pre-wetting transition. These interesting and exotic phenomena are likely to be as relevant to polymers as to other systems and indeed there are arguments suggesting that polymers may form ideal model systems in which to search for these effects. However, studies of these fundamental aspects of wetting in polymers have been few indeed and so we shall not consider them further.

Instead we can take a cruder approach that will give us some feel for the likely location of wetting transitions in polymer mixtures. It turns out that, after some approximations, one can use the square gradient theory outlined above to construct a universal wetting phase diagram for polymer mixtures, subject to the restriction that the components' relative molecular masses are equal (Jones 1994). At this level of approximation the location of the wetting transition is controlled by the balance of two parameters, the degree of incompatibility, χN, and the difference in surface energy between the two polymers divided by the square root of the Flory–Huggins interaction parameter χ. Because we know that the interfacial tension between incompatible polymers scales as the square root of χ (see chapter 4), we can see that this reduced surface energy parameter reflects the relative importance of the surface energy difference and the interfacial tension.

The resulting surface phase diagram is plotted in figure 5.27. What is striking from this plot, particularly if one considers the range of parameter values that will be encountered in practice, is that quite large degrees of incompatibility

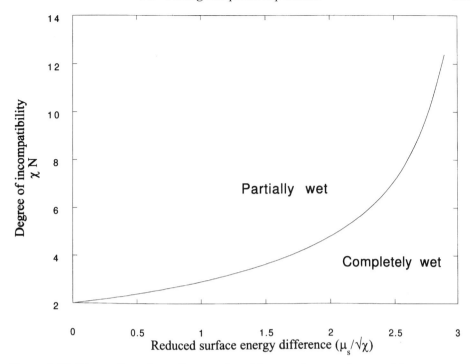

Figure 5.27. The surface phase diagram for a binary polymer mixture with components of equal relative molecular mass. After Jones (1994).

are required in order to yield partial wetting. This is in contrast to the situation for small molecules, for which complete wetting is usually obtained only for temperatures rather close to the critical point. The reason for this difference lies in the fact that interfacial tensions between even quite strongly immiscible polymers are rather small.

Experimental studies concerning polymer wetting over a wide range of parameters have been rather rare. Part of the reason for this is that equilibrium is reached rather slowly in polymer systems; the corollary of this, however, is that the kinetic aspects of wetting, which are themselves of considerable intrinsic interest, are particularly relevant to polymers. We can consider two routes to achieving equilibrium at the surface of a polymer blend. We might create a polymer mixture in the one-phase part of the phase diagram and then change the temperature in order to reach the coexistence curve. In this case we would expect diffusion of the species of lower surface energy from the bulk to the surface and the formation of a wetting layer that grows with time. This idea is the basis of a theory of wetting-layer growth due to Lipowsky and Huse (1986). An outline of the idea behind this is sketched in figure 5.28; near the surface is a layer of material depleted in the phase of low surface energy which

(a)

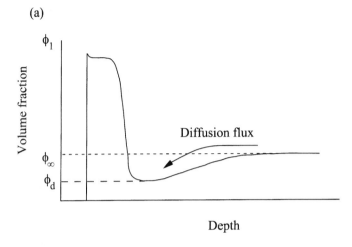

(b)

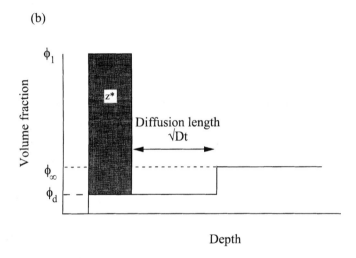

Figure 5.28. The diffusion limited model for growth of a wetting layer. (a) The true situation; a wetting layer exists in local equilibrium with the depleted concentration ϕ_d. Growth of the wetting layer is driven by diffusion of material from the bulk. (b) A schematic 'box model' for the same situation.

is in local equilibrium with the wetting layer, while diffusion of material from the bulk down the concentration gradient into the depleted region feeds the growth of the wetting layer. The problem can be formally solved by setting up a diffusion equation with a boundary condition reflecting the local equilibrium of the wetting layer and the depleted region; a cruder but instructive solution is obtained by neglecting the concentration gradients in the depleted region and observing that the size of the depletion zone must scale with time t as $(Dt)^{1/2}$, where D is the mutual diffusion coefficient for the polymer mixture. If the depleted concentration is ϕ_d, then the condition for local equilibrium is obtained by writing for the time-dependent adsorbed amount $z^*(t)$

$$z^*(t) = f(\phi_d(t)), \tag{5.3.3}$$

where the function f defines the equilibrium adsorption isotherm

$$z^*(\infty) = f(\phi_\infty). \tag{5.3.4}$$

We can also write an equation expressing conservation of matter,

$$z^*(t) = (Dt)^{1/2}(\phi_\infty - \phi_d), \tag{5.3.5}$$

and if we use (5.3.3) to eliminate the unknown ϕ_d we get an equation for $z^*(t)$ involving the inverse of the adsorption isotherm. The square gradient theory given above and indeed any mean-field theory where the surface interaction is short-ranged, will predict an adsorption isotherm that diverges logarithmically as the bulk volume fraction ϕ_∞ approaches its value on the binodal ϕ_b; thus

$$z^*(\infty) = -A \ln(1 - \phi_d/\phi_\infty). \tag{5.3.6}$$

In these circumstances growth of the wetting layer will be logarithmic in time – that is to say, the growth is much slower than the square root of time growth that one might expect for a diffusion-limited process. This has been observed in experiments using model olefin copolymers, where the growth of the wetting layer could be followed directly using nuclear reaction analysis (Steiner *et al.* 1992). The results are shown in figure 5.29. These characteristically slow growth kinetics lead to considerable experimental difficulties in measuring the equilibrium properties of wetting layers in polymer systems.

Another method of creating a wetting layer would involve creating a bilayer of one pure material on top of another. This then would be the situation corresponding to processes like coating and co-extrusion. If one creates a continuous film of polymer in this way, if the system falls in the partially wetting region of the surface diagram of figure 5.27, then at equilibrium the film will break up into droplets with a finite contact angle. Figure 5.27 indicates that this will most likely be the case for situations in which polymer and substrate have a moderate to large unfavourable thermodynamic interaction. Nonetheless, such a polymer coating layer may well be metastable, because the dynamics by which such a film will de-wet is a strong function of the viscosities both of the film and of the substrate polymers. For polymers of high relative molecular mass these may be very large and effectively prevent de-wetting on laboratory time scales. Experiments that probe the dynamics by which a hole in a coating layer, once it has been nucleated, subsequently grow have been carried out by Faldi *et al.* (1995) and Lambooy *et al.* (1996).

Our discussion so far has been couched in terms of polymer blends, but the

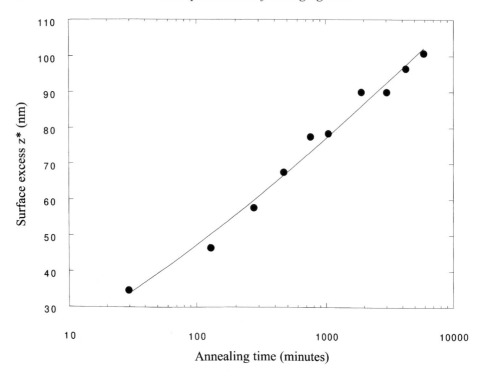

Figure 5.29. The growth of a wetting layer at the surface of a blend of ethyl–ethyl ethylene copolymers, measured by nuclear reaction analysis. After Steiner *et al.* (1992).

same concepts of wetting and critical adsorption should be equally applicable to the case of polymers in solution. In fact, rather little systematic work has been done on polymer adsorption at temperatures below the theta temperature, for which these issues are likely to become important. Some calculations from the theoretical point of view have been made; Klein and co-workers used a square gradient approach to calculate both the concentration profiles and the force profiles to be expected when mica plates are brought together under poor solvent conditions (Klein and Pincus 1982), whereas Johner and Joanny have carried out a scaling analysis (Johner and Joanny 1991).

Experimental studies have been few. Perhaps the most extensive study and certainly one of the few whose authors discussed their results in terms of wetting ideas was carried out by Johnson and Grannick (1991). They used Fourier transform infra-red spectroscopy in the attenuated total reflection mode to study adsorption of polystyrene from cyclohexane on to the native oxide of silicon. For relatively low relative molecular mass (43 900) they found an increase in the adsorbed amount as the temperature was reduced towards the cloudpoint temperature; their results (shown in figure 5.30) are consistent with

(a)

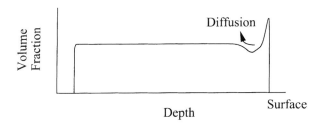

(b)

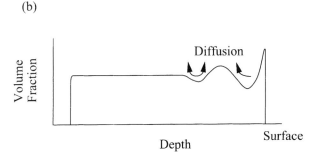

Figure 5.30. Surface-directed spinodal decomposition. In (a) material near the surface has segregated to the surface, leading to the formation of a depletion layer. In (b) material has diffused from regions of low concentration to regions of high concentration ('uphill diffusion'), which deepens the concentration minimum and leads to a secondary maximum.

a logarithmic divergence of the form of equation (5.3.6). For a higher relative molecular mass (355 000) the results were less clear; as the temperature was reduced towards the cloudpoint temperature the amount adsorbed did increase, but as the layer became thicker it became impossible to reach equilibrium on an experimental time scale. This may have been simply a reflection of the intrinsically slow dynamics of growth of wetting layers mentioned above. However a considerable complication for polymers of high relative molecular mass in solution is that, as the phase boundary is approached, not only do the chains undergo phase separation in a macroscopic way, but also each individual chain itself collapses. This is likely to have an effect both on the kinetics of formation of the wetting layer and on the equilibrium adsorption isotherm. However, the implications of this complication have not yet been worked out fully.

An interesting twist to wetting behaviour in polymer solutions arises in those cases in which a polymer solution will form a gel under certain conditions of temperature and concentration. Typically these conditions will involve high concentrations; in these circumstances a solution may be well below the concentration to form a gel, but, if a wetting layer of polymer-rich material is formed at the system's surface, then this surface layer may be above the critical concentration for gelation. Thus a thin polymer skin with quite different viscoelastic properties from those of the bulk may be formed at the surface of the solution. The term 'surface gelation' has been coined for this phenomenon, that has been identified in solutions of polystyrene sulphonate (Kim *et al.* 1984).

We began this section by considering what happened at the surface of a polymer blend as one approached either the critical point or the coexistence curve. The contrasting situation in which the system is quenched deep into the unstable, two-phase part of the phase diagram is also interesting. In the bulk, of course, we would expect the mixture to undergo phase separation by the mechanism of spinodal decomposition, as discussed in section 4.4. The presence of the surface profoundly modifies the mechanism of phase separation. Qualitatively, we might expect the first stage of the process to resemble the situation in figure 5.28 – the material of lower surface energy is attracted to the surface and leaves behind a depleted region. Because we are inside the spinodal line, the mutual diffusion coefficient is negative: material diffuses from regions of low concentration to regions of high concentration (figure 5.30). Thus the depleted region is made deeper and a secondary layer of higher concentration is formed below the surface. In this way an oscillatory concentration profile is formed near the surface. Another way of thinking about this process is to recall that the process of spinodal decomposition may be considered to lead to a superposition of composition waves of given wavelength but random directions and phases; the effect of the surface, then, is to fix the direction and phase of the waves at the surface, leading to the real-space concentration wave near the surface. The term surface-directed spinodal decomposition has been coined to describe this phenomenon.

Surface-directed spinodal decomposition was first observed in an isotopic polymer blend (Jones *et al.* 1991); thin films of a mixture of poly(ethylene–propylene) and its deuterated analogue were annealed below the upper critical solution temperature and the depth profiles measured using forward recoil spectrometry, to reveal oscillatory profiles similar to those sketched in figure 5.30. Similar results have now been obtained for a number of other polymer blends, including polystyrene with partially brominated polystyrene (Bruder and Brenn 1992), polystyrene with poly(α-methyl styrene) (Geoghegan *et al.* 1995) and polystyrene with tetramethylbisphenol-A polycarbonate (Kim *et al.* 1994), suggesting that the phenomenon is rather general.

Theoretically, we can understand this effect in the same spirit as bulk spinodal decomposition. If the Cahn–Hilliard equations are supplemented by boundary conditions expressing the fact that one component is favoured at the surface, numerical solutions reveal oscillatory concentration profiles very similar to those observed in experiment (Marko 1993, Puri and Binder 1994). The initial wavelength of the composition wave is essentially the same as the wavelength of the fastest growing concentration fluctuation in the corresponding bulk process, as observed experimentally. The coherence of the composition wave is progressively lost as one goes further away from the surface; of course, deep in the bulk one must recover the normal, isotropic spinodal decomposition pattern. The depth to which the surface-directed pattern persists is determined by the competition between the strength of the surface field – i.e. the degree of preference for one component at the surface – and the thermal noise strength. It appears that, for typical experimentally realisable situations, only one or perhaps two layers will be formed near the surface before isotropic phase separation is recovered.

In these circumstances, it is possible to be relatively certain about the mechanism of growth of the surface layer. If, at a late stage in the phase-separation process, one is left with essentially a single complete layer at the surface, with an essentially isotropic morphology in the bulk, then we can use the arguments given in section 4.5 to show that the growth of the surface layer must depend on time as $t^{1/3}$; growth of the surface layer is driven by the process of Ostwald ripening, with the surface layer, with its zero curvature, growing at the expense of the smaller droplets in the bulk. This type of growth has been very clearly observed in the PEP/dPEP system (Krausch *et al.* 1993).

To summarise this section, then, we can see that the presence of a surface or an interface can dramatically change the mechanism of phase separation. For relatively shallow quenches into the two-phase part of the phase diagram, close to the coexistence curve in the metastable part of the phase diagram, a wetting layer will grow at the surface. In some ways this is the analogue of the bulk mechanism of nucleation and growth, with the important difference that, as long as the system is on the wetting side of the wetting transition, no activation energy is required to commence the process. For deeper quenches, into the unstable part of the phase diagram, the bulk process of spinodal decomposition has its analogue in surface-directed spinodal decomposition. During the early stages of phase separation both mechanisms produce structures characterised by a length that is the same in both cases. During late stages of phase separation, no matter by which route it is initiated, when a surface layer is coexisting with bulk isotropic phase separated domains, the characteristic size of the surface layer grows with time according to the classical Lifshitz–Slyozov cube root of time law.

Before leaving this subject, it is worth asking whether these interesting phenomena can be translated into ways of making useful structures in polymer thin films. It is not difficult to imagine circumstances in which one might want to make a coating, for example, whose surface properties were different from those of the bulk. Surface-driven phase separation seems an attractive route to such structures, because in some sense the structure is 'self-assembled'. However, we should remember the point made in section 4.5; there are many routes to achieve a phase separated morphology in a polymer mixture and the method of changing the temperature to move the sample from the one-phase region of the phase diagram to the two-phase region, that is the focus of almost all academic study, is likely to be one of the least important in practice.

One of the simplest ways to create a film from an immiscible polymer blend, in a laboratory at least, is to cast it from a common solvent. There have been several experimental studies in which the surface composition of a film created in such a way has been studied, with rather complicated results. The reason for the complexity is understood if we consider what might happen in the casting process in more detail. The starting solution is in the one-phase region of the phase diagram for the ternary system comprising the two polymers and the solvent. During the casting process the solvent is removed by evaporation and, as the solvent concentration decreases, the system will cross into the two-phase part of the phase diagram. At this point phase separation between solution phases rich in each of the two polymers will be initiated. At this point it is likely that a layer of one phase will form at the surface, either due to the growth of a wetting layer or by the process of surface-directed spinodal decomposition. Any tendency towards layering due to differences in surface energy between the two phases will be enhanced by the fact that the composition of solvent through the film will not be uniform – this also breaks the symmetry of the phase-separation process promoting layering. Note, though, that the driving force for layering at this stage is not the surface energy of the pure polymers, but rather that of the respective polymer-rich solutions. The growth of the surface layer will be determined by the complex interplay of the transport of molecules in the film and the evaporation of solvent. During this process two things are happening; as the solvent leaves the system will be becoming less mobile and indeed a glass transition may be encountered, that will have the effect of completely freezing the morphology in a non-equilibrium state. On the other hand as the system becomes more concentrated, the phases will become more immiscible and the driving force for separation will become greater; it is also possible that one will cross a wetting transition, with the result that the layered structure is unstable with respect to breaking up into droplets at the surface.

The detailed theory of this situation is perhaps impossibly complex, though it is possible to imagine progress being made with computer models. However, there is no doubt about the richness of the possible morphologies that could be achieved in this way; many studies have indicated a variety of possibilities, ranging from the formation at the surface of a complete layer of one component, to cases in which there was no difference in composition at all between the surface and the bulk. For example, depth profiles deduced from ion beam analysis and neutron reflectivity for spin-cast films of mixtures of polystyrene and polybutadiene (Geoghegan *et al.* 1994) showed that almost complete layering had taken place during the casting process. On the other hand, a confocal microscope study of films cast from polystyrene/poly(methyl methacrylate) mixtures (Li *et al.* 1994) showed that the surface was decorated by a rather regular array of evenly spaced PMMA droplets, whereas in the bulk there was a much broader range of sizes of PMMA domains, all of which were much smaller than the surface drops. In this case it is tempting to interpret the results as suggesting that a layer of PMMA transiently formed during the casting process and subsequently broken up into droplets when the condition for complete wetting was no longer satisfied. In a final example, in which casting produced a layered morphology that had practically useful properties, a gas separation membrane was produced from casting a mixed solution of cellulose acetate and PMMA (Bikson *et al.* 1994). The membrane had at its surface a thin skin of PMMA that had the effect of improving its performance.

5.4 References

Bhatia, Q. S., Pan, D. H. *et al.* (1988). *Macromolecules* **21**, 2166.

Bikson, B., Nelson, J. K. *et al.* (1994). *Journal of Membrane Science* **94**, 313.

Bruder, F. and Brenn, R. (1992). *Physical Review Letters* **69**, 624.

Bruder, F. and Brenn, R. (1993). *Europhysics Letters* **22**, 707.

Cahn, J. W. (1977). *Journal of Chemical Physics* **66**, 3367.

Carmesin, I. and Noolandi, J. (1989). *Macromolecules* **22**, 1689.

Chen, Z. Y., Noolandi, J. *et al.* (1991). *Physical Review Letters* **66**, 727.

Cosgrove, T. (1990). *Journal of the Chemical Society: Faraday Transactions* II **86**, 1323.

Cosgrove, T., Crowley, T. L. *et al.* (1990). *Colloids and Surfaces* **51**, 255.

Cosgrove, T., Heath, T. G. *et al.* (1991). *Macromolecules* **24**, 94.

Cotton, J. P., Nierlich, M. *et al.* (1976). *Journal of Chemical Physics* **65**, 1101.

Daoud, M., Cotton, J. P. *et al.* (1975). *Macromolecules* **8**, 804.

Daoud, M. and Jannink, G. (1975). *Journal de Physique* **37**, 973.

de Gennes, P. G. (1979). *Scaling Concepts in Polymer Physics*. Ithaca, NY, Cornell University Press.

de Gennes, P. G. (1981). *Macromolecules* **14**, 1637.

de Gennes, P. G. (1983). *Journal de Physique Lettres* **44**, 1241.

de Gennes, P. G. (1985). *Reviews of Modern Physics* **57**, 827.

des Cloizeaux, J. and Jannink, G. (1990). *Polymers in Solution: Their Modelling and Structure*. Oxford, Oxford University Press.

Dietrich, S. (1987). Wetting Phenomena. *Phase Transitions and Critical Phenomena*. C. Domb and J. L. Lebowitz. San Diego, Academic Press. **12**.

Edwards, S. F. (1966). *Proceedings of the Physical Society* **88**, 265.

Edwards, S. F. and Jeffers, E. (1979). *Journal of the Chemical Society: Faraday Transactions* II **75**, 1020.

Faldi, A., Composto, R. J. *et al.* (1995). *Langmuir* **11**, 4855.

Fisher, M. E. and de Gennes, P. G. (1978). *Comptes Rendus de l' Académie des Sciences (Paris)* B **287**, 207.

Fleer, G. J., Stuart, M. A. C. *et al.* (1993). *Polymers at Interfaces*. London, Chapman and Hall.

Flory, P. J. (1953). *Principles of Polymer Chemistry*. Ithaca, NY, Cornell University Press.

Frantz, P. and S. Granick (1992). *Langmuir* **8**, 1176.

Frantz, P. and S. Granick (1995). *Macromolecules* **28**, 6915.

Fredrickson, G. H. and Donley, J. P. (1992). *Journal of Chemical Physics* **97**, 8941.

Freed, K. F. (1987). *Renormalisation Group Theory of Macromolecules*. New York, Wiley-Interscience.

Genzer, J., Faldi, A. *et al.* (1994). *Physical Review E* **50**, 2273.

Geoghegan, M., Jones, R. A. L. *et al.* (1994). *Polymer* **35**, 2019.

Geoghegan, M., Jones, R. A. L. *et al.* (1995). *Journal of Chemical Physics* **103**, 2719.

Guiselin, O., Lee, L. T. *et al.* (1991). *Journal of Chemical Physics* **95**, 4632.

Hariharan, A., Kumar, S. K. *et al.* (1993). *Journal of Chemical Physics* **98**, 4163.

Israelachvili, J. (1991). *Intermolecular and Surface Forces*. San Diego, Academic Press.

Jenkel, E. and Rumbach, B. (1951). *Zeitschrift für Elektrochemie* **55**, 612.

Johner, A., Bonet-Avalos, J. *et al.* (1996). *Macromolecules* **29**, 3629.

Johner, A. and Joanny, J. F. (1991). *Journal de Physique* II **1**, 181.

Johnson, H. E. and Grannick, S. (1991). *Macromolecules* **24**, 3023.

Jones, R. A. L. (1993). *Physical Review* E **47**, 1437.

Jones, R. A. L. (1994). *Polymer* **35**, 2160.

Jones, R. A. L., Kramer, E. J. *et al.* (1989a). *Physical Review Letters* **62**, 280.

Jones, R. A. L., Kramer, E. J. *et al.* (1989b). *Materials Research Society Symposium Proceedings* 133.

Jones, R. A. L., Norton, L. J. *et al.* (1990). *Europhysics Letters* **12**, 41.

Jones, R. A. L., Norton, L. J. *et al.* (1991). *Physical Review Letters* **66**, 1326.

Kent, M. S., Bosio, L. *et al.* (1992). *Macromolecules* **25**, 6231.

Kim, E., Krausch, G. *et al.* (1994). *Macromolecules* **27**, 5927.

Kim, M. W., Peiffer, D. G. *et al.* (1984). *Journal de Physique Lettres* **45**, L953.

Klein, J. and Pincus, P. (1982). *Macromolecules* **15**, 1129.

Krausch, G., Dai, C.-A. *et al.* (1993). *Physical Review Letters* **71**, 3669.

Krigbaum, W. R. and Flory, P. J. (1952). *Journal of Chemical Physics* **20**, 873.

Kumar, S. K., Yethiraj, A. *et al.* (1995). *Journal of Chemical Physics* **103**, 10 332.

Kurata, M. (1982). *Thermodynamics of Polymer Solutions*. London, Harwood Academic.

Lambooy, P., Phelan, K. C. *et al.* (1996). *Physical Review Letters* **76**, 1110.

Lee, E. M., Thomas, R. K. *et al.* (1990). *Europhysics Letters* **13**, 135.

Li, L., Sosnowski, S. *et al.* (1994). *Langmuir* **10**, 2495.

Linden, C. C. van der and F. A. M. Leermakers (1992). *Macromolecules* **25**, 3449.

Linden, C. C. van der and van Leemput, R. (1978). *Journal of Colloid and Interface Science* **67**, 48.

Lipowsky, R. and Huse, D. A. (1986). *Physical Review Letters* **57**, 353.

Lu, J. R., Su, T. J. *et al.* (1996). *Polymer* **37**, 109.

Marko, J. F. (1993). *Physical Review* E **48**, 2861.

Napper, D. H. (1983). *Polymeric Stabilisation of Colloidal Dispersions.* London, Academic Press.

Norton, L. J., Kramer, E. J. *et al.* (1995). *Macromolecules* **28**, 8621.

Pan, D. H. and Prest, J. W. M. (1985). *Journal of Applied Physics* **58**, 15.

Patel, A., Cosgrove, T. *et al.* (1991). *Polymer* **32**, 1313.

Puri, S. and Binder, K. (1994). *Physical Review* E **49**, 5359.

Richards, R. W., Maconnachie, A. *et al.* (1978). *Polymer* **19**, 266.

Richards, R. W., Maconnachie, A. *et al.* (1981a). *Polymer* **22**, 147.

Richards, R. W., Maconnachie, A. *et al.* (1981b). *Polymer* **22**, 153.

Richards, R. W., Maconnachie, A. *et al.* (1981c). *Polymer* **22**, 158.

Schaefer, D. W. (1984). *Polymer* **25**, 387.

Scheutjens, J. M. H. M. and Fleer, G. J. (1979). *Journal of Physical Chemistry* **83**, 1619.

Scheutjens, J. M. H. M. and Fleer, G. J. (1980). *Journal of Physical Chemistry* **84**, 178.

Schmidt, I. and Binder, K. (1985). *Journal de Physique* **46**, 1631.

Semenov, A. N. and Joanny, J.-F. (1995). *Europhysics Letters* **29**, 279.

Shull, K. R. (1991). *Journal of Chemical Physics* **94**, 5723.

Steiner, U., Klein, J., *et al.* (1992). *Science* **258**, 1126.

Wilson, K. G. and Kogut, J. (1974). Physical Reports C **12**, 75.

Yamakawa, H. (1971). *Modern Theory of Polymer Solutions.* New York, Harper and Row.

Yethiraj, A. (1995). *Physical Review Letters* **74**, 2018.

Yethiraj, A., Kumer, S. *et al.* (1994). *Journal of Chemical Physics* **100**, 4691.

Zhao, H., Penninckx-Sans, A. *et al.* (1995). *Physical Review Letters* **75**, 1977.

6

Tethered polymer chains in solutions and melts

6.0 Introduction

We saw in the last chapter that we can modify the properties of interfaces using polymers; for example, we can reduce the surface energy of a polymer melt by blending in a polymer of lower surface energy that segregates to the surface or we can modify a solid/liquid interface by the adsorption of a polymer at that interface. So far we have only considered simple homopolymers, but for a long time it has been realised that we can use cleverer polymer chemistry to design materials whose architecture leads to especially desirable interfacial properties. The basic idea of a surface-active polymer is the same as that for a small-molecule surfactant: we create a molecule in which various parts of the molecule have different affinities for the various parts of the interface. For example, to make a molecule that would modify the interface between two immiscible polymers we might make a diblock copolymer, composed of a sub-chain of each polymer type linked togther covalently (see figure 6.1). Such a molecule would at equilibrium segregate to the polymer/polymer interface, whose properties would thereby be modified. As we shall see, if they are carefully designed, such molecules are able to make huge changes to properties such as the interfacial energy, which can in some cases be reduced to zero, and the interfacial fracture toughness, which can be increased by more than an order of magnitude. The use of such 'compatibilisers' has become very important in improving the properties of multi-phase polymer blends. Useful modifications of interfacial properties are just as important in solutions and suspensions, for which it is often found that chains end-grafted to the solid/liquid interface of particles are more efficient at conferring colloidal stability than are adsorbed homopolymers.

  In this chapter we consider the physics which determines the interfacial properties and conformations of molecules whose architecture gives them

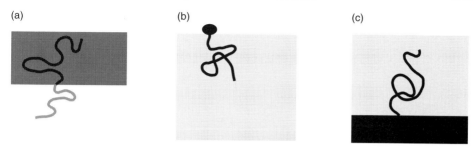

Figure 6.1. Interfacially active polymer architectures: (a) a diblock copolymer at a polymer/polymer interface, (b) a chain with surface active end-group at a polymer surface and (c) a chain covalently bound at a liquid/solid interface.

useful interfacial activity. The essence of most of these architectures (see figure 6.1) is to constrain the position of a definite part of the molecule, perhaps one end, or a segment in the middle, leaving a tail to dangle freely. A central theme in the physics of these molecules, then, is the idea of a polymer 'brush' – an array of polymer tails with one end constrained to lie in a plane and the other end free to move. Properties of such brushes are discussed in the context both of a polymer solution and of a polymer melt. Finally, we consider what happens when these intrinsically surface-active materials are present in neat form, whereupon they form microphase-separated materials whose properties are dominated by interfaces.

6.1 Polymer brushes in solution

The prototypical polymer brush is formed by grafting a set of polymer chains on to a planar solid/liquid interface. We assume that the grafting density is uniform and conveniently expressed in a dimensionless way as σ, the number of chains grafted in an area the square of the segment size a^2[§], and that the chains are monodisperse, with degree of polymerisation N. We assume that the liquid is a good solvent for the polymer, so if the statistical segment length is a, then in dilute solution the chains would have a radius of gyration given by the Flory formula $R_F \sim aN^{3/5}$. We can immediately distinguish two regimes for the grafted chains (figure 6.2); if the grafting density is such that the average distance between grafting points is greater than the radius of gyration, then each chain will be isolated from its neighbours. The result will be an array of

---

[§] In this section we make the assumption, for the sake of simplicity, that the statistical segment length b is the same as the cube root of the monomer volume a. This is not generally exactly true; it is relatively straightforward to lift this restriction but it makes the formulae less clear.

little 'mushrooms' whose dimensions must be comparable to the radius of gyration of the free chains R_F (see figure 6.2a).

As the grafting density increases the chains will start to interact with one another. Excluded volume interactions between neighbouring chains mean that they will start to stretch, until the free energy gain thus obtained, by reducing excluded volume interactions, is offset by the entropic penalty associated with stretching the chains out from their random walk configurations (figure 6.2b). The resulting array of interacting chains is known as a 'brush'.

At relatively low grafting densities, conditions inside the brush are semi-dilute. Here we can use a scaling argument (Alexander 1977) to construct the dependence of the brush height on the grafting density and degree of polymerisation of the grafted chains. The fundamental distance in the problem is the average distance between grafting points D;

$$D = a\sigma^{-1/2};$$ (6.1.1)

this gives us the size of the 'blobs' into which we can divide the grafted chains (see figure 6.3). On length scales smaller than the blob size the sub-chains follow excluded volume statistics, so we can say that each blob contains N_b subunits, where

$$D = aN_b^{3/5}.$$ (6.1.2)

Since space is filled by the blobs the polymer volume fraction ϕ_b inside the brush must have the form

$$\phi_b \sim a^3 N_b / D^3,$$ (6.1.3)

which gives us a relation between the volume fraction and the grafting density:

$$\phi_b \sim \sigma^{2/3}.$$ (6.1.4)

From this we can obtain an expression for the thickness of the grafted chain h by noting that the volume of one chain, which contains N monomers, is hD^2, from which we find

$$h \sim Na\sigma^{1/3}.$$ (6.1.5)

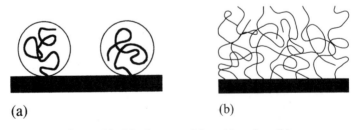

(a)   (b)

Figure 6.2. Mushrooms (a) and brushes (b).

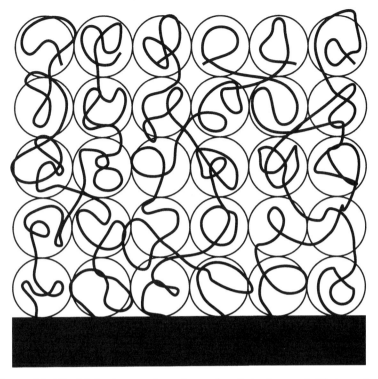

Figure 6.3. The blob picture of a polymer brush in the semi-dilute regime.

The important feature of this expression is the fact that the brush height depends on N in a linear way, rather than as the square root, as for an ideal chain, or the $3/5$ power, as it does for an isolated chain in a good solvent. This means that the chains are strongly stretched; this is most easily seen by rewriting (6.1.5) in terms of the Flory radius of gyration of an isolated chain R_F and the average distance between grafting points D. This yields

$$\frac{h}{R_F} = \left(\frac{R_F}{D}\right)^{2/3}; \qquad (6.1.6)$$

the degree of stretching is simply the ratio of the average distance between grafting points and the Flory radius raised to the power $2/3$.

A fully close-packed brush, with the whole area of the interface taken up by the grafting sites, would have a value of σ of unity and thus a volume fraction inside the brush of unity. However, this invalidates our assumption that the concentration inside the brush is in the semi-dilute regime; recall that, for volume fractions of polymer greater than a crossover value $\phi^{**} \approx v/a^3$, where v is the excluded volume parameter, we have a concentrated solution, for which concentration fluctuations are weak and mean-field theories should work. To get a feel for the range of concentrations involved, note that, in Flory–Huggins theory, the excluded volume parameter is related to χ as

$$v = a^3(1 - 2\chi); \tag{6.1.7}$$

the value of χ for a good solvent ranges from its value at the theta point of $1/2$ to its value in the athermal limit of zero. Taking as a typical polymer–good solvent combination polystyrene in toluene at 22 °C, we find a value of 0.4; thus, as a rule of thumb, when the volume fraction is in the range of tens of per cent, we can usually assume that we are in the concentrated regime.

In the concentrated regime we can use an alternative approach to finding the brush height, which emphasises the physical origin of the strong stretching phenomenon (de Gennes 1980). This is an energy balance argument similar in spirit to the one employed by Flory to estimate the Flory exponent for a swollen chain in a good solvent. We can estimate the free energy cost of stretching a single chain to the height of the brush h from Gaussian chain statistics as

$$\frac{F_{\text{stretch}}}{kT} = \frac{h^2}{a^2 N}; \tag{6.1.8}$$

the free energy due to the excluded volume interaction has the form

$$\frac{F_{\text{vol}}}{kT} = \tfrac{1}{2} v \frac{\phi^2}{a^6} kT \tag{6.1.9}$$

per unit volume. The volume associated with a single chain is ha^2/σ, so we can write down the form of the total free energy per chain as a function of the brush height as

$$\frac{F_{\text{chain}}}{kT} \sim \frac{h^2}{2a^2 N} + \frac{v N^2 \sigma}{2ha^2}, \tag{6.1.10}$$

which we can minimise with respect to h to find

$$h \sim N(v\sigma)^{1/3}. \tag{6.1.11}$$

So we find the same, linear dependence on N as emerges from the scaling argument and likewise a similar, cube-root dependence on the grafting density. However, this expression emphasises the role of solvent quality. Essentially, osmotic pressure drives solvent into the brush, stretching out the chains until the excluded volume interaction and the stretching energy are in balance. The better the solvent for the polymers making up the brush the more the brush will be swollen.

What evidence is there for these scaling laws? The first clear confirmation of equation (6.1.11) came from small-angle neutron scattering work by Auroy *et al.* (1991), who studied poly(dimethyl siloxane) chains terminally attached by a covalent bond to porous silica. Figure 6.4 shows their results, confirming the

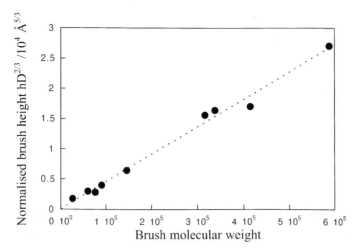

Figure 6.4. The characteristic size of a poly(dimethyl siloxane) brush grafted on to porous silica in dichloromethane, a good solvent, measured by small-angle neutron scattering. The brush height h is normalised by multiplying it by the $2/3$ power of the average distance between anchoring points D to verify equation (6.1.11). After Auroy *et al.* (1991).

linear scaling of the brush height with the relative molecular mass for relative molecular masses changing by more than an order of magnitude.

Earlier experiments had used the surface forces apparatus to measure the force profiles between mica sheets to which polystyrene had been end-grafted (Taunton *et al.* 1988). Because the height of an isolated polymer brush minimises its free energy, if we bring two brushes into contact and compress them we must always increase the free energy. Thus there is always a repulsive force between plates coated by polymer brushes, whose dependence on separation could be obtained by differentiating the expression for free energy (for example equation (6.1.10) for a concentrated solution) with respect to h. An excellent measure of the height of the brush is obtained from the value of the plate separation which marks the onset of significant repulsive force. These experiments apparently revealed a different scaling of brush height with the relative molecular mass, with an exponent closer to 0.6 than to unity. This apparent discrepancy disappears when one notes that, in these experiments, the polystyrene chains were physically, rather than chemically, grafted; in this case the grafted layer is in equilibrium with free chains in solution. In these circumstances the surface coverage is not fixed, but will itself change when the relative molecular mass is changed. A rough argument allows us to estimate this effect for the mean-field case. From equations (6.1.10) and (6.1.11) we can write down the free energy per brush chain as

$$\frac{F_{\text{chain}}}{kT} \sim \frac{N}{a^2}(v\sigma)^{2/3}. \tag{6.1.12}$$

If we equate this to the energy of attachment Δ we find that the coverage itself has a dependence on the relative molecular mass;

$$\sigma \sim \frac{1}{v}\left(\frac{a^2\Delta}{N}\right)^{3/2}. \tag{6.1.13}$$

Combining this with equation (6.1.11) leads to the expectation that the brush height for a physically grafted chain with constant sticking energy should scale as the square root of the relative molecular mass. A similar argument using scaling laws appropriate to the semi-dilute case yields a predicted relative molecular mass exponent of $3/5$, close to that which is observed experimentally. Of course, these arguments are too crude; to calculate the adsorption isotherm of physically end-grafted polymers we need properly to equate the chemical potentials of free and grafted chains, taking into account the non-ideality of the polymer solution and the translational entropy of chains both in the grafted layer and in the free solution (Ligoure and Leibler 1990). Nonetheless, it captures an important point – in a highly stretched brush long polymer chains are in a state of high free energy and they need a correspondingly large sticking energy to keep them in place.

As remarked above, both the scaling and the mean-field approach can be extended to provide a prediction for the force between the plates as a function of their separation under conditions of fixed grafting density. For the scaling case, appropriate to the semi-dilute concentration regime, de Gennes found (de Gennes 1987)

$$f(z) = \frac{kT}{\sigma^3}\left[\left(\frac{2h_0}{z}\right)^{9/4} - \left(\frac{z}{2h_0}\right)^{3/4}\right] \tag{6.1.14}$$

for $z < 2h_0$, where h_0 is the equilibrium height of the brush. Quite reasonable agreement is obtained between this expression and experimental data; figure 6.5 shows the experimental force profile obtained for a polystyrene brush, of relative molecular mass 140 000, attached to the mica plates via a zwitterionic group (Taunton *et al.* 1990), with good agreement except at very high and very low separations.

In many cases the anchoring point is provided not by a single chemical group (in the experiments described above this was a zwitterionic end-group) but by one half of a block copolymer, which may be much less soluble in the solvent than is the half that forms the brush. In this situation the less soluble part of the diblock that adsorbs on the wall is known as the 'anchor', while the soluble part that forms the brush is known as the 'buoy'. The amount of adsorbed

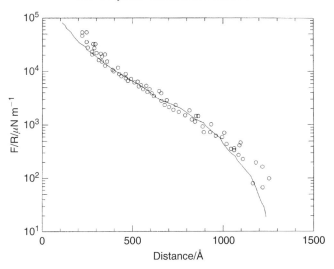

Figure 6.5. A force–distance profile obtained using the surface forces apparatus for the force between two mica surfaces each bearing a physically adsorbed polystyrene brush, of relative molecular mass 140 000, in toluene. The solid line is a fit to equation (6.1.14). After Taunton *et al.* (1990).

copolymer and the resulting extension of the brush are found by offsetting the free energy cost of stretching the 'buoy' chains against the free energy gain of having a dense wetting layer of the adsorbing 'anchor' chains at the interface (Marques *et al.* 1988) to find an effective chemical potential for the chains in the adsorbed layer as a function of the amount adsorbed, which in turn can be compared with the chemical potential of chains in the micellar solution.

We will return to this situation in more detail in chapter 8, when we consider block copolymers spread at a liquid surface; in the context of the surface forces apparatus an extensive study was made by Watanabe and Tirrell (1993). They measured forces between surfaces bearing adsorbed block copolymers of 2-vinyl pyridine and isoprene; earlier work by the same group had concerned 2-vinyl pyridine/isoprene block copolymers (Hadzioannou *et al.* 1986). In addition to interactions between chemically identical brush polymers of equal relative molecular mass, three different asymmetries were investigated: chemical asymmetry using PS and PI layers; asymmetry of relative molecular mass, using PS layers of different relative molecular mass; and structural asymmetry using two PI layers with different relative molecular masses and surface grafted densities. In common with the results of Taunton *et al.* discussed above, interaction between symmetrical layers was noted at separations 5–10 times the radius of gyration of the brush polymer in solution. Although the grafting density on the surface was greater than that required for overlap, Watanabe and

Tirrell concluded that the non-adsorbing brush-like PS block layers were not highly overlapping. Since toluene is a good solvent both for PS and for PI blocks, excluded volume effects need to be taken into account because such intramolecular osmotic effects may be of similar magnitude to intermolecular osmotic influences. Results for symmetrical adsorbed layers were compared in detail with the predictions of two models. The first due to Patel, Tirrell and Hadzioannou (Patel *et al.* 1988), referred to as PTH, is an extension of the Alexander–de Gennes scaling approach discussed above, whereas the second is the self-consistent field approach due to Milner, Witten and Cates (Milner *et al.* 1988) which is discussed below. PTH were able to derive an expression for the force as a function of distance in terms of reduced variables. All the constants involved may be independently measured, with the exception of an adjustable parameter, β, which accounts for an apparent discrepancy in the relationship between the osmotic pressure and the concentration observed for free chains and deduced for brushes. Satisfactory data collapse is observed for a wide range of data and, if the parameter β is allowed to vary to a value of 0.58, quite good agreement is achieved with the theory of PTH (see figure 6.6). However, as for the data of Taunton *et al.*, there seems to be some departure between theory and experiment at large separations.

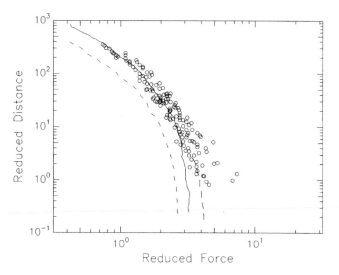

Figure 6.6. Reduced force as a function of reduced distance for the close approach of two brush-like layers. The data points are for a number of 2-vinyl pyridine/isoprene and 2-vinyl pyridine/styrene diblocks with various degrees of polymerisation. The solid line is the prediction of PTH theory with $\beta = 0.58$, the short dashed line PTH theory with $\beta = 1$ and the long dashed line the prediction of self-consistent field theory. After Watanabe and Tirrell (1993).

At the root of the discrepancy between the experimental force–distance curves and the predictions of the simple scaling theory is the fact that the latter says nothing about the true shape of the concentration profile of the brush layers. When the brushes are compressed this is not terribly important, because in these circumstances the segment concentration between the plates will be relatively uniform. At larger separations, however, the force–distance curve will be different if the density of brush segments falls to zero gradually rather than in a very abrupt fashion. It should prove possible to obtain information about the shape of the concentration profile near the interface from small-angle neutron scattering experiments of the type discussed above or from neutron reflectivity measurements.

To obtain useful theoretical results for the concentration profile, we need to go beyond these simple scaling arguments. Luckily, at least for the situation of relatively dense, strongly stretched, brushes, we can expect self-consistent field theories to work rather well; in such a dense brush the basic mean-field assumption that any polymer chain will interact with its neighbours more than it will with itself should be well obeyed. Numerical mean-field theories of the kind described in chapter 5 are very well suited to this kind of calculation; the earliest results, due to Hirz (these results are still unpublished but some were reproduced by Milner *et al.* (1988)) showed profiles very different in character from those found for adsorbed chains. Rather than a concave concentration profile, the curves were notably convex, with the concentration dropping rather abruptly to zero on the outside of the brush. In fact it turns out that the profiles are rather well described by a parabolic form (see figure 6.7). It soon turned out that there was a remarkably good analytical solution to the self-consistent mean-field equations which provided an explanation for these parabolic profiles.

As a starting point for the analytical self-consistent mean field theory it is convenient to take, not the diffusion equation approach introduced in section 4.3, but an equivalent formulation in which we write down a partition function for a single chain in the field due to all the other chains, which of course is inititially unknown and needs to be determined self-consistently. This partition function takes the form of a sum of the Boltzmann factors corresponding to each possible trajectory of the grafted chain. If we represent the trajectory of the chain in continuum language, as the function $r(t)$, where r represents the position in space of monomer unit number t, we find that the free energy has two parts. The first is a stretching energy and the second is the effective, mean-field potential $U(r)$. Thus we write the single-chain partition function Z_{sc} as

$$Z_{sc} = \sum_{\{r(t)\}} \exp(-S_k), \tag{6.1.15}$$

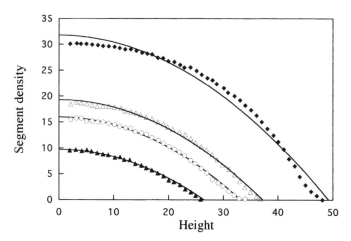

Figure 6.7. Segment density profiles for end-grafted chains in a good solvent calculated by Hirz using self-consistent mean-field theory, for four values of the grafting density σ ($\blacklozenge$, 0.25; $\triangle$, 0.12; $\circ$, 0.08; and $\blacktriangle$, 0.04). The solid lines are the parabolic profiles predicted by equation (6.1.15). After Milner *et al.* (1988).

where the sum is over all possible chain configurations $r(t)$ and S_k is given for each possible chain configuration as

$$S_i = \int dt \left[\frac{1}{2} \left(\frac{d\boldsymbol{r}_i}{dt} \right)^2 - U(\boldsymbol{r}(t)) \right]. \tag{6.1.16}$$

The key insight of this analytical approximation to the SCF is that the sum of equation (6.1.15) is, under conditions of strong stretching, dominated by the configuration that minimises the equation (6.1.16). That this saddle-point approximation is a good one is easiest to see by a physical argument. Just as the diffusion equation formulation of the self-consistent field theory, equation (4.3.2), is analagous in form to Schrödinger's equation, with the monomer index t taking the role of time, so equations (6.1.15) and (6.1.16) are analogous to the path integral formulation of quantum mechanics. The stretching energy corresponds to the particle's kinetic energy and the function S_i corresponds to the action. A strongly stretched configuration corresponds to the trajectory of a particle with a large momentum, which can be described by the classical limit of quantum mechanics. In the classical limit the path taken by a particle is the one which minimises the action, so the 'equation of motion' for the trajectories of our polymer chains is the analogue of Newton's second law of motion

$$\frac{d^2\boldsymbol{r}}{dt^2} = -\nabla U. \tag{6.1.17}$$

(This classical approximation for the strong stretching case was first introduced by Semenov (1985)). Together with this differential equation we need boundary conditions. We know that one end of every chain must be located at the wall; we do not know the location of the free end, but we do know that the local stretching must be zero there in the absence of an applied force. So, in our mechanical analogy, each chain corresponds to a particle dropped from rest some distance away from the wall, which moves through the potential in such a way that it arrives at the wall after a time corresponding to the degree of polymerisation of the chains.

Surprisingly, we now have enough information to write down the form of the self-consistent potential. By symmetry, for chains grafted on to a planar surface it is a function only of the distance away from the wall, z. If we can assume that all the chains have the same length, we need a so-called equal time potential – one in which particles dropped from any position will arrive at the origin at the same time. The form of the potential that has this property is a quadratic one; to put this another way, we know that the period of a simple harmonic oscillator is independent of its amplitude. Thus we can write the potential in the form

$$U(z) = Bz^2 - A. \tag{6.1.18}$$

The constants B and A are easily found: the time taken for the particle to fall to the origin is one quarter of the period of the simple harmonic oscillator and this time corresponds to the chain length N. Thus we find

$$B = \frac{\pi^2}{8N^2}. \tag{6.1.19}$$

To find A we need the relationship between the potential and the volume fraction. The simplest choice, which is valid for concentrated solutions, is $U(\phi) = -v\phi$, where v is, as above, the excluded volume parameter. Now we can convert the potential of (6.1.18) into a volume fraction profile and fix A by the requirement that the integrated volume fraction profile gives us the coverage. This gives as our final result for the volume fraction profile

$$\phi(z) = \frac{B}{v}(h^2 - z^2), \tag{6.1.20}$$

where the brush height h is given by

$$h = \left(\frac{12}{\pi^2}\right)^{1/3} N(\sigma v)^{1/3}. \tag{6.1.21}$$

So, in the analytical self-consistent field theory, the brush height has the same functional dependence as that predicted by the simple energy balance argument that led to equation (6.1.11).

Several experimental results that appear to confirm this picture for dense end-grafted polymers in good solvents have now appeared. Auroy and co-workers (Auroy *et al.* 1992) used small-angle neutron scattering from porous silica, with polystyrene chemically grafted by one end on to the solid surface of the pores, and found that the data were best fitted by a parabolic profile of the form equation (6.1.20), though with a modification far from the wall that we will discuss below. Neutron reflectivity experiments provided similar results (Cosgrove *et al.* 1991, Field *et al.* 1992).

One difference between the profiles deduced experimentally and the parabolic profile of equation (6.1.20) is at the tip of the profile; rather than an abrupt change in gradient where the concentration profile goes to zero, one finds that the experimental data are better accounted for if there is a more gradual, exponential decay of concentration to zero in the outermost region of the tip. This is to be expected; the classical approximation is likely to fail close to the tip because the ends of each chain are no longer strongly stretched and fluctuations around the most probable chain conformation contribute significantly to the partition function of equation (6.1.15). Numerical self-consistent field calculations do not have this limitation (Wijmans *et al.* 1992); figure 6.8 shows a comparison between the numerical calculations and the analytical solution, revealing a deviation in the tail region that becomes less important for increasing relative molecular mass. It should also be clear that these fluctuation corrections to the classical limit become more important as the solvent quality is reduced and the chains are less strongly stretched.

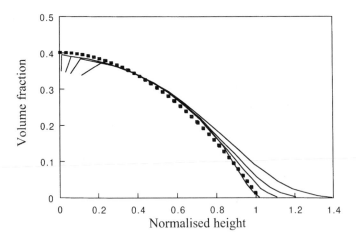

Figure 6.8. A comparison of the numerical solution of the self-consistent field equations with the classical approximation. The dashed curve is the parabolic, classical profile, whereas the solid curves are for grafted chains with degrees of polymerisation of 600, 100, 50 and 25, in order of increasing deviation from the parabolic profile. The solvent is assumed athermal (i.e. $\chi = 0$). After Wijmans *et al.* (1992).

Deviations from the parabolic profile are also found close to the wall. Experimentally one often finds a region close to the wall depleted in polymer. The details of the profile in this region depend on the nature of the interaction between the wall and the backbone segments of the grafted polymer; for a repulsive wall there will be a depleted region, whereas if there is a favourable interaction between the wall and segments, the concentration will actually be enhanced. Only for a precisely neutral wall–polymer interaction will the concentration gradient at the wall be zero, as predicted by the simple classical theory.

The effect of solvent quality on the brush structure is expressed through equation (6.1.21) for the situation in which the solvent still remains a good one. If the solvent quality is reduced while keeping the grafting density constant, the brush becomes less extended. At the theta temperature this type of treatment must break down – here the excluded volume parameter v becomes equal to zero. The Flory energy balance analysis can be extended through the theta point into the bad solvent regime by including three-body interactions, as well as modifying the stretching energy term to include the energy cost of compressing the brush; the result may be conveniently expressed in terms of the reduced temperature $t \equiv (T - T_\theta)/T_\theta$, where T_θ is the theta temperature, as

$$h = N\sigma^{1/2} f(t/\sigma^{1/2}), \tag{6.1.22}$$

where the function $f(x)$ has the properties

$$f(x) \sim \begin{cases} x^{1/3} & \text{for } x \gg 1 \\ |x|^{-1} & \text{for } x \ll -1. \end{cases} \tag{6.1.23}$$

Note that, in the limit of low temperatures, for which the solvent is very bad, the scaling of the brush height with the degree of polymerisation and grafting density is exactly the same as would obtain for a dry polymer layer; as one might expect, in the limit of a very bad solvent all the solvent is simply expelled from the brush to leave an undiluted polymer film.

The analytical self-consistent field theory can also be applied to theta and poor solvent conditions; as in the good solvent case the scaling relations for the brush height of equation (6.1.22) are preserved, but we also find expressions for the volume fraction profile. At the theta temperature, the profile once again takes a simple, analytical form – an ellipse (Zhulina *et al.* 1991). However, both numerical solutions of the self-consistent field theory and computer simulations suggest that, for finite relative molecular masses, fluctuation corrections at the foot of the profile are large. Within the poor solvent regime, the SCF profiles are characterised by discontinuities, corresponding to the regions of concentration that would fall into the two-phase region of a bulk phase diagram (Shim and Cates 1989). Once again, at finite relative molecular mass these discontinuities are unphysical and will be smoothed out by the fluctuation corrections (Wijmans *et al.* 1992).

The collapse of a polymer brush as the temperature is dropped through theta conditions is nicely illustrated by a neutron reflection study of end-grafted polystyrene in cyclohexane (Karim *et al.* 1994). The results are shown in figure 6.9; note that the brush height varies in a smooth way through the theta temperature, with no sign of a sharp transition.

The question of the degree to which the bulk phase separation and collapse phenomena that occur in solutions below the theta temperature are reflected in the behaviour of brushes is an interesting one. An essential parameter must be the grafting density; for very low grafting densities the behaviour of an isolated grafted chain must be rather similar to the behaviour of a single chain in a very dilute solution. As the grafting density increases, however, chains begin to interact with their neighbours; since this interaction is favourable this may lead to a kind of lateral phase separation or clustering phenomenon, as sketched in figure 6.10. This effect was first predicted using self-consistent field theory by Yeung, Balazs and Jasnow (Yeung *et al.* 1993). Such clustering has also been predicted by scaling arguments (Tang and Szleifer 1994) and has been seen in

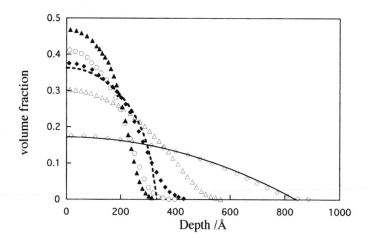

Figure 6.9. Volume fraction profiles of an end-grafted polystyrene brush, of relative molecular mass 105 000, under various solvent conditions ($\Diamond$, toluene at 21 °C; and cyclohexane at $\triangle$, 53.4 °C; $\blacklozenge$, 31.5 °C; $\circ$, 21.4 °C; and $\blacktriangle$, 14.6 °C), deduced from neutron reflectivity measurements (all the solvents are deuterated). Toluene at 21 °C is a good solvent and the solid line is the classical parabolic profile. The theta temperature for d-cyclohexane is 34 °C and the dashed line is the elliptical profile predicted by analytical self-consistent field theory for theta conditions. After Karim *et al.* (1994).

Mushrooms

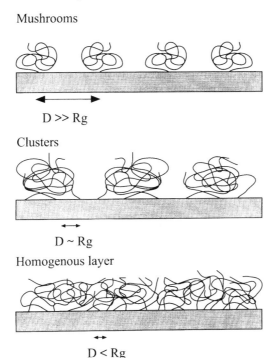

D >> Rg

Clusters

D ~ Rg

Homogenous layer

D < Rg

Figure 6.10. Possible structures of grafted polymer layers in poor solvents, showing isolated collapsed coils (mushrooms) laterally phase-separated clusters or bundles and a homogeneous collapsed layer. After Tang and Szleifer (1994).

Monte Carlo simulations (Lai and Binder 1992). Figure 6.11 shows scanning force microscopy images of an end-grafted layer of a water-soluble polymer, dextran, which in water appears as a uniform layer, but, after the addition of a poor solvent, propanol, develops lateral structure (Frazier *et al.* 1997). As the grafting density is increased yet further the homogeneous, layer structure is expected once again to become stable.

Of course, this discussion of clustering phenomenon in brushes assumes that the brushes are permanently grafted and in contact with pure solvent. If the grafting points are not permanent then large-scale lateral phase separation may take place, whereas if the brush is in equilibrium with a polymer solution, then, in addition to grafting of the brushes by their anchoring groups, a wetting layer of polymer-rich solution may form in addition to the end-grafted material.

The ability of a polymer brush to respond to external conditions, by changing its conformation in response to a change in temperature, suggests intriguing uses for brush systems, perhaps as the lining for micro-channels that can be made to open or close according to the surrounding conditions (Israels *et al.*

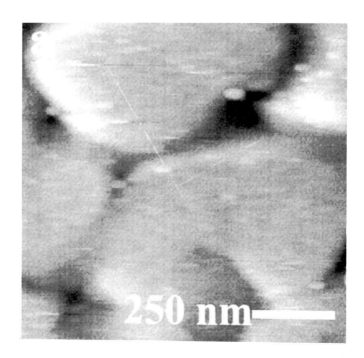

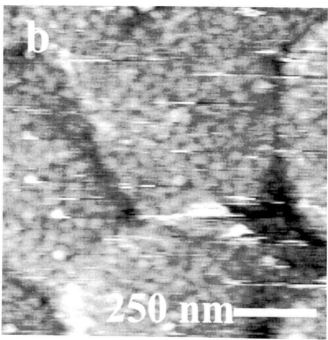

Figure 6.11. (a) A scanning force micrograph of a layer of dextran end-grafted into a gold-coated mica surface, immersed in water. The large-scale structure is composed of large, atomically flat gold islands. (b) A similar layer after the water had been replaced by propanol, a poor solvent for dextran. Small-scale lateral structure of the brush is now visible. Micrographs courtesy of R. A. Frazier and M. C. Davies; see Frazier *et al.* (1997).

1994). Considerably more flexibility is possible if we move beyond simple polymers in organic solvents to situations in which the grafted chains carry charged groups and the solvent is water. Now electrostatic interactions play an important role and two new variables in principle can enter; the salt concentration, which affects the range of the electrostatic force, and the pH, which in the case of a polyelectrolyte involving weak acids or bases will control the degree of dissociation of the acid groups and thus the charge state of the brush. It is convenient to distinguish between strong polyelectrolytes, for which the degree of charging of the chains is essentially fixed, and weak electrolytes, for which the degree of dissociation depends on the local pH.

These extra variables lead to a considerable extra richness of possible behaviours, even when considered at the level of scaling theory for a strong electrolyte brush (Borisov *et al.* 1994). A key factor in determining the behaviour of the brush is the disposition of the counterions, which is determined by the solution of the Poisson–Boltzmann equation near the grafting surface (Israelachvili 1991). When the counterions are localised within the brush, the brush thickness is given by a balance between the elastic energy of the brush chains and the confinement entropy of the counterions; since the brush in this situation is overall charge neutral, electrostatic forces play no explicit role. On the other hand, if the counterions extend well beyond the brush layer then the brush will be charged and electrostatic repulsion will lead to stretching of the brush even at relatively low grafting densities. The situation is even more complicated for weak electrolytes (Zhulina *et al.* 1995), for which it turns out that the local pH may in some circumstances vary within the brush layer itself. This theoretical work has up to now been supported by little in the way of systematic experimental work. In view of the importance of polyelectrolyte systems both industrially and in biology and biochemistry, this is an area in which we should expect progress in the future.

6.2 Polymer brushes in melts

We now consider the case in which a layer of grafted polymer chains is in contact not with a solvent, but with a dense polymer melt. This situation is of intrinsic interest as a much simplified model for the interphase between a polymer matrix and the reinforcing particles or fibres of a filled polymer or polymer composite; it is also closely related to the situation encountered when a block copolymer is used to modify an interface between immiscible polymers.

There are two major differences between the case of a brush in a solution and that of a brush in a melt; firstly, the effect of the swelling of the brush by the solvent is much reduced due to the effect of the screening of excluded volume interactions. This is the same phenomenon that results in a labelled chain in a polymer melt having ideal random walk dimensions, rather than being swollen, as it is in a good solvent. The result of this is that brushes in melts will usually be much less strongly stretched than are their counterparts in solution. The second difference is that, when the solvent is itself polymeric, we cannot impose on it an arbitrary concentration profile without incurring free energy costs due to forcing the solvent chains to take up non-ideal configurations. In particular, even in situations in which the brush might tend to take up a configuration with a steep concentration gradient at its tip, this would incur excessive gradient free energy penalties for the free chains and in the equilibrium situation the brush profile will adjust itself to present a broader interface between brush and free chains.

We can investigate the consequences of screening at the level of a Flory energy balance argument (de Gennes 1980) by expanding the Flory–Huggins expression for the free energy of mixing of two polymers in the limit of a small volume fraction. This yields a term in the free energy proportional to the square of the volume fraction which is the counterpart of equation (6.1.9);

$$\frac{F_{\text{vol}}}{kT} = \frac{1}{2}\left(\frac{1}{P} - 2\chi\right)\frac{\phi^2}{a^3}kT, \qquad (6.2.1)$$

where P is the degree of polymerisation of the 'solvent' or 'matrix' chains. For the case of $\chi = 0$, corresponding to the situation in which the brush and the matrix chains are chemically identical, the excluded volume parameter is reduced by a factor of P. The effect of this reduction is to decrease greatly the degree of stretching obtained for a given grafting density in situations in which the relative molecular mass of the matrix is relatively high.

For $P = 1$, of course, we return to the case of a brush in a good solvent. For low coverages we have a mushroom whose size scales like that of a swollen single chain in a good solvent, whereas when the coverage increases to the point at which the mushrooms overlap, we recover a brush, which, as the grafting density increases still further, is progressively stretched. The scaling of the brush height is given by equation (6.1.11) with the excluded volume parameter v replaced by a^3/P. We can estimate how much larger than unity the degree of polymerisation of the matrix chains must be before the excluded volume parameter is too small to cause the chains to swell by using the Flory argument to find the linear dimensions of mushrooms when the grafting density is low. Writing the free energy per chain as a sum of stretching and excluded volume energies as

$$\frac{F_{\text{chain}}}{kT} \sim \frac{h^2}{a^2 N} + \frac{N^2 a^3}{Ph^3} \qquad (6.2.2)$$

and minimising the free energy we find

$$h \sim aN^{3/5} P^{-1/5}. \qquad (6.2.3)$$

This corresponds to the classical Flory result for the dimensions of a swollen chain in a good solvent for the case $P = 1$; the chain remains swollen but the degree of swelling decreases as P increases, until the mushroom reaches the ideal, unswollen random walk dimensions $aN^{1/2}$, which occurs when $P = \sqrt{N}$.

So, if the degree of polymerisation of the matrix chains is less than the square root of the degree of polymerisation of the brush chains, we expect the system to behave essentially like a brush in a good solvent. For longer matrix chains, the effect of screening is to introduce qualitative differences in behaviour. As we increase the grafting density of such a layer, even when the unswollen mushrooms characterising the layer at low grafting density start to overlap, the interaction between them is screened in such a way that no stretching occurs; we have a brush, in the sense that neighbouring grafted chains can substantially overlap, but it is unswollen and still characterised by ideal random walk dimensions.

The grafting density cannot increase indefinitely with the brush height remaining constant; when $\sigma > N^{-1/2}$ the volume fraction in an unstretched brush would start to exceed unity. At this point the brush must stretch, driven simply by space-filling considerations; the scaling of the layer height is that which would be expected for a grafted layer that is not mixed with any solvent chains. This situation is referred to as a 'dry brush', because the brush layer is substantially unpenetrated by matrix chains. However, as we shall see, this does not mean that the interface between the brush and the matrix is abrupt; it is in this regime that the free energy penalties associated with the distortion of the matrix chains are particularly important.

These predictions for different regimes for melt brushes are summarised in figure 6.12 (Aubouy *et al.* 1995). It should be stressed that in reality the sharp boundaries between regimes will be replaced by broader regions of crossover; indeed, if the degree of polymerisation of the brush chains is modest, most of the accessible parameter space will be in one or other of these crossover regimes and it will be difficult to establish the asymptotic power law behaviour.

This problem is avoided if we resort to self-consistent field methods. These should provide reliable predictions for the concentration profile over virtually the whole range of parameter space, though of course this comes at the expense of the loss of physical transparency which accompanies the need for numerical

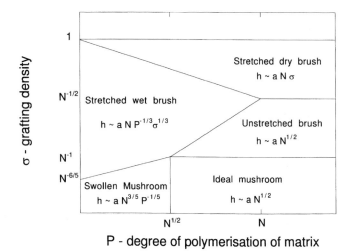

Figure 6.12. Various regimes for a brush formed of chains of degree of polymerisation N with a grafting density σ, exposed to a matrix of a chemically identical polymer melt with degree of polymerisation P. Included is the scaling of the layer thickness h in each regime; for more details see the text. Both axes are logarithmic. After Aubouy *et al.* (1995).

calculations. Shull has implemented the self-consistent field method for the case of end-grafted chains in the melt (Shull 1991); a comparison of the prediction of this approach with the scaling theory is shown in figure 6.13, which shows rather clearly that in reality the sharp boundaries between regimes assumed by the scaling theory are replaced by broad regions of crossover when calculations are done with self-consistent field theory. One interesting prediction that emerges from these calculations is that, when the degree of polymerisation of the matrix is more than a few times larger than that of the brush, a universal dry brush regime is entered wherein both the degree of stretching of the brush and the shape of the brush are universal functions of the normalised grafting density.

Experiments have been performed to probe the conformation of brushes in melt matrices (Jones *et al.* 1992, Clarke *et al.* 1995). Just like in the solution case, it is useful to distinguish between the case of chemically grafted layers, for which the grafting density is essentially fixed, and physically grafted layers, for which the grafting density is set by an equilibrium between free and grafted chains. Figure 6.14 shows examples of volume fraction profiles obtained by neutron reflectivity for a chemically grafted case (Clarke *et al.* 1995); here the brush chains are deuterium-labelled polystyrene of relative molecular mass around 100 000, terminally attached to a silicon oxide surface by a silane linkage. The matrix polymer is normal polystyrene; for matrices of relatively

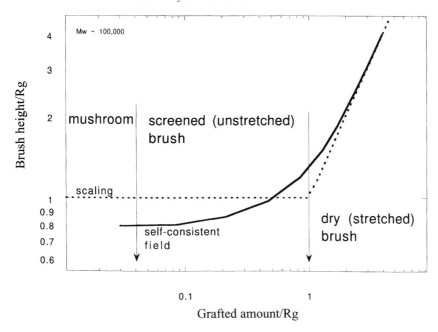

Figure 6.13. The height of a polymer brush in a chemically identical matrix of equal degree of polymerisation 1000 as predicted by the scaling theory and calculated by the numerical self-consistent mean-field method. The SCF calculations were done using a program written by K. Shull.

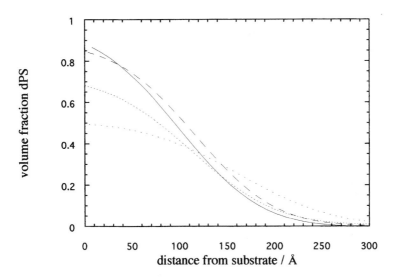

Figure 6.14. Segment density profiles of a deuterium-labelled polystyrene brush of relative molecular mass 100 000 in contact with polystyrene matrices of a variety of relative molecular masses (———, 156 000; – –, 52 000; ·····, 28 500; · · ·, 7000) measured by neutron reflectivity. After Clarke *et al.* (1995).

high relative molecular mass the profiles are qualitatively similar, but the profile corresponding to the lowest relative molecular mass, 7000, does start to look significantly stretched. This relative molecular mass does correspond to a degree of polymerisation approaching the square root of the degree of polymerisation of the brush, so this is qualitatively in accord with the scaling expectations. Detailed comparison with the results of numerical self-consistent field theory is possible and the agreement is quantitative.

The case of physically grafted brushes has perhaps been more widely studied. The anchoring entity may be a single chemical group that interacts with the surface or a short block of an immiscible species; the latter arrangement has the useful property that one can use a short block of a highly surface-active material, which will then form a layer at the surface of a polymer film that is firmly connected to the rest of the polymer by virtue of the entanglement of the brush and matrix. Examples of anchoring groups have included, for polystyrene brushes, carboxy groups (Zhao *et al.* 1991) and short butadiene blocks (Jones *et al.* 1992) at silicon interfaces, short isoprene blocks both at silicon interfaces and surfaces (Budkowski *et al.* 1995), and short fluorocarbon blocks at the free surface (Affrossman *et al.* 1994, Schaub *et al.* 1996, Hopkinson *et al.* 1997). The case of a fluorocarbon end-group is particularly interesting because of the very favourable surface properties of fluorocarbons for a number of applications. What is interesting is that even for a very short fluorocarbon group the large surface energy difference and strong bulk incompatiblity between styrene and fluorocarbon segments combine to produce a significant driving force for segregation. In one example a perfluorohexane end-group yielded an effective sticking energy of about $1.9k_BT$ (Hopkinson *et al.* 1997).

Rather like the solution case, in these physically anchored systems the amount adsorbed is given by the balance between a 'sticking energy' Δ_i, which is gained when the attaching unit is stuck to the surface or wall, and the energy cost associated with brush formation. We can calculate this energy balance at the level of the Flory approach, following Leibler (1988), by writing the free energy per grafted chain g as

$$\frac{g}{kT} = \frac{h}{a\sigma}\frac{1}{P}(1-\phi)\ln(1-\phi) + \ln(N\sigma) + \frac{3}{2}\frac{h^2}{Na^2} + \Delta_i. \qquad (6.2.4)$$

Here the first term represents the entropy of mixing of the matrix chains with the brush, the second term accounts for the translational entropy of the brush chains in the plane (note that this term would be absent for chemically grafted brushes) and the third term is the familiar stretching energy. Knowing the relations between the brush height h and N, P and σ, we can evaluate the chemical potential of brush chains in the dry-brush and wet-brush regimes and, by equating these with the chemical potential of the chains in the bulk, we can construct the adsorption isotherms.

As an example of such adsorption isotherms, figure 6.15 shows the amount of a polystyrene chain with a short isoprene block on one end segregated to a polystyrene/silicon interface and the free surface, as measured by nuclear reaction analysis (Budkowski *et al.* 1995). By fitting these kinds of data with adsorption isotherms derived by the Leibler approach, it is possible to estimate the sticking energy Δ_i. In the case of these copolymers there is another driving force for segregation in addition to the sticking energy of the group; there will also generally be an unfavourable energy of interaction between the sticky group and the bulk polymer when the copolymer chain is in the bulk phase; this unfavourable energy can be avoided if the chain joins the brush. Taking these two factors together we can define the nett driving force for segregation as $\beta' = \chi N_{anchor} - \Delta_i$, where N_{anchor} is the degree of polymerisation of the anchoring block and χ is the Flory–Huggins parameter for the interaction between the anchoring block and the surrounding matrix. For the example shown in figure 6.15, the values of β' deduced are 5.2 for the surface and 4.52 for the silicon substrate.

The numerical self-consistent field method can also be used to construct adsorption isotherms, from which estimates of sticking energies may be

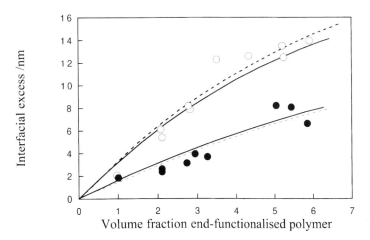

Figure 6.15. The interfacial excess of a deuterated styrene–isoprene diblock, with styrene block relative molecular mass 1 000 000 and isoprene block relative molecular mass 10 000. The matrix was polystyrene of relative molecular mass 2 890 000: (○), the excess at the surface of the film; and (●), the excess at the silicon substrate. Solid lines are the predictions of the Leibler theory, whereas the dashed lines are calculated from self-consistent field theory. After Budkowski *et al.* (1995).

obtained (Shull 1991). Figure 6.15 shows adsorption isotherms thus calculated; the values of β' deduced, 5.2 for the surface and 4.4 for the silicon, are in agreement with those found from the Leibler approach.

It is worth commenting on the smallness of these values; a single group with an interaction energy only of order kT or less causes what is quite a substantial degree of segregation even when it is attached to a very long polymer chain. This can be understood in general terms as another consequence of the smallness of the entropy of mixing of polymers; rather small perturbations are sufficient to overcome the very weak tendency of polymers to mix. Thus polymers with any kind of chemical heterogeneity are intrinsically surface active and their inclusion in a polymer system, whether intentional or not, will profoundly modify properties such as the surface energy and adhesion.

6.3 Block copolymers at polymer/polymer interfaces

The addition of a block copolymer to an incompatible polymer mixture can have a profound effect both on the morphology and on the mechanical properties of that mixture. For example, the extensive experiments of Teyssié, Jerome, Fayt and co-workers (Fayt *et al.* 1989, Harrats *et al.* 1995) have shown that the addition of block copolymers of styrene and hydrogenated polybutadiene to blends of polystyrene and polyethylene significantly modifies both the morphology and mechanical properties, properties such as the ultimate tensile strength and the elongation at breakage showing substantial improvements. Changes in morphology must be mainly related to the effect of the block copolymers on interfacial tension, which leads to a reduction in phase size and in some circumstances a stabilisation of co-continuous morphologies against coalescence. Improvements in mechanical properties will then derive both from these morphological changes and from an intrinsic toughening of the interfaces. The toughening effect of block copolymers at interfaces will be discussed in chapter 7; here we will discuss their interfacial activity. The degree to which added block copolymers can reduce interfacial tension in real processing situations must depend both on the equilibrium adsorption isotherm of a given architecture and on the kinetics by which the block copolymer reaches equilibrium at the interface in the processing time available. In a real blend disentangling all these effects can be complex, so much progress has been made by the study of more idealised systems.

An experiment that clearly shows the interfacial activity of block copolymers at polymer/polymer interfaces may be done by making a bilayer sample of two incompatible polymers, which we shall refer to as A and B, and incorporating in one of the polymers some AB block copolymer that has been labelled in some way. On annealing such a sample, one may expect the block copolymer to accumulate at the interface. A very convenient approach is to use deuterium labelling on the block copolymer, so that its build-up at the interface may be studied using forward recoil spectrometry.

A nice example of this technique in action is shown in figure 6.16 (Shull *et al.* 1990). Here a bilayer is prepared by spin-casting a 2000 Å film of poly(vinyl pyridine); on top of this a 4000 Å film of polystyrene containing a small amount of styrene/vinyl pyridine block copolymer was deposited. The styrene block of the copolymer was labelled with deuterium and forward recoil spectrometry revealed that in the as-prepared sample the block copolymer was uniformly distributed throughout the polystyrene film. On annealing above the glass transition temperature, however, substantial amounts of the block copolymer can be seen to segregate to the interface between the incompatible polymers. This technique, then, allows one to quantify the equilibrium relation between the amount of block copolymer adsorbed at the interface and the amount which remains in solution in the homopolymers – the adsorption isotherm. Figure 6.17 shows an example of such an isotherm.

Such adsorption isotherms may be predicted using numerical self-consistent field theory. This involves a relatively straightforward extension of the ap-

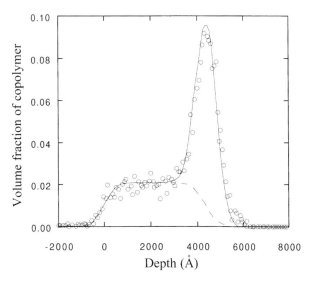

Figure 6.16. Segregation of a deuterium-labelled styrene–vinyl pyridine block copolymer to an interface between polystyrene and poly(vinyl pyridine), revealed by forward recoil spectrometry. The block copolymer was initially uniformly distributed in the upper, polystyrene film; after annealing for 8 h at 178 °C an interfacial excess of 100 Å has developed. After Shull *et al.* (1990).

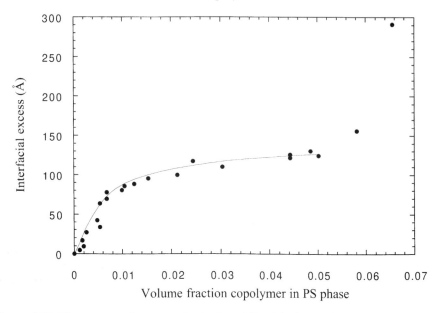

Figure 6.17. The excess of styrene–2-vinyl pyridine block copolymer at a polystyrene/poly(2-vinyl pyridine) interface, determined by forward recoil spectrometry. The degrees of polymerisation of the styrene and vinyl pyridine blocks were 391 and 68, respectively. The solid line is the prediction of self-consistent field theory, assuming a value for the Flory–Huggins interaction parameter χ_{ps-pvp} of 0.11. After Shull *et al.* (1990).

proach due to Edwards and Helfand outlined in section 4.3; this was done for the case of a diblock copolymer at a polymer/polymer interface in the presence of solvent by Hong and Noolandi (1982, 1984) and for an unswollen polymer interface by Shull and Kramer (1990). As shown in figure 6.17, the theory works very well, at least at relatively low copolymer concentrations. We will discuss what happens at higher copolymer concentrations below.

The Gibbs adsorption equation (equation (2.2.7)) implies that segregation of a block copolymer to an interface is associated with a reduction in interfacial tension. This can conveniently be expressed in the form

$$\gamma = \gamma_0 - \frac{\rho}{N_c} \int_{-\infty}^{\mu} z_i^* \, d\mu, \qquad (6.3.1)$$

where γ_0 is the interfacial tension of the bare interface, ρ is the segment density of the copolymer and z^* is the interfacial excess. For example, the predicted interfacial tension for the system whose interfacial excess is shown in figure 6.17 is shown in figure 6.18.

In this case direct measurements of the interfacial tension were not made,

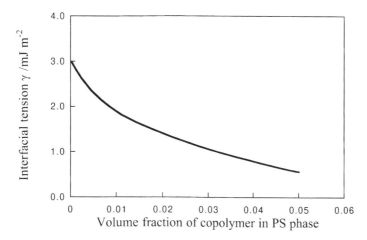

Figure 6.18. Interfacial tension between polystyrene and poly(vinyl pyridine) in the presence of styrene–2-vinyl pyridine block copolymer, as calculated by self-consistent field theory for the system whose interfacial excess is shown in figure 6.17. The value of the Flory–Huggins interaction parameter χ_{ps-pvp} was taken as 0.11, which provides a good fit to the adsorption isotherm below the CMC. After Shull *et al.* (1990).

because of the high viscosity of the homopolymer phases. Direct measurements have been made for the case of interfaces between polystyrene of low relative molecular mass and 1,2-polybutadiene (Anastasiadis *et al.* 1989a, b) and poly-(dimethyl siloxane) (Hu *et al.* 1995). Results are shown in figure 6.19.

The self-consistent field theory also allows one to calculate the segment density profiles of each homopolymer and each block of the copolymer. Forward recoil spectrometry is unable to resolve the details of these concentration profiles – the apparent finite width of the copolymer layer shown in figure 6.6 is entirely due to the instrumental resolution – but from neutron reflectivity measurements on a series of differently labelled samples one is able to extract all four segment density profiles. Figure 6.20 shows an example of this, for a styrene/methyl methacrylate copolymer at an interface between polystyrene and poly(methyl methacrylate).

From this extremely instructive data set we can draw a number of conclusions. Firstly, it is clear that the block copolymer is localised at the interface between the two homopolymers and is organised with the styrene block at the polystyrene side of the interface and the methyl methacrylate block at the PMMA side. The joints between the two halves of the copolymer are localised to a relatively narrow region, while each half of the copolymer penetrates rather deeply into the corresponding homopolymer. The overall interface between all styrene segments and all methyl methacrylate segments is broadened to 75 Å from the value for the homopolymer interface of 50 Å (see section 4.3).

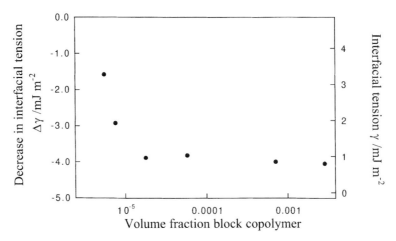

Figure 6.19. Interfacial tension between polystyrene (relative molecular mass 4000) and poly(dimethyl siloxane) (relative molecular mass 4500) as a function of the copolymer volume fraction in the PDMS phase. The relative molecular mass of the styrene block of the copolymer is 5500 and that of the dimethyl siloxane block is 7500. The interfacial tension in the absence of block copolymer was $4.85 \pm 0.05 \ \mathrm{mJ \, m^{-2}}$. After Hu *et al.* (1995).

Thus our picture of a block copolymer at a strongly immiscible polymer/polymer interface is as sketched in figure 6.21. The joints between the two halves of the copolymer are localised to a well-defined plane, from which each half of the copolymer protrudes into the corresponding copolymer. In effect, we have two polymer brushes joined back to back at the interface. Armed with this insight it is possible to construct a scaling theory for block copolymers at interfaces very much in the spirit of the scaling theories for melt brushes discussed in section 6.2 (Leibler 1988).

Any theory predicting the adsorption isotherm works by comparing the chemical potential of copolymer chains at the interface with the chemical potential of copolymer chains in the bulk, μ_c. If the two homopolymers are highly immiscible we can write this in the form (Dai *et al.* 1992)

$$\frac{\mu_c}{kT} = \ln \phi_c - \phi_c + 1 - \frac{N_c}{N_{ha}} + \chi_{ab} N_{cb}, \qquad (6.3.2)$$

where N_c is the total degree of polymerisation of the copolymer, N_{cb} is the degree of polymerisation of the b block of the copolymer and N_{ha} is the degree of polymerisation of the homopolymer in which the copolymer is dissolved at a volume fraction ϕ_c. The first four terms in this expression arise from the entropy of mixing of the copolymer with its surroundings, whereas the last

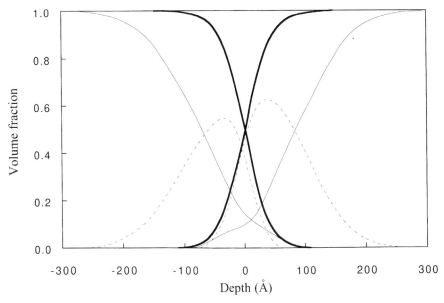

Figure 6.20. Segment distributions of a styrene–(methyl methacrylate) block copolymer (relative molecular masses of each block were in the range 48 000–65 000) at an interface between polystyrene (relative molecular mass in the range 110 000–127 000) and poly(methyl methacrylate) (relative molecular mass in the range 107 000–146 000), revealed by a series of neutron reflection experiments in which various parts of the copolymer and/or one of the homopolymers was labelled with deuterium. The bold lines are the segment density profiles for all styrene and methyl methacrylate segments, summed over both the homopolymer and the copolymer; the solid lines are the homopolymers, and the dotted lines are the styrene and methyl methacrylate blocks of the copolymer. After Russell *et al.* (1991).

term arises from the interaction between each block and its surroundings and is expressed in terms of the Flory–Huggins interaction parameter χ_{ab}.

As we discussed in section 6.2, we can distinguish between two limits for the behaviour of polymer brushes in the melt – the 'dry brush' case, in which the brushes are not significantly penetrated by the homopolymer melt, and the 'wet brush' case, in which they are. The conditions for a 'dry brush' are $N_{hi}^{3/2} > N_{ci}$ – the homopolymer chains are long compared with the copolymer block forming the brush – and the areal copolymer density $\sigma > N_{ci}^{-1/2}$ – the copolymers are packed densely enough to be forced to stretch. In these circumstances it is the stretching energy that dominates the free energy of the copolymers at the interface and one finds the following expression for the surface excess:

$$z_i^* = \frac{a}{3}\left(2N_c\frac{\mu_c}{kT}\right)^{1/2},$$ (6.3.3)

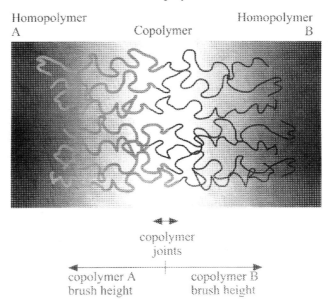

Figure 6.21. A sketch of the arrangement of a block copolymer at an immiscible polymer/polymer interface.

where a is the statistical segment length. Similarly, one can obtain for the interfacial tension

$$\gamma = \gamma_0 - \frac{\rho_0 a}{9} \left[\frac{8}{N_c} \left(\frac{\mu_c}{kT} \right)^3 \right]^{1/2} kT, \qquad (6.3.4)$$

where ρ_0 is the segment density.

The cases in which one or both of the copolymer blocks form a wet brush are somewhat more complicated and are discussed by Dai *et al.* (1992). Our discussions in section 6.2 should lead us to suspect that this scaling approach is not likely to be completely accurate – we know from neutron reflectivity measurements (for example those shown in figures 6.14 and 6.20) that, even in the dry-brush regime, there is substantial penetration of the brush by the homopolymer. Nonetheless, the theory does succeed in capturing much of the physics and allows at least semi-quantitative predictions of the interfacial excess and interfacial tension.

Our expression relating the chemical potential of a bulk copolymer to the total volume fraction, equation (6.3.2), is applicable only when the copolymer does not associate into micelles. As the volume fraction increases such association occurs at a critical micelle concentration (or CMC). This also can be found by a scaling approach; for a spherical micelle the chemical potential at which micelles will form is given by (Leibler 1988)

$$\frac{\mu_{\text{cmc}}^{\text{spheres}}}{kT} = 1.72(\chi_{ab}N_c)^{1/3}g^{4/9}(1.74g^{-1/3} - 1)^{1/3}, \qquad (6.3.5)$$

where g is the ratio of the degree of polymerisation of the block which forms the core of the micelle to the total degree of polymerisation of the copolymer. As this asymmetry parameter changes one expects micelles of different geometry – for example cylinders or lamellae – to be formed; but in all cases the scaling of the chemical potential at the CMC with $(\chi_{ab}N_c)^{1/3}$ still holds.

Comparing equations (6.3.4) and (6.3.5) shows us that the critical micelle concentration is essentially an exponential function of $\chi_{ab}N_c$, so for copolymers of high relative molecular mass the CMC will tend to be rather low. What effect will this have on the adsorption of copolymer at the interface? The analogy with small-molecule surfactants would suggest that the adsorption isotherm would reach a plateau at the CMC; in the absence of strong inter-action between the micelles above the CMC we would expect the chemical potential to be essentially independent of concentration. In fact what is found experimentally for copolymers is that the interfacial excess rapidly increases at the CMC, as shown in figure 6.17 by the steep rise in excess for volume fractions greater than 5%. This steep rise in interfacial excess is mirrored by a rise in the surface excess of copolymer, even in cases in which below the CMC no adsorption of copolymer to the surface at all takes place.

This large rise in interfacial excess above the CMC is not caused by the incorporation of more copolymer chains into the brush layer, but by the segregation of whole micelles at the interface. This has been confirmed by electron microscopy (Shull *et al.* 1991), showing micelles segregated both at a polymer/polymer interface and at a free surface, as sketched in figure 6.22. The driving force for this separation is entropic; two polymer brushes in a melt attract each other, because by approaching closer they expel homopolymer chains that were initially present in non-ideal, low-entropy configurations.

The effect of the micelles on the interfacial tension is to limit the possible reduction. This is shown in figure 6.18, in which the interfacial tension has fallen to less than one third of its initial value by the time the copolymer volume fraction has reached 0.05. But the adsorption isotherm, figure 6.17, reveals that this is the critical micelle composition and above this composition the chemical potential becomes a very weak function of the copolymer volume fraction and very little further reduction in interfacial tension is to be expected.

An interesting question to ask is that of whether, and in what circumstances, it would be possible to achieve an interfacial tension of zero. A vanishing

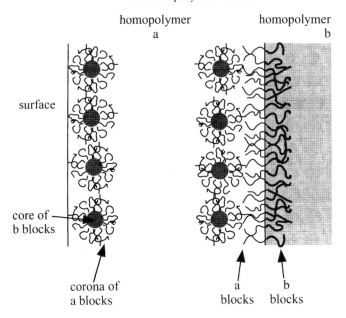

Figure 6.22. Segregation of micelles to a surface and an immiscible polymer/polymer interface in a case in which block copolymer is present in excess of the critical micelle concentration. Segments of a have a lower surface energy than do segments of b. After Shull *et al.* (1991).

interfacial tension would be required for the formation of a microemulsion phase – a stable microscopic dispersion of one immiscible polymer in another, which might well have fascinating and useful properties. At the crudest scaling level, we see that the chemical potential at the CMC scales like $(\chi_{ab} N_c)^{1/3}$; putting this into equation (6.3.4) for the dry-brush case, for example, reveals that the maximum decrease in interfacial tension should scale like $\rho_0 a k T \sqrt{\chi}$, which is exactly the same scaling as that for the interfacial tension between homopolymers. This suggests that the question of whether one can achieve a vanishing interfacial tension using block copolymers is likely to be rather delicate and probably not usefully approached by such crude scaling arguments.

The question has been considered more carefully by Israels *et al.* (1995), who have used numerical self-consistent field calculations both in one and in two dimensions to evaluate the relative stabilities of a microemulsion phase compared with an interface with non-vanishing tension coexisting either with micelles or with multi-lamellar phases. Their calculations suggested that vanishing interfacial tension was possible for symmetrical copolymers with a much larger degree of polymerisation than the homopolymers and they also found some experimental indications that the interfacial tension can at least become very small in these circumstances.

Vanishing interfacial tension and the associated polymer microemulsion phases have been directly and unambiguously observed in a slightly more complicated situation than the one discussed so far, in which at least one block of the copolymer is chemically different from either of the homopolymers and has a strong favourable interaction with one of them (Shull *et al.* 1992, Xu *et al.* 1995). Specifically, these studies looked at an interface between polstyrene and a copolymer of styrene and parahydroxystyrene (PHS) in the presence of a styrene/vinyl pyridine block copolymer. The vinyl pyridine block of the block copolymer strongly interacts with the PHS component of the random copolymer via hydrogen bonding; this highly favourable interaction is characterised by a negative value for the Flory–Huggins parameter for the interaction between poly(vinyl pyridine) and the styrene/PHS copolymer, estimated as $\chi \approx -0.16$ when the random copolymer PHS mole fraction was 0.71 (Shull *et al.* 1992). The signature of the vanishing interfacial tension is a large increase in the amount of copolymer adsorbed at compositions well below the CMC, coupled with the development of a convoluted interface layer with a thickness greater than 100 nm, whose structure can be directly observed by electron microscopy.

6.4 Other interfacially active species at polymer/polymer interfaces

The majority of academic studies of interfacially active species at polymer/ polymer interfaces have involved block copolymers. These are not the only materials that are interfacially active, however, and indeed they are of relatively minor importance in the practical production of polymer blends with useful properties. In industrial applications it is graft copolymers, whether synthesised separately from the blend components or produced *in situ* at the polymer/ polymer interface, that are the most widely used compatibilising agents. The huge diversity of compatibilised blends has been systematically reviewed by Datta and Lohse (1996). Academic studies in this area have lagged far behind, but some recent studies have begun to suggest some fascinating insights.

Perhaps the simplest interfacial agent one can imagine is a small molecule. It is easy to see that, if a solvent that is soluble in both polymers is present, then, by segregation at the interface between polymers 'a' and 'b' it decreases the number of a–b interactions, albeit at the cost of a loss of translational entropy. Calculations of this effect have been performed using self-consistent field theory (Hong and Noolandi 1981). For a situation in which two homopolymers have an interaction parameter of 0.005, they found that adding a volume fraction ϕ_s of a neutral solvent with interaction parameter 0.4 for both solvents reduced the interfacial tension as $(1 - \phi_s)^{1.5}$ and increased the interfacial width as $(1 - \phi_s)^{-0.51}$.

These laws are, to within error, what one expects if the role of the solvent is simply to dilute the interactions between the polymers throughout the system, without appreciably segregating at the interface (this approximation also assumes that the polymer concentration is high enough for the mean-field assumption to be valid – that is to say we must be in a concentrated solution rather than a semi-dilute one). For more immiscible polymers, with $\chi_{ab} = 0.1$, the interfacial tension scaled as $(1 - \phi_s)^{1.6}$ and the interfacial width as $(1 - \phi_s)^{-0.62}$; in this case the solvent reduces the interfacial tension and increases the interfacial width to a greater extent than would be expected by a simple volume dilution of unfavourable interactions; the solvent in this case is segregating to the interface significantly. This effect has also been investigated in the context of a square gradient theory appropriate for the semi-dilute regime (Broseta *et al.* 1987), where the conclusion is similarly reached that solvent segregation becomes more important as the degree of immiscibility of the two polymers increases.

If solvent molecules are expected to segregate to polymer/polymer interfaces, then by the same token so should the low relative molecular mass fraction of a polydisperse polymer, also lowering the interfacial tension. This effect has been investigated in the context of square gradient theory (Broseta *et al.* 1990). For relatively small polydispersities they find that the interfacial tension is controlled by the number average degree of polymerisation. In fact, in one experiment in which the effect of polydispersity on interfacial tension was measured directly, the effect was more dramatic than this (Nam and Jo 1995). Figure 6.23 shows the interfacial tension between polybutadiene and polystyrene, with the polystyrene made artificially polydisperse by blending narrow relative molecular mass fractions in proportions that maintain the same number average relative molecular mass. The interfacial tension of the sample with the highest polydispersity is less than half the interfacial tension of the narrow relative molecular mass fraction.

The same idea – of having a third species that has less unfavourable interactions with either of the polymers forming the interface, in the hope that the third species will segregate to the interface and lower the interfacial energy – can also be applied to a homopolymer additive. Thus, if one has a homopolymer C such that χ_{AC} and χ_{BC} are less than χ_{AB}, one can expect the homopolymer C to segregate to the A/B interface. One case that has been studied both theoretically (Helfand 1992) and experimentally (Faldi *et al.* 1995) occurs when the polymer C is completely miscible with A but not with

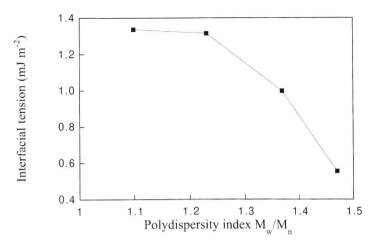

Figure 6.23. The effect of polydispersity on the interfacial tension. The interfacial tension between polybutadiene ($M_n = 4100$ and $M_w/M_n = 1.40$) and polystyrene samples made by blending three fractions of narrow relative molecular mass to achieve a constant number average relative molecular mass ($M_n = 5500$) and various polydispersity indices was measured at 150 °C. After Nam and Jo (1995).

B. If one prepares an A/C blend and makes an interface with B then C will be adsorbed at the interface with a resultant lowering of the interfacial tension. The adsorbed amount will increase if the interaction parameter χ_{AB} is increased – just as, in the case of the small-molecule additives discussed above, segregation of an interfacially active species becomes more marked as the degree of incompatibility between the polymers forming the interface is increased. On the other hand, as polymers C and A become more miscible – that is χ_{AC} becomes more negative – the amount of C adsorbed at the interface decreases as the free energy cost of maintaining a region of material at a different composition from the stable, bulk composition increases. The discussion closely parallels the case of surface segregation from a miscible polymer blend discussed in section 5.1. Continuing the parallel with surface segregation, we would expect that, in the case in which A and C became immiscible, interfacial segregation should still occur. However, we will now need to consider whether a C-rich phase will completely wet the A/B interface, interposing a thick layer between A and B, or whether lenses of the C-rich phase with a finite contact angle will coexist with regions of A/B interface with a microscopic degree of enrichment of C. Just like for the case of a surface discussed in section 5.3, the two situations will be separated by a wetting transition. This so-called internal wetting transition has been studied in the framework of numerical self-consistent field theory (Yeung *et al.* 1994).

A practically convenient way of arranging a polymer C that is less immiscible with each of the blend components A and B than they are with each other is to use a random copolymer of A and B. Self-consistent field calculations (Gersappe and Balazs 1995) reveal that a random copolymer of A and B will reduce the interfacial tension at an A/B interface in exactly the same way as a homopolymer C with the same interaction parameters χ_{AC} and χ_{BC}. The efficiency of a random copolymer in reducing interfacial tension turns out to be a strong function of its composition, with by far the best results obtained for symmetrical compositions, for which $\chi_{AC} = \chi_{BC} = \chi_{AB}/4$. The great advantage of a random copolymer over a homopolymer as an interfacial agent, then, is the fact that one can easily and precisely tune χ_{AC} and χ_{BC} for maximum effect in a way that would not be practicable for a homopolymer. There may be other advantages too; as we shall see in chapter 7 random copolymers prove to be suprisingly effective in promoting adhesion at immiscible interfaces. One factor that is likely to be important in practice is that random copolymers are generally compositionally heterodisperse; this in fact will make them more efficient at reducing interfacial tension than if all chains have the same composition (Gersappe and Balazs 1995). Moreover, the sequence of monomers along the chain may well not be ideally random; some composition drift along the chain will give the copolymers some of the character of a block copolymer and improve their interfacial activity further. This is approaching the case of a so-called tapered diblock copolymer, in which A and B blocks are separated by a length of A–B random copolymer. Such tapered diblocks have been shown to produce a polymer blend with a finer blend morphology and superior mechanical properties than the corresponding pure diblock (Fayt *et al.* 1989).

Although solvent molecules, homopolymers and random copolymers can be surprisingly effective in modifying polymer/polymer interfaces, using an amphiphilic molecule – in which chemically distinct chemical units are covalently bonded – has advantages in terms of flexibility of design. The obvious architecture for an amphiphilic polymer is a block copolymer, but the major limitations preventing the widespread use of these materials are their cost and the limited number of chemical types of copolymer that can be synthesised. Other architectures may be less intrinsically efficient as interfacial agents, but their cost advantages may make them more useful practically. This is certainly true of graft copolymers, which in a direct comparison with diblock copolymers were found to give polymer blends with poorer mechanical properties (Fayt *et al.* 1989). There has been much less theoretical work predicting the interfacial properties of graft copolymers than there has been for diblocks; however, a direct comparison of the relative efficacies of the two architectures in reducing interfacial tension has been made by Lyatskaya *et al.* (1995) using self-consistent field theory. Their results, shown in figure 6.24, show that, in a situation in which the total relative molecular mass is kept constant, a diblock has a greater effect on the interfacial tension than does a molecule with a single graft, which in turn is more efficient than a two-toothed comb.

Another scheme for compatibilisation combines the idea of modifying an A/B interface with a homopolymer C with the use of graft copolymers (Gersappe *et al.* 1994). Both the A phase and the B phase have graft copolymers added; the A phase by a graft copolymer consisting of C chains on an A backbone and the B phase by a graft of C chains on a B backbone. Dynamic SIMS measurements revealed that both graft copolymers segregate to an A/B interface, where their respective C chains form a single mixed layer, whereas mechanical measurements on blends thus compatibilised show increases in strength and modulus.

In many practical cases the interfacial agent – usually a graft copolymer – is not synthesised separately and added to the blend, but is formed *in situ* at the interface, by using homopolymers with complementary reactive groups (Hobbs *et al.* 1983, Datta and Lohse 1996). This has the great advantage that segregation to the interface is not limited by competition with micellisation. The kinetics of the reaction are, however, limited by diffusion of chains bearing reactive groups to the interfacial region and potentially, at high degrees of reaction, by the effective barrier to further reaction formed by the dense brush

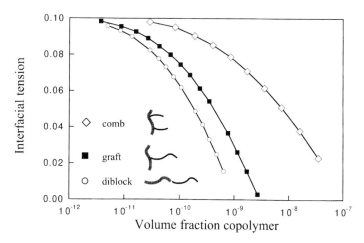

Figure 6.24. The reduction in interfacial tension as a function of the bulk volume fraction of a copolymer for three different architectures, calculated by self-consistent field theory. The total degree of polymerisation of the copolymers was 600 and their composition was symmetrical. The homopolymer degrees of polymerisation were 100 and the interaction parameter $\chi = 0.1$. After Lyatskaya *et al.* (1995).

layer that the preformed copolymers present. These reaction kinetics have been studied theoretically in the simplest case, in which the two polymers forming the interface have complementary end-groups (Fredrickson 1996, O'Shaughnessy and Sawhney 1996). Assuming that the reactivity of the groups is rather high and the polymers are entangled, one expects at early times a diffusion-controlled regime with a rate constant inversely proportional to the degree of polymerisation of the chains, whereas at later times, when the interface is saturated, further reaction becomes exponentially suppressed. This is in fact closely related to the problem of building up a brush at a solid/polymer interface, where the entropic barrier presented by chains grafted earlier drastically slows down the rate of further grafting as soon as the brush approaches the stretched regime.

In one of the few experiments to measure the time dependence of an interfacial grafting reaction (Norton *et al.* 1995), rather more complex behaviour was observed, no doubt reflecting the difference between the simple, ideal situation discussed in the theories and the more complicated chemistry that occurs in practice. Evidently a substantial effort to understand the kinetics of these interfacial reactions is still required. The effect on the properties of immiscible polymers is clear enough, however. Measurements have been made of the interfacial tension of polybutadiene (PBD) and poly(dimethyl siloxane) (PDMS) as a function of the concentration of amine end-functionalised PDMS (Fleischer *et al.* 1994); the PBD was mixed with PBD end-functionalised with a carboxy group, leading to the formation of an acid–base complex at the interface. Reductions in interfacial tension of up to 70% were achieved. A reduction in interfacial tension would be expected to lower the average dispersed particle size in melt-blended polymers; a direct study of a blend of polystyrene and poly(ethylene propylene) with a maleic anhydride graft (Sundararaj and Macosko 1995) showed that introducing reactive groups on the polystyrene reduces the phase size by a factor of five for a 20% blend. The interfacial tension is reduced by a factor of two as a result of the interfacial reaction; interestingly this is not a large enough factor to account for the magnitude of the reduction in phase size. The conclusion that is drawn from this is that, in this kind of blend, with a relatively high concentration of dispersed phase, the effect of the interfacial graft copolymers is primarily to prevent particle coalescence by providing steric stabilisation.

6.5 Phase behaviour of neat block copolymers in bulk and in thin films

We discussed in section 6.3 the tendency of block copolymers dispersed in a homopolymer to form micelles. This is an example of self-assembly – the tendency of amphiphilic molecules to form, at equilibrium, complex, micro-phase-separated structures. Consider an A–B diblock copolymer, in which there is an unfavourable interaction between A and B – that is to say, the Flory–Huggins interaction parameter χ_{AB} is large and positive. If the A and B blocks were not covalently linked the system would macroscopically undergo phase separation, as described in chapter 4. In the diblock case, free energy will still be minimised by keeping A and B in separate phases as far as possible, but the covalent links mean that phase separation can only occur on a microscopic length scale – we have microphase separation. If we now imagine raising the temperature there will come a point at which the entropy gain from allowing A and B segments to mix will outweigh the unfavourable interaction, and we will have a transition to a disordered fully mixed state, the order–disorder transition (ODT).

In the microphase-separated state the structure will have some characteristic length scale that arises from a competition between surface energy – leading to a tendency to larger domains – and the free energy associated with the stretching of the chains. If we assume that the A and B blocks are highly immiscible – the 'strong segregation limit' (SSL) – we can find the way in which this characteristic length depends on the relative molecular mass and the interaction parameter by the following scaling analysis. Suppose that the system adopts a simple lamellar morphology (figure 6.25) – we shall see that this is the morphology found for symmetrical diblocks.

If we consider an area A in the plane of the interfaces, we can write the free energy per unit volume as the sum of a term depending on the inter-

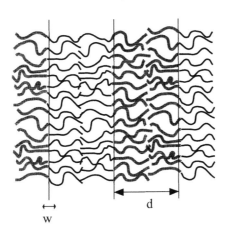

Figure 6.25. A lamellar microphase-separated block copolymer in the strong segregation limit – the domain size d is very much bigger than the interfacial width w.

facial energy γ and a term expressing the free energy cost of stretching the chains:

$$F \sim \frac{\gamma}{d} + \frac{1}{Na^3} \frac{d^2 kT}{Na^2}. \qquad (6.5.1)$$

On minimising this with respect to the lamellar spacing d, we find that the equilibrium value is given by

$$d \sim \left(\frac{\gamma a^5}{kT} \right)^{1/3} N^{2/3}. \qquad (6.5.2)$$

In this strong segregation limit the interfacial tension γ scales in just the same way as the interfacial tension between immiscible homopolymers,

$$\gamma \sim \frac{kT}{a^2} \sqrt{\chi}, \qquad (6.5.3)$$

so we find for the lamellar spacing d

$$d \sim a\chi^{1/6} N^{2/3}. \qquad (6.5.4)$$

It is worth noticing that the condition for this argument assuming strong stretching is that d is much bigger than the random walk size of the block $a\sqrt{N}$; combining this condition with equation (6.5.4) gives us the equivalent condition that the degree of incompatibility $\chi N \gg 1$. Thus the strong stretching condition is equivalent to the strong segregation limit.

Our assumption that the system will adopt a lamellar phase will clearly fail when the block copolymers become very asymmetrical. If one block is much shorter than the other, a lamellar morphology will impose unequal stretching constraints on each block. The asymmetry is conveniently parametrised by the ratio of the length of one block to the total length of the copolymer, f; as this changes away from a value of 0.5 an intrinsic interfacial curvature is imposed on the system, leading to different shapes of the microdomains.

Helfand applied the self-consistent field method described in section 4.3 to the block copolymer phase diagram in the strong segregation limit (Helfand and Wasserman 1982). This approach yielded a derivation of the prediction of domain size of equation (6.5.4), as well as a prediction of the regions of stability of the series of classical morphologies shown in figure 6.26. In the strong segregation limit, it was found that the limits of stability of the different morphologies were essentially independent of the degree of incompatibility χN and depended solely on the composition parameter f.

As the degree of incompatibility decreases, the assumptions of the strong segregation theory break down. The interfacial width between the domains starts to increase and the compositions of the domains are themselves increasingly mixed. In this regime (the weak segregation limit) Leibler applied the

Classical Structures

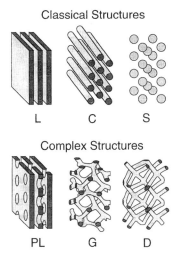

L   C   S

Complex Structures

PL   G   D

Figure 6.26. Block copolymer morphologies. L, C and S (lamellae, cylinders and spheres, respectively) are the classical morphologies whose stability limits were determined in the strong segregation limit by Helfand and Wasserman (1982). PL, G and D (perforated lamellae, gyroid and bi-continuous double diamond) are complex, non-classical phases that have subsequently been identified. Diagram courtesy of M. Matsen.

square gradient theory based on the random phase approximation that we discussed in section 4.2, to find that the critical value of the degree of incompatibility marking the order–disorder transition for the symmetrical case was $(\chi N)_{\mathrm{crit}} = 10.495$. Note that this is much larger than the value for the case of a homopolymer blend, for which the critical degree of incompatibility is 2 (section 4.1). Unsurprisingly, the effect of covalently linking two unfavourably interacting homopolymers is to increase the range of their miscibility substantially.

The classical picture of spheres, cylinders and lamellae was first challenged by the discovery of a new morphology in styrene/isoprene diblocks (Hasegawa *et al.* 1987) and styrene/isoprene star diblocks (Thomas *et al.* 1986). In the latter case the morphology has subsequently been identified as a so-called gyroid phase (Hajduk *et al.* 1995); this is a bicontinous nework with cubic symmetry that has now been identified in a number of different diblock systems (Schulz *et al.* 1994, Hajduk *et al.* 1995). Other non-classical phases have also been identified; these include a perforated layer phase and a modulated lamellar phase (Bates *et al.* 1994).

The gyroid phase may be accounted for theoretically by lifting the approximations used in the strong segregation limit and the weak segregation limit and solving the self-consistent field equations in full (Matsen and Schick

1994). For a diblock copolymer that is structurally symmetrical this results in the phase diagram shown in figure 6.27. The meaning of structural symmetry in this context is that each block of the diblock, although chemically different, is assumed to have identical conformational properties. For real block copolymers this is not so; one block is likely to be stiffer than the other and this will make the phase diagram asymmetrical. Even if the diblock is compositionally symmetrical, with $f = 0.5$, the chain-stretching free energy penalty will be different for each block, leading to a tendency towards a mean curvature (Bates *et al.* 1994).

It is now not clear whether any of the other non-classical phases shown in figure 6.27 represent stable equilibrium phases. In particular, the perforated lamellar phase is most likely to be a long-lived non-equilibrium state that occurs because of the inability of the lamellar phase to convert directly to the gyroid phase when the system is brought into the gyroid region of the phase diagram (Hajduk *et al.* 1997).

The self-consistent field theory phase diagram is also likely to be inaccurate at low relative molecular mass, because, like any mean-field theory, it neglects fluctuations. The effect of fluctuations is to stabilise the disordered phase somewhat (Fredrickson and Helfand 1987); in addition the second-order transition predicted for the symmetrical diblock is replaced by a first-order transition and, for asymmetrical diblocks, there are first-order transitions directly from the disordered into the hexagonal and lamellar phases. In addition it seems likely that fluctuations tend to stabilise high symmetry states such as the gyroid (Bates *et al.* 1994).

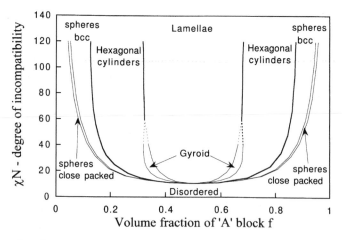

Figure 6.27. The phase diagram for a conformationally symmetrical diblock melt, as calculated by self-consistent field theory. After Matsen and Bates (1996).

As we have seen, even for the relatively simple case of a pure diblock copolymer the phase behaviour is very rich. Naturally, when one goes beyond this simple case to consider different architectures, blends of copolymers with other copolymers and homopolymers, the effects of solvent and possible non-equilibrium effects the range of behaviours possible is huge and the effort involved in cataloguing them is likely to be very great, even if few fundamentally new principles emerge.

We now move on to consider the effect of external surfaces and interfaces on pure block copolymer phases. Let us first consider a block copolymer melt in the disordered state. We saw in chapter 5 that the effect of a surface that favours one component on a simple two-phase blend is to lead to segregation of the component of lower surface energy, with the surface composition perturbation decaying back to the bulk value over a length scale of order the bulk correlation length for composition fluctuations. One can think of this as the stabilisation of a more ordered phase by the surface, with the range of that stabilisation set by the correlation length. For block copolymers we would expect something similar. For a symmetrical composition the ordered phase is a lamellar one and, in the disordered part of the phase diagram, such a material will have fluctuating regions of incipient lamellar ordering, the size of such fluctuations once again being characterised by some correlation length. If one of the blocks of the copolymer has a lower surface energy this incipient lamellar ordering will be stabilised near the surface and one expects a concentration profile consisting of an oscillation damped over a distance given by the correlation length.

Calculations in the framework of square gradient theory (Fredrickson 1987) and self-consistent field theory (Shull 1992) confirm this expectation. Figure 6.28 shows the results of such calculations for three values of the degree of incompatibility χN, for a symmetrical block copolymer. For a value of $\chi N = 5$, well below the critical value for microphase separation of $(\chi N)_{\text{crit}} = 10.495$, we find enhancement of one component at the surface, with a region depleted of that component between about one and two radii of gyration below the surface, but beneath that the system is fully disordered. As the value of χN increases towards the critical value we see that the composition has a damped sinusoidal form, with order penetrating more deeply into the bulk as the microphase separation temperature is approached.

Surface-induced ordering in a block copolymer that is disordered in the bulk has been detected using neutron reflectometry for styrene/methyl methacrylate block copolymers, in which the styrene block is favoured at the surface and the methacrylate block is favoured at a silicon oxide substrate (Anastasiadis *et al.* 1989a, b). Figure 6.29 shows how the decay length characterising the oscilla-

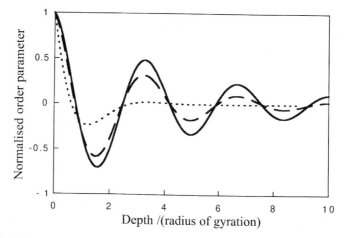

Figure 6.28. Order parameter profiles for a symmetrical block copolymer near a surface that attracts one component of the block, calculated from the theory of Fredrickson (1987). The order parameter ψ is defined as the departure of the volume fraction ϕ from its average value and is here normalised with respect to its value at the surface, where the volume fraction is ϕ_1 – thus for this symmetrical block copolymer $\psi = (\phi - 1/2)/(\phi_1 - 1/2)$. The value of the degree of incompatibility χN is 5 (dotted line), 10 (dashed line) and 10.3 (solid line).

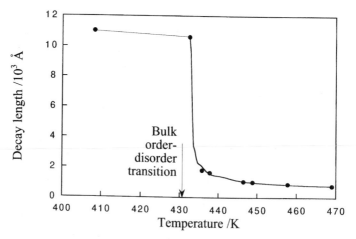

Figure 6.29. The decay length characterising the damping of concentration oscillations near the surface and substrate of block copolymer thin films. The data were obtained using neutron reflectivity for a symmetrical styrene–methyl methacrylate diblock with relative molecular mass 29 700. After Menelle *et al.* (1992).

tory profile increases as the temperature is decreased towards the bulk order–disorder transition. As expected, it rapidly increases as the transition temperature is approached. However, in this experiment the decay length cannot ultimately diverge. This is because the sample used was a thin film; when the decay length is comparable in size to the film thickness the whole sample is effectively completely ordered.

Thus in the ordered part of the phase diagram the effect of a surface or interface on a thin block copolymer film can be very dramatic. If the surface favours one or other component of the block, the film may become essentially completely ordered with lamellae parallel to the plane of the film. This has been confirmed both by dynamic SIMS (Russell *et al.* 1989) and by neutron reflectivity (Anastasiadis *et al.* 1989a, b) studies of annealed block copolymer films. Figure 6.30 shows this effect very clearly; on annealing prominent Bragg peaks appear in the reflectivity, indicating that a multilayer structure has formed in the direction perpendicular to the interface, as discussed in section 3.1 (see in particular figures 3.6 and 3.7).

This tendency towards surface-induced ordering of a lamellar phase is so strong that it may result in a lamellar phase being stabilised close to the surface even when the equilibrium bulk morphology is not lamellar (Turner *et al.* 1994). To a first approximation, in the strong segregation limit, if the difference in surface tensions between the two components of the copolymer is greater than the interfacial tension between them, then the lowering in surface energy obtained by having a pure layer of the lower surface tension component at the surface will outweigh the free energy cost of having the lamellar rather than hexagonally packed cylinders. This will frequently be the case in practice and indeed complete coverage of the surface of a styrene/isoprene block copolymer by isoprene has been observed by electron microscopy for bulk morphologies consisting both of spheres and cylinders (Hasegawa and Hashimoto 1992).

One interesting consequence of the ordering effect of surfaces on lamellar block copolymers is that the thickness of thin, lamellar, block copolymer films is essentially quantised. We saw above (equation (6.5.4)) that the lamellar spacing d (see figure 6.25) is determined by the relative molecular mass of the copolymer and the Flory–Huggins interaction parameter. If, for example, one component of the block copolymer is favoured at the substrate and the other component is favoured at the free surface, then the total film thickness must be given by $2nd$, where n is an integer. If a film is prepared by spin-casting with an arbitrary thickness, on annealing it will develop an incomplete top layer – taking the form of a pattern of terrraces and islands – to accommodate the material that cannot be fitted into an integral number of lamellae (Coulon *et al.* 1989). These effects are easily (and strikingly)

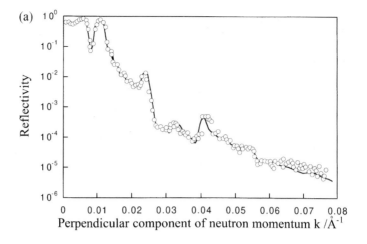

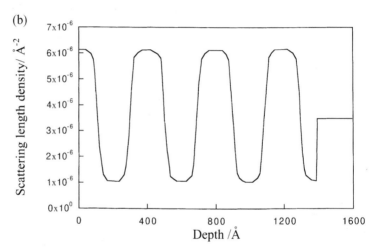

Figure 6.30. (a) Neutron reflectivity from a 130 nm thick film of a styrene–methyl methacrylate copolymer, relative molecular mass 100 000, after annealing for 37 h at 443 K. The solid line is a fit to the profile shown in (b). After Anastasiadis *et al.* (1990).

observed using light interference microscopy or scanning force microscopy (Carvalho and Thomas 1994).

6.6 References

Affrossman, S., Hartshorne, M. *et al.* (1994). *Macromolecules* **27**, 1588.
Alexander, S. (1977). *Journal de Physique, Paris* **38**, 983.
Anastasiadis, S. H., Gancarz, I. *et al.* (1989a). *Macromolecules* **22**, 1449.
Anastasiadis, S. H., Russell, T. P. *et al.* (1989b). *Physical Review Letters* **62**, 1852.

Anastasiadis, S. P., Russell, T. P. *et al.* (1990). *J. Chem. Phys.* **92**, 5677.

Aubouy, M., Fredrickson, G. H. *et al.* (1995). *Macromolecules* **28**, 2979.

Auroy, P., Auvray, L. *et al.* (1991). *Physical Review Letters* **66**, 719.

Auroy, P., Mir, Y. *et al.* (1992). *Physical Review Letters* **69**, 93.

Bates, F. S., Schulz, M. F. *et al.* (1994). *Faraday Discussions* **98**, 7.

Borisov, O. V., Zhulina, E. B. *et al.* (1994). *Macromolecules* **27**, 4795.

Broseta, D., Fredrickson, G. H. *et al.* (1990). *Macromolecules* **23**, 132.

Broseta, D., Leibler, L. *et al.* (1987). *Journal of Chemical Physics* **87**, 7248.

Budkowski, A., Klein, J. *et al.* (1995). *Macromolecules* **28**, 8571.

Carvalho, B. L. and Thomas, E. L. (1994). *Physical Review Letters* **73**, 3321.

Clarke, C. J., Jones, R. A. L. *et al.* (1995). *Macromolecules* **28**, 2048.

Cosgrove, T., Heath, T. G. *et al.* (1991). *Macromolecules* **24**, 94.

Coulon, G., Russell, T. P. *et al.* (1989). *Macromolecules* **22**, 2581.

Dai, K. H., Kramer, E. J. *et al.* (1992). *Macromolecules* **25**, 220.

Datta, S. and Lohse, D. J. (1996). *Polymeric Compatibilisers*. Munich, Hanser.

de Gennes, P. G. (1980). *Macromolecules* **13**, 1069.

de Gennes, P. G. (1987). *Advances in Colloid and Interface Science* **27**, 189.

Faldi, A., Genzer, J. *et al.* (1995). *Physical Review Letters* **17**, 3388.

Fayt, R., Jérôme, *et al.* (1989). *Journal of Polymer Science: Part B: Polymer Physics* **27**, 775.

Field, J. B., Toprakcioglu, C. *et al.* (1992). *Macromolecules* **25**, 434.

Fleischer, C. A., Morales, A. R. *et al.* (1994). *Macromolecules* **27**, 379.

Frazier, R. A., Davies, M. C. *et al.* (1997). *Langmuir* **13**, 4795.

Fredrickson, G. H. (1987). *Macromolecules* **20**, 2535.

Fredrickson, G. H. (1996). *Physical Review Letters* **76**, 3440.

Fredrickson, G. H. and Helfand, E. (1987). *Journal of Chemical Physics* **97**, 697.

Gersappe, D. and Balazs, A. C. (1995). *Physical Review* E **52**, 5061.

Gersappe, D., Irvine, D. *et al.* (1994). *Science* **265**, 1072.

Hadzioannou, G., Patel, S. *et al.* (1986). *Journal of the American Chemical Society* **108**, 2869.

Hajduk, D. A., Harper, P. E. *et al.* (1995). *Macromolecules* **28**, 2570.

Hajduk, D. A., Takenouchi, H. *et al.* (1997). *Macromolecules* **30**, 3788.

Harrats, C., Blacher, S. *et al.* (1995). *Journal of Polymer Science: Part B: Polymer Physics* **33**, 801.

Hasegawa, H. and Hashimoto, T. (1992). *Polymer* **33**, 455.

Hasegawa, H., Tanaka, H. *et al.* (1987). *Macromolecules* **20**, 1651.

Helfand, E. (1992). *Macromolecules* **25**, 1676.

Helfand, E. and Wasserman, Z. R. (1982). *Developments in Block and Graft Copolymers – 1*. J. Goodman. New York, Applied Science. **1**, 99.

Hobbs, S. Y., Bopp, R. C. *et al.* (1983). *Polymer Engineering and Science* **23**, 380.

Hong, K. M. and Noolandi, J. (1981). *Macromolecules* **14**, 736.

Hong, K. M. and Noolandi, J. (1982). *Macromolecules* **15**, 482.

Hong, K. M. and Noolandi, J. (1984). *Macromolecules* **17**, 1531.

Hopkinson, I., Kiff, F. T. *et al.* (1997). *Polymer* **38**, 87.

Hu, W., Koberstein, J. T. *et al.* (1995). *Macromolecules* **28**, 5209.

Israelachvili, J. (1991). *Intermolecular and Surface Forces*. San Diego, Academic Press.

Israels, R., Gersappe, D. *et al.* (1994). *Macromolecules* **27**, 6679.

Israels, R., Jasnow, D. *et al.* (1995). *Journal of Chemical Physics* **102**, 8149.

Jones, R. A. L., Kramer, E. J. *et al.* (1992). *Macromolecules* **25**, 2359.

Karim, A., Satija, S. K. *et al.* (1994). *Physical Review Letters* **73**, 3407.
Lai, P.-Y. and Binder, K. (1992). *Journal of Chemical Physics* **97**, 586.
Leibler, L. (1988). *Makromoleculare Chemie, Makromoleculare Symposia* **16**, 1.
Ligoure, C. and Leibler, L. (1990). *Journal de Physique* **51**, 1313.
Lyatskaya, Y., Gersappe, D. *et al.* (1995). *Macromolecules* **28**, 6278.
Marques, C., Joanny, J. F. *et al.* (1988). *Macromolecules* **21**, 1051.
Matsen, M. W. and Bates, F. S. (1996). *Macromolecules* **29**, 1091.
Matsen, M. W. and Schick, M. (1994). *Physical Review Letters* **72**, 2660.
Menelle, A., Russell, T. P. *et al.* (1992). *Physical Review Letters* **68**, 67.
Milner, S. T., Witten, T. A. *et al.* (1988). *Macromolecules* **21**, 2610.
Nam, K. H. and Jo, W. H. (1995). *Polymer* **36**, 3727.
Norton, L. J., Smigolova, V. *et al.* (1995). *Macromolecules* **28**, 1999.
O'Shaughnessy, B. and Sawhney, U. (1996). *Physical Review Letters* **76**, 3444.
Patel, S., Tirrell, M. *et al.* (1988). *Colloids and Surfaces* **31**, 157.
Russell, T. P., Anastasiadis, S. H. *et al.* (1991). *Macromolecules* **24**, 1575.
Russell, T. P., Coulon, G. *et al.* (1989). *Macromolecules* **22**, 4600.
Schaub, T. F., Kellogg, G. J. *et al.* (1996). *Macromolecules* **29**, 3982.
Schulz, M. F., Bates, F. S. *et al.* (1994). *Physical Review Letters* **73**, 86.
Semenov, A. N. (1985). *Soviet Physics – Journal of Experimental and Theoretical Physics* **61**, 733.
Shim, D. F. K. and Cates, M. E. (1989). *Journal de Physique (Paris)* **50**, 3535.
Shull, K. R. (1991). *Journal of Chemical Physics* **94**, 5723.
Shull, K. R. (1992). *Macromolecules* **25**, 2122.
Shull, K. R., Kellock, A. J. *et al.* (1992). *Journal of Chemical Physics* **97**, 2095.
Shull, K. R. and Kramer, E. J. (1990). *Macromolecules* **23**, 4769.
Shull, K. R., Kramer, E. J. *et al.* (1990). *Macromolecules* **23**, 4780.
Shull, K. R., Winey, K. I. *et al.* (1991). *Macromolecules* **24**, 2748.
Sundararaj, U. and Macosko, C. W. (1995). *Macromolecules* **28**, 2647.
Tang, H. and Szleifer, I. (1994). *Europhysics Letters* **28**, 19.
Taunton, H., Toprakcioglu, C. *et al.* (1988). *Nature (London)* **332**, 712.
Taunton, H. J., Toprakcioglu, C. *et al.* (1990). *Macromolecules* **23**, 571.
Thomas, E. L., Alward, D. B. *et al.* (1986). *Macromolecules* **19**, 2197.
Turner, M. S., Rubinstein, M. *et al.* (1994). *Macromolecules* **27**, 4986.
Watanabe, H. and Tirrell, M. (1993). *Macromolecules* **26**, 6455.
Wijmans, C. M., Scheutjens, J. M. H. M. *et al.* (1992). *Macromolecules* **25**, 2657.
Xu, Z., Jandt, K. D. *et al.* (1995). *Journal of Polymer Science: Part B: Polymer Physics* **33**, 2351.
Yeung, C., Balazs, A. C. *et al.* (1993). *Macromolecules* **26**, 1914.
Yeung, C., Desai, R. C. *et al.* (1994). *Macromolecules* **27**, 55.
Zhao, X., Zhao, W. *et al.* (1991). *Europhysics Letters* **15**, 725.
Zhulina, E. B., Birshtein, T. M. *et al.* (1995). *Macromolecules* **28**, 1491.
Zhulina, E. B., Borisov, O. V. *et al.* (1991). *Macromolecules* **24**, 140.

7

Adhesion and the mechanical properties of polymer interfaces at the molecular level

7.0 Introduction

Many applications of polymers rely on their adhesion, between different pieces of identical polymer, between chemically different polymers or between a polymer and a non-polymer. Examples of situations in which identical polymers are joined include the wide variety of packaging materials that are sealed by heat, the welding of plastic pipes and, less obviously, the integrity of weld lines in injection moulded objects. Adhesion between different polymers is important in determining the mechanical properties of polymer mixtures and of coextruded polymer sheets, whereas polymer/non-polymer adhesion determines the mechanical properties of filled polymers and composites, as well as being the basis of the widespread use of polymers in glues and adhesives. Given the increasing degree of understanding of the microscopic structure of polymer interfaces that we have discussed in earlier chapters, can we predict a macroscopic property such as the strength of such an interface?

A natural starting point would be to enumerate the bonds crossing an interface, whether chemical or physical, multiplying the average force needed in order to break each bond by the number of bonds per unit area to yield a maximum stress that the interface can sustain. In considering the bulk strength of materials, however, one rapidly finds that this is not always the most useful approach; flaws and cracks lead to values of the local stress that are greatly in excess of the average stress by the mechanism of stress concentration; this usually leads to brittle failure at values of the applied stress much lower than the calculated theoretical ultimate tensile stress (see for example Ashby and Jones (1980)). The same considerations apply to interfaces, which will also contain cracks and flaws. Being guided by the experience of bulk materials, what we should try to do is to consider the circumstances under which a pre-existing crack will grow. This is the subject of fracture mechanics; the basic insight of this is that a crack will grow under conditions in which the strain energy released by the failure of the interface more than compensates for the energy required to create two new surfaces. It is the magnitude of this fracture energy, then, that determines whether a crack of given shape and dimensions will be stable or not when the joint is under a given loading. Thus it is the interfacial fracture energy, rather than the maximum interfacial stress, that will determine the strength of a joint in practical circumstances.

A starting point for estimating the interfacial fracture energy, which we write as G, is based on simple thermodynamics. Supposing that we start off with an interface between materials A and B, with an interfacial tension σ_{AB}. If we were able to separate these two surfaces in a thermodynamically reversible way the total free energy change would be $W_{AB} = \sigma_A + \sigma_B - \sigma_{AB}$, where σ_A and σ_B are the surface tensions of A and B, respectively. We shall refer to W_{AB} as the thermodynamic work of adhesion. Note that, for an interface between two identical materials, we expect the thermodynamic work of adhesion to be simply twice the surface tension; $W_{AA} = 2\sigma_A$. In the light of the numerical values of surface tensions given in chapter 2, then, we would expect the thermodynamic work of adhesion to be no more than $0.1 \, \mathrm{J\,m^{-2}}$. However, experimentally it is found that actual measured fracture energies may be orders of magnitude greater than this.

The key to this discrepancy is that interfacial fracture takes place under conditions very far from thermodynamic reversibility. Associated with the propagation of an interfacial crack are processes that result in substantial energy dissipation and it is this that produces usefully large values of the interfacial fracture energy. This dissipation may be relatively localised, close to the crack tip; it may in some cases, however, take place over macroscopic volumes. The latter situation is common for polymer melts and concentrated solutions, as anyone who has pulled their finger out of a pot of glue will attest.

The key to understanding the energy of polymer interfaces, then, lies in identifying the mechanisms of energy dissipation active during crack growth. Conversely, in order to strengthen an interface one needs to ensure that effective energy dissipation mechanisms are available. This emphasis on fracture energy rather than interfacial stress does not mean that the maximum interfacial stress is irrelevant, though. Many highly effective energy dissipation mechanisms – notably crazing in glassy polymers – only become active above a certain minimum stress. In these circumstances what one must do to ensure a tough interface is ensure that it can sustain enough stress to bring these energy dissipation mechanisms into play.

7.1 The strength of interfaces involving glassy polymers

In comparison with inorganic glasses, which are the classical brittle materials, glassy polymers are quite tough. The fracture energy of polystyrene of high relative molecular mass, for example, is around $1000\,\mathrm{J\,m^{-2}}$; four orders of magnitude greater than the value of twice the surface energy expected for an ideal brittle material. This relatively high degree of energy indicates that mechanisms are available during fracture that can dissipate large amounts of energy as a crack is propagated. As the relative molecular mass of the polystyrene decreases, so too does the fracture energy, until at a relative molecular mass that corresponds quite closely to the critical relative molecular mass for entanglement, determined from melt viscosity measurements, the fracture energy has fallen to around $1\,\mathrm{J\,m^{-2}}$. The connection between the fracture energy and entanglement arises because a major mechanism of energy dissipation in the fracture of a glassy polymer such as polystyrene is crazing, the microscopic mechanisms of which rely on the polymers being entangled.

Thus in bulk glassy polymers the ultimate source of toughness is entanglement. Similarly, at an interface between two glassy polymers, it is the degree of entanglement across the interface that determines how strong the interface is. This degree of entanglement may be less at an interface than it is in the bulk in one of two circumstances. If two identical polymers are brought together and annealed above the glass transition temperature for a limited time, the full degree of entanglement will not have been reached for kinetic reasons. This is the case of auto-adhesion or welding. On the other hand, if two dissimilar, immiscible, polymers are brought together, we know from the arguments made in chapter 4 that the equilibrium interface width will be relatively small, so even at equilibrium only limited entanglement will take place.

In discussing interfacial strength from a fracture mechanics point of view, it is important first to note that there are a number of different modes by which an interface will fail and the energy dissipation measured in any fracture test will depend on which mode of failure dominates. A situation in which a crack at the interface is stressed in a purely tensile way (figure 7.1a) is known as mode I; however, it is also possible for shear stresses to be applied, either in the plane of the crack or perpendicular to it. These situations involving shear are referred to as mode II and mode III, respectively (figures 7.1b and c). In general the fracture energy will depend on the relative weight of each of these modes in the actual loading situation. Pure mode I fracture is generally the smallest value, as well as being the most important both in practical terms and in the degree of theoretical understanding. The fracture energy for pure mode I fracture is referred to as G_{Ic} and a major part of the design of fracture tests is to

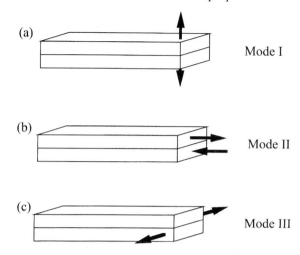

Figure 7.1. Modes of loading an interface.

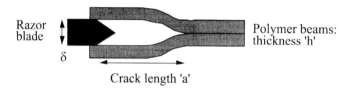

Figure 7.2. The double cantilever beam test for measuring interfacial fracture energy. Two welded polymer bars are driven apart by a razor blade of width δ and the length of the crack ahead of the blade is measured.

ensure that the stress conditions are as close to pure mode I as possible, which is not always easy, particularly when the two materials forming the interface are very different in mechanical properties.

Many testing methods are available with which to measure the interfacial fracture energy and the choice of an appropriate test is dictated by several factors, including particularly the degree to which the mode of crack opening is controlled. These methods are extensively discussed by, for example, Kinloch (1987), and we will not reproduce this discussion here. However, it is worth mentioning one of the most convenient and reliable tests, the double cantilever beam test. The test pieces are rectangular bars of uniform thickness, which are joined together to form the interface. In the test (figure 7.2) a razor blade is driven into the interface at constant speed and the length of the crack ahead of the razor blade is measured. From this crack length a we can calculate the fracture energy. If the two beams have identical Young's moduli E and thicknesses h, then failure is pure mode I and the interfacial fracture energy G_{Ic} is given by (Kanninen 1973)

$$
G_{Ic} = \frac{3\delta^2 E h^3}{16 a^4 \left[1 + 0.64\left(\dfrac{h}{a}\right)\right]^4} .
$$

The situation is more delicate when the two materials have different moduli. In this case, if the beams are of identical thickness the failure will no longer be purely mode I. In these circumstances the crack will deviate from the interface into the material with the lower deformation resistance, leading to additional energy dissipation. In these circumstances the measured values of the interfacial fracture energy will be larger than G_{Ic}. This problem can be overcome by using an asymmetrical test, in which the thicknesses of the two beams are unequal. At a particular ratio of thicknesses the measured fracture energy will be a minimum and this may be taken as the true value of G_{Ic}.

Using such tests it is possible to measure the interfacial fracture energy between immiscible polymers. For example, between polystyrene and poly-(methyl methacrylate) (PMMA) a value of fracture energy of about 20 J m^{-2} is found (Brown *et al.* 1993). This value of course is substantially greater than the ideal work of adhesion, indicating that additional loss mechanisms are at work in the fracture of such an interface; on the other hand, it is substantially less than the bulk fracture energy either of polystyrene or of PMMA. Thus the extent of entanglement of chains across the interface must be very much smaller than the extent of entanglement in the bulk materials. We discussed in chapter 4 the microscopic structure of the interface between immiscible polymers. These interfaces are diffuse on a microscopic scale, with interfacial widths typically of order tens of ångström units. Specifically, for the polystyrene/PMMA interface we expect an interfacial width of some 30 Å (neglecting the effect of capillary waves). We know that the distance between entanglements in a polymer such as polystyrene is of order 90 Å, so this gives us at least a qualitative understanding of why entanglement at such interfaces is very incomplete, leading to rather poor adhesion at immiscible polymer/polymer interfaces.

One might wonder whether it is possible to correlate the interfacial fracture energy of an incompatible polymer pair more precisely to the width of the interface. Such a correlation clearly exists at a qualitative level. For example, polystyrene is substantially less miscible with poly(2-vinyl pyridine) (PVP) than it is with PMMA. This is reflected via equation (4.2.4) in the width of the interface, which is around 17 Å for PS/PVP interfaces (Dai *et al.* 1994b), neglecting capillary waves, in contrast to the value of 30 Å for PS/PMMA. The measured interfacial fracture energy for PS/PVP is about $1.5\,\mathrm{J\,m^{-2}}$ (Creton *et al.* 1992), very much smaller than the $20\,\mathrm{J\,m^{-2}}$ for PS/PMMA. Another illustration is provided by a study in which the strength of interfaces between polycarbonate and a styrene/acrylonitrile copolymer was studied as a function of the acrylonitrile content of the copolymer (Willett and Wool 1993). For this pair of polymers the Flory interaction parameter has a minimum at a particular copolymer content. We expect from equation (4.2.4) that the interfacial width is inversely proportional to the square root of the interaction parameter and indeed it turns out that the maximum value of the interfacial fracture energy is found at the copolymer content corresponding to the smallest interaction parameter. At present, then, we can say that broader interfaces will lead to higher values of interfacial energy. Can we make this relationship quantitative? De Gennes has argued (de Gennes 1992) that the crucial quantity is the probability that a chain which crosses the interface is entangled. Extending the argument of section 4.2, we can estimate this probability as

$$p_e = \exp(-N_e\chi). \tag{7.1.1}$$

Brown has shown (Brown 1991b), as will be discussed in much more detail below, that the fracture energy of an interface that fails by the formation and subsequent breakdown of a craze is proportional to the square of the number of effectively entangled chains crossing the interface. Thus we would expect the interfacial fracture energy to vary like

$$G_{IC} = G_0 \exp(-2N_e\chi). \tag{7.1.2}$$

Schnell *et al.* (1998) have made a direct correlation between the interfacial width, as measured by neutron reflectivity, and the interfacial fracture energy, using bilayers of polystyrene and poly(paramethylstyrene), in which the interfacial width can be varied by varying the annealing temperature. Some of their results are shown in figure 7.3. These results are for pairs with high enough relative molecular masses for the bulk fracture energy to be independent of the relative molecular mass; the functional form of equation (7.1.2) does not hold very well, but the data for different pairs do appear to lie on the same curve, suggesting that the interfacial fracture energy might be some other universal function of the ratio of the interfacial width to the distance between mechanically effective entanglement points (which is about 9.3 nm for polystyrene).

This weakness of the interfaces between most immiscible polymers explains why the mechanical properties of two-phase polymer mixtures are generally so poor. However, as we saw in chapter 6, we can modify the interfaces between immiscible polymers using block copolymers. In section 6 we concentrated on the role of such so-called compatibilisers in lowering the energy of the interface between the immiscible phases. However, in addition to this effect

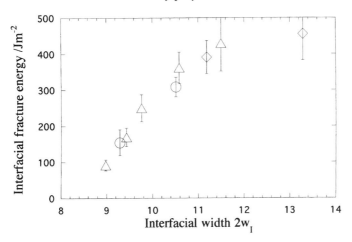

Figure 7.3. Fracture energies of interfaces between polystyrene and poly(p-methyl styrene) of various relative molecular masses ($\triangle$, PS 1 250 000 and PpMS 570 000; $\circ$, PS 310 000 and PpMS 570 000; and $\diamond$, PS 862 000 and PpMS 157 000) as functions of their interfacial widths, measured by neutron reflectivity. After Schnell *et al.* (1998).

these molecules can dramatically increase the fracture energy of the interfaces, promoting efficient stress transfer between the phases and making the materials potentially useful.

Several careful studies have been carried out that have greatly clarified the mechanisms underlying the reinforcing action of block copolymers at interfaces (Brown 1989, 1991a, 1993, Brown *et al.* 1993, Char *et al.* 1993, Creton *et al.* 1992, 1994a, Kramer *et al.* 1994, Washiyama *et al.* 1994). These experiments are important not merely in suggesting ways in which interfaces can be optimally reinforced, but also in the light they throw on failure mechanisms in general at polymer/polymer interfaces.

Creton and co-workers (Creton *et al.* 1992) studied the way that an interface between polystyrene (PS) and poly(2-vinyl pyridine) (PVP) could be reinforced with block copolymers of styrene and 2-vinyl pyridine. These two polymers are highly incompatible – more incompatible than polystyrene and PMMA – and the fracture energy of the unreinforced interface is only of the order of 1 J m^{-2}. However, the fracture energy of the interface may be dramatically increased when a block copolymer of relatively high relative molecular mass is located at the interface. Figure 7.4 shows fracture energies as a function of the amount of block copolymer at the interface for these circumstances. This shows that orders of magnitude increases in fracture energy are possible with block copolymer reinforcement. Over a wide range of copolymer coverages the fracture energy rises in proportion to the square of the copolymer coverage; at

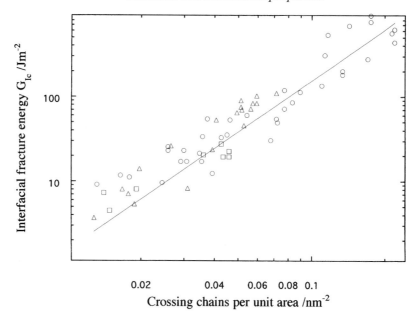

Figure 7.4. Fracture energies of interfaces reinforced by block copolymers as a function of the effective areal density of chains crossing the interface. Triangles and squares are for polystyrene/poly(2-vinyl pyridine) interfaces reinforced with styrene–2-vinyl pyridine block copolymers (Creton *et al.* 1992); circles are for poly(xylenyl ether)/poly(methyl methacrylate) interfaces reinforced with styrene–methyl methacrylate block copolymers (Brown 1991a, b). After Creton *et al.* (1992).

lower coverages the increase in G_{Ic} is closer to linear in the coverage, whereas at very high coverages a plateau in the fracture energy is reached. Qualitatively these results are easily understood in terms of entanglement; one half of each block copolymer molecule is able to entangle completely with its corresponding homopolymer, but the two halves remain joined by a strong covalent bond, effectively stitching the interface together.

The importance of entanglement is highlighted by experiments in which the relative molecular masses of the blocks are varied. Figure 7.5 shows the results of such an experiment. For short blocks, which are not long enough to be fully entangled with their corresponding homopolymer, only modest increases in energy are achieved. However, as soon as the block length exceeds the entanglement length the energy increases rapidly.

If this sort of experiment is supplemented by surface analysis of the fracture surfaces, in order to determine on which side of the joint the block copolymer ends up after failure, it becomes possible to identify the main possible mechanisms of failure of the interface (Kramer *et al.* 1994). There are essentially three possibilities, which are illustrated in figure 7.6. Which mechanism is operative

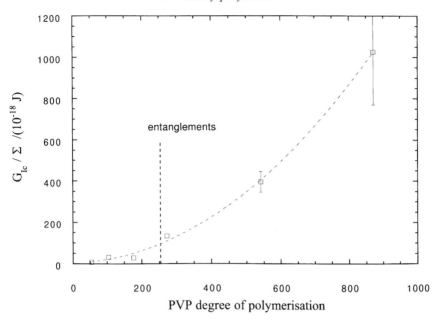

Figure 7.5. Fracture energy per reinforcing chain for polystyrene/poly(2-vinyl pyridine) interfaces reinforced with styrene–2-vinyl pyridine block copolymers, as a function of the degree of polymerisation of the 2-vinyl pyridine block. After Creton *et al.* (1992).

depends on how much stress the interface is able to transfer before failing. The fracture energy is determined by the work done during any plastic deformation as the crack is propagated and is very roughly determined by the stress multiplied by the distance the crack opens as it deforms. We can thus see that the scission mechanism (figure 7.6b) will lead to a very brittle interface, because very little plastic deformation takes place. More dissipation will occur during the process of pulling out (figure 7.6a), whereby the interface will maintain stress while the crack is opened a distance comparable to the stretched out length of the blocks, but it is the process of crazing (figure 7.6c) that will lead to much the biggest values of the fracture energy. The consequence of the fact that these mechanisms have such different fracture energies associated with them is that, if a parameter such as the interfacial coverage of copolymer chains is gradually changed, there may be rather abrupt and large changes in fracture energy when the failure mechanism in operation changes.

The pull-out mechanism of failure (figure 7.6a) is most likely to occur for block copolymers in which one block is relatively short. The fracture energy can then be estimated as the work done against the frictional force which resists the chain being pulled out. If the block is not fully entangled we might expect

(a)

(b)

(c)

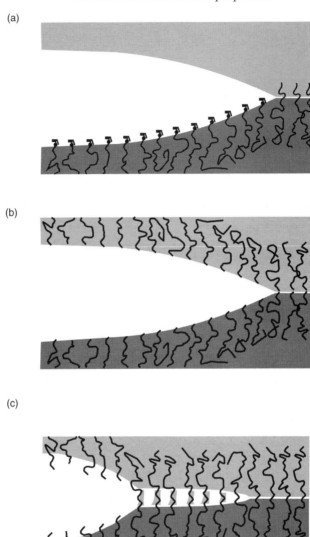

Figure 7.6. Possible modes of failure of a block copolymer reinforced interface. (a) Pull-out. (b) Scission. (c) Crazing.

the frictional force to be proportional to the block length N_b, in which case we arrive at the prediction

$$G_{Ic} \propto \Sigma N_b^2, \tag{7.1.3}$$

where Σ is the number of block copolymer chains per unit area. In general one should expect the frictional force to be a function of the velocity at which the chain is pulled out, but the nature of this dependence is unknown and experimentally the fracture energy in this regime does not seem to be a strong function of the velocity over the range of velocities at which experiments are carried out.

On the other hand, if both blocks of the copolymer are relatively well anchored, the force on an individual chain may exceed the force required to break chemical bonds along the backbone of the chain (figure 7.6b). This is a situation that is likely to lead to a relatively low fracture energy, because the size of any zone of plastic deformation at the crack tip must be rather small, of order the monomer size. It is difficult to estimate the value of the fracture energy precisely, but if the force required in order to break a bond is f_b, then the stress the interface can sustain is

$$\sigma_i \sim \Sigma f_b. \tag{7.1.4}$$

As the coverage of the interface by block copolymers increases, the interface will transfer increasingly large values of stress, until this stress exceeds the stress at which the material on one or other side of the interface will craze (figure 7.6c). The onset of interfacial crazing will be associated with a large and rather abrupt increase in the fracture energy, because of the large volume of material that undergoes plastic deformation in a craze. In fact, it is now becoming clear that crazing is the only mechanism of adhesive failure at glassy polymer/polymer interfaces that can lead to usefully large values of the interfacial fracture energy.

The fracture energy of an interface that fails by crazing has been calculated by Brown (1991b). Here is a highly simplifed version of his argument (see also de Gennes (1992)).

The stress through most of the craze has a constant value σ_{craze}; this crazing stress is essentially a material constant for a given type of polymer. This stress is carried primarily by the main craze fibrils which run perpendicular to the craze/bulk interface. However, there are in addition to the main craze fibrils cross-tie fibrils, which connect the main craze fibrils laterally. These permit some transfer of stress in the lateral direction (see figure 7.7), with the result that there is a stress concentration at the crack tip. It is this stress concentration that causes the breakdown of the last load-bearing fibril and the growth of the crack. At the crudest level we can model the craze as an elastic continuum and write the stress as a function of the distance x from the end of the crack as

$$\sigma(x) = \sigma_{craze} \left(\frac{h}{x}\right)^{1/2}, \tag{7.1.5}$$

where h is the width of the craze. If the spacing between craze fibrils is d this means that the stress on the last load-bearing craze fibril is

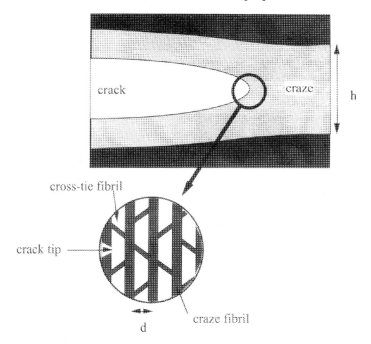

crack

craze

h

cross-tie fibril

crack tip

d

craze fibril

Figure 7.7. The geometry of a craze breaking down to form a crack. After Sha *et al.*
(1995).

$$\sigma_{\text{fibril}} \approx \sigma_{\text{craze}} \left(\frac{h}{x}\right)^{1/2}. \tag{7.1.6}$$

The craze will break down when the stress on the last load-bearing fibril
reaches a value

$$\sigma_{\text{fibril}} = \Sigma f_{\text{b}}, \tag{7.1.7}$$

where f_{b} is the force to break or pull out a single one of the reinforcing chains,
which are present at an areal density of Σ (in fact this is an overestimate
because some of the chains are broken or pulled out when the craze forms).
This allows us to write an expression for the craze width h:

$$h = \frac{\Sigma^2 f_{\text{b}}^2 d}{\sigma_{\text{craze}}^2}. \tag{7.1.8}$$

Now, in the Dugdale model of craze micro-mechanics we can estimate the
fracture energy as

$$G_{\text{c}} = \delta_{\text{c}} \sigma_{\text{craze}}, \tag{7.1.9}$$

where δ_{c}, the critical crack opening displacement, is related to h via

$$\delta_{\text{c}} = h(1 - \nu_{\text{f}}), \tag{7.1.10}$$

where v_f is the volume fraction of fibrils. Thus finally we are led to an expression for the fracture energy:

$$G_{Ic} \approx \frac{\Sigma^2 f_b^2 d(1 - v_f)}{\sigma_{craze}}. \qquad (7.1.11)$$

This correctly accounts for the observed quadratic dependence of the fracture energy on the areal density of reinforcing chains. A more sophisticated version of the same argument would treat the craze as an anisotropic elastic continuum (Brown 1991b). For weak interfaces the craze is rather narrow, and the approximation of treating the crazed material as a continuum breaks down. A discrete treatment of craze breakdown is now available (Sha *et al.* 1995); for such weak interfaces the quadratic dependence of the fracture energy on the areal coverage breaks down and the fracture energy rapidly drops off as the coverage is reduced.

The way in which transitions between these mechanisms can occur is shown in figure 7.8. In figure 7.8a, we have a transition between chain scission and crazing as the coverage of block copolymers increases. This corresponds to a situation in which the diblocks are well anchored on both sides. Figure 7.9 shows such a transition observed experimentally for a polystyrene/poly(vinyl pyridine) interface reinforced by a high relative molecular mass styrene/vinyl pyridine diblock, both of whose blocks are expected to be well entangled with their respective homopolymers. Surface analysis confirms that, as expected, in the scission regime, after the failure of the interface, all the styrene part of the diblock is to be found on the polystyrene part of the interface. Fracture energies in this regime are rather low.

It is also possible to observe transitions from failure by pull-out to failure by crazing, though the conditions for this to occur are rather more restrictive. At low coverages by a copolymer with one rather short block the interface will fail by pull-out. A transition to failure by crazing will be observed if the saturation value of the areal density exceeds the pullout–crazing critical areal chain density $\Sigma^\dagger$, as illustrated in figure 7.8b.

Block copolymers are efficient reinforcers of weak interfaces, but they are costly and difficult to get to the interface. Other reinforcing strategies, which may in some cases be more practical, do exist. Two examples are the use of reactive grafting at interfaces and the use of random copolymers. In both of these methods one uses the same basic principle of reinforcement as that which operates in block copolymers: a single, covalently bonded molecule straddles the interface in such a way that it is well entangled with the homopolymers on both sides of the interface.

In a reactively grafted interface copolymers are effectively created *in situ* at

(a)

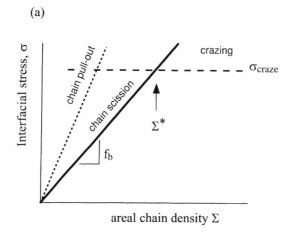

(b)

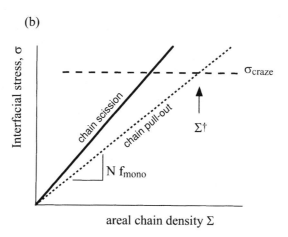

Figure 7.8. Failure transitions in block copolymer reinforced interfaces. The critical stress versus the areal chain density is sketched for chain pull-out (dotted lines) and chain scission (solid). The dashed lines show the crazing stress, which is independent of the areal chain density. For large N (a) there is a transition from failure by chain scission to failure by crazing at a critical areal chain density Σ^*, whereas for a smaller value of N (b) chains are pulled out before they break and there is a transition from failure by chain scission to failure by crazing at a critical areal chain density $\Sigma\dagger$. After Kramer *et al.* (1994).

a polymer/polymer interface (Hobbs *et al.* 1983); at least some of each of the polymers have complementary reactive groups, which, when they meet at the interface, can produce a covalent bond between polymer strands from each side of the interface (figure 7.10). This method of controlling interfacial properties is effective and often used in practice, though there have been few studies at the microscopic level of the way such interfaces are modified by the reaction.

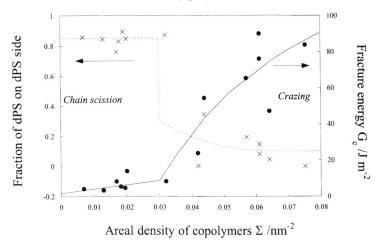

Figure 7.9. Interfacial reinforcement of a polystyrene/poly(vinyl pyridine) interface by a high relative molecular mass deuterated styrene–vinyl pyridine block copolymer, with degrees of polymerisation of each block 800 and 870, respectively. Circles (right-hand axis) show the measured interfacial fracture energy as a function of the areal chain density of the block copolymer Σ, whereas crosses show the fraction of dPS found on the polystyrene side of the interface after fracture. The discontinuity in the curves at $\Sigma = 0.03$ nm^{-2} is believed to reflect a transition from failure by chain scission to failure by crazing. After Kramer *et al.* (1994).

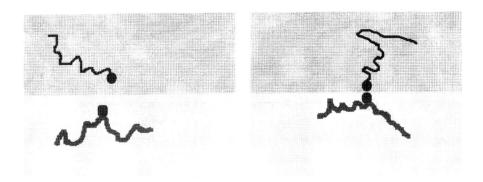

Figure 7.10. A reactively grafted interface. Some polymers on each side of the interface have reactive groups that allow a graft copolymer to be formed *in situ* at the interface.

The formation of such grafts must modify the interfacial tension in a similar way to block copolymers, as discussed in chapter 6. The mechanical effect of grafting has been studied by Norton *et al.* (1995), who looked at the effect on fracture energy of grafting polystyrene with a carboxy end-group at an interface between polystyrene and a thermosetting epoxy resin. The unmodified interface had a fracture energy of $4 \, J \, m^{-2}$. When relatively short polystyrene chains with functional end-groups were introduced into the system, with relative molecular masses below the entanglement relative molecular mass, very little enhancement of the fracture energy was observed. Surface analysis of the failure surface confirmed that the polystyrene chains had become anchored to the epoxy network, but that they had been pulled out of the polystyrene homopolymer. However, for functionalised polystyrene chains longer than the entanglement relative molecular mass substantial increases in the fracture energy were observed, with maximum fracture energies of more than $100 \, J \, m^{-2}$. For these long functionalised chains a sharp transition was noted as a function of the coverage of grafted chains; for low coverages the interface failed by scission of the functionalised chains and the fracture energies were rather low, but when sufficient grafted chains were present to allow the interfacial stress to reach the crazing stress for polystyrene the interface failed by crazing and very large increases in fracture energy were observed.

Our final example of an interfacial agent that can cause significant increases in fracture energy is perhaps the most surprising; these are random copolymers. Brown and co-workers (Brown *et al.* 1993) discovered that a random styrene/methyl methacrylate copolymer could increase the fracture energy of a polystyrene/poly(methyl methacrylate) interface up to a value of about $80 \, J \, m^{-2}$. This is of course a less dramatic increase in strength than is possible with diblock copolymers, but it is still a respectable value, given the potential overwhelming economic advantage of random copolymers over diblocks. Subsequent work (Dai *et al.* 1994a) has demonstrated that the best reinforcing effect is obtained when the copolymer is most symmetrical in its composition. The suggestion is that the copolymer chain makes multiple crossings of the interface (Yeung *et al.* 1992), each loop being well entangled with the homopolymer. Any sub-strand of the random copolymer will have a composition that differs from the average purely as a result of the statistical nature of the copolymer; in an A–B random copolymer some sub-strands will be richer in component A than average and will tend to form loops into the A homopolymer, whereas other sub-strands will be richer in component B and will form loops into the B homopolymer. The result is an interface that is stitched together by multiple crossings in a way that is highly effective at enhancing its fracture energy.

Let us now return to the apparently simpler problem of the way in which interfacial strength develops with time when two identical glassy polymers are brought together and annealed above their glass transition temperature – the case of auto-adhesion. Several authors have studied the time dependence of this process (Wool 1995) and the picture that emerges is that, at early times, the interfacial fracture energy grows proportionally to the square root of the annealing time.

Understanding the origin of this relation is less straightforward than it might appear at first sight. For glassy polymers of high relative molecular mass failure of the imperfectly formed interface will be by crazing for all but the very earliest times, so we would expect a relation of the form of equation (7.1.11) to hold, the fracture energy being proportional to the square of the number of effectively entangled crossing chains. Thus we need to be able to predict the time dependence of this number of entangled crossing chains during the early stages of diffusion, invoking those ideas about interface formation at very early times that we discussed in section 4.4. Progress along these lines is reported by Schnell *et al.* (1998) while earlier work is reviewed in the monograph by Wool (1995).

7.2 The strength of interfaces involving rubbery polymers

Adhesion involving polymers above their glass transition temperature – polymer melts or elastomers with relatively light crosslinking – has some similarities with adhesion involving glassy polymers. Stress is transferred across an interface by physical bonds, chemical bonds or entanglements. If the stress transferred is sufficiently high, yielding processes within the polymer will dissipate substantial amounts of energy, resulting in usefully high values of the fracture energy that are many times greater than the ideal, thermodynamic work of adhesion.

One difference between glassy polymers and melts and elastomers, however, is that the stress required to bring these dissipative mechanisms into play can be much lower for melts than it is for glassy polymers; physical bonds are quite often sufficient. This is familiar to anyone who has put their finger into a pot of glue and accounts for the sticky feel of polymer melts and concentrated solutions. A second difference is that the dissipation mechanisms in melts and elastomers are in many cases dominated by visco-elastic losses. These are characterised by a complex modulus that is strongly dependent on the rate of deformation and consequently adhesion involving melts and elastomers itself has a strong dependence on the rate of deformation. A third difference is that, whereas in glassy polymer the dissipative processes are relatively localised – to craze fibrils and active zones, for example – in polymer melts the dissipation sometimes occurs over macroscopic volumes.

The rate of energy input needed to advance a crack unit area, the adhesive fracture energy *G*, can be written in a form that emphasises the contribution of viscoelastic losses. This was proposed by Gent and Schulz (1972) and Andrews and Kinloch (1973); they wrote G as the product of two factors:

$$G = G_0[1 + f(R, T)]. \tag{7.2.1}$$

Here G_0 is the limiting value of the fracture energy at zero rate of crack growth or the adhesion energy. In the absence of physical or chemical bonds across the interface this will be equivalent to the thermodynamic work of adhesion W. The effect of viscoelastic losses is expressed through the function f, which depends both on the rate at which the crack grows R and on the temperature. In the limit of very low rates and high temperatures, f tends to zero. We know that all time scales in a viscoelastic material have the same temperature dependence; as discussed in chapter 4.4 this leads to the principle of time–temperature superposition, whereby the time and temperature dependences of all viscoelastic properties can be expressed in a single universal curve by the use of temperature-dependent shift factors a_T (Ferry 1980). Thus $f(R, T) = f(Ra_T)$, where the shift factor takes a form given by the WLF equation (this is equivalent to the Vogel–Fulcher equation, equation (4.4.1)):

$$\log_{10}(a_T) = \frac{C_1^g(T - T_g)}{C_2^g + T - T_g}, \tag{7.2.2}$$

where T_g is the glass transition temperature and C_1^g and C_2^g are constants, typical values of which are 17.6 and 52, respectively (Ferry 1980).

This approach is illustrated very well by data obtained by Gent and Lai (1994). They performed peeling tests as a function of the rate and temperature on thin sheets of elastomers that had been partially crosslinked before being bonded together, after which the crosslinking process had been completed. This procedure gave rise to an interface crossed by a controllable number of covalent bonds; the limiting fracture energy G_0, is taken as being proportional to the number of crossing bonds and could thus be systematically varied by varying the crosslinking time before contact. Results are shown in figure 7.11.

The data show that there are strong dependences of the fracture energy both on the rate and on the temperature, the fracture energy increasing with increasing peeling rate and decreasing temperature. The data can be reduced to a single curve by plotting the fracture energy against the effective peeling rate reduced to $-20\,°C$ using the shift factors of equation (7.2.2); this is shown in figure 7.12.

The importance of viscoelastic effects in adhesion as measured by methods such as peeling tests means that it is quite difficult to measure the limiting fracture energy G_0. One method that has been used successfully for elastomers was developed by Johnson, Kendall and Roberts (Johnson *et al.* 1971) and is commonly known as the JKR experiment. The experimental arrangement is

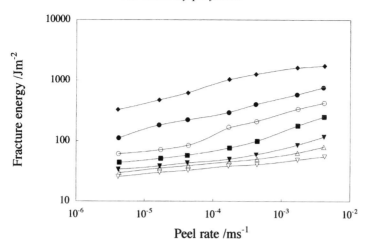

Figure 7.11. The fracture energy for partially cured crosslinked polybutadiene sheets as a function of the rate of peeling and the temperature ($\blacklozenge$, $-40\,°C$; $\bullet$, $-20\,°C$; $\circ$, $0\,°C$; $\blacksquare$, $25\,°C$; $\blacktriangledown$, $50\,°C$; $\triangle$, $80\,°C$; and $\triangledown$, $130\,°C$), as measured in a peeling test. Lines are to guide the eye. After Gent and Lai (1994).

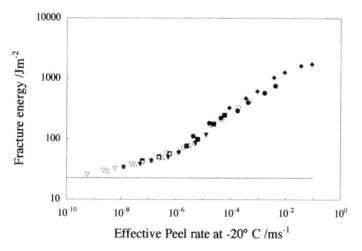

Figure 7.12. The fracture energy versus the effective rate of peeling at $-20\,°C$, with data from a number of different temperatures ($\blacklozenge$, $-40\,°C$; $\bullet$, $-20\,°C$; $\circ$, $0\,°C$; $\blacksquare$, $25\,°C$; $\blacktriangledown$, $50\,°C$; $\triangle$, $80\,°C$; and $\triangledown$, $130\,°C$) reduced using the WLF shift factors (equation (7.2.2)). The line is an estimate of the low-peeling-rate limit of the fracture energy G_0. After Gent and Lai (1994).

illustrated in figure 7.13, for the case in which we are seeking to measure the adhesion energy between an elastomeric sphere and a rigid substrate (the method is easily generalised to the more general situation in which we wish to find the adhesion energy between two elastomers). A hemisphere of the

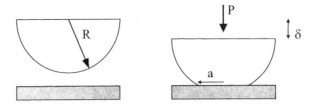

Figure 7.13. The JKR experiment for measuring the ideal adhesion energy between a rigid substrate and an elastomer. A hemispherical cap of the elastomer of radius R is brought into contact with the substrate and loaded with a force P. This results in a displacement of the cap δ and the formation of a circular area of contact radius a.

elastomer is brought into contact with the substrate and loaded with a force P. This results in an elastic distortion of the hemisphere and the formation of a circular area of contact of radius a. In the absence of any surface forces there is a relation between this radius and the load that takes the form, for small loads,

$$a^3 = \tfrac{3}{4}kRP, \tag{7.2.3}$$

where k is an elastic constant related to Young's Modulus E and Poisson's ratio ν of the elastomer by $k = (1 - \nu^2)/(\pi E)$. Likewise the displacement of the hemisphere δ is related to the load P by

$$\delta^3 = \tfrac{9}{16}\pi^2 k^2 \frac{P^2}{R}. \tag{7.2.4}$$

These equations, originally due to Hertz, are modified in the presence of interactions between the surfaces, which even in the absence of an external load P will produce a distortion of the hemisphere. An approximate argument illustrates the way this works. An adhesion energy G_0 makes a contribution to the energy of the system $-\pi a^2 G_0$; this gives rise to an extra effective force pushing the hemisphere into contact with the substrate of magnitude $(\mathrm{d}/\mathrm{d}x)(\pi a^2 G_0)$. Here x is the displacement of the hemisphere; this is approximately related to the contact radius a and the radius of the hemisphere R by $x \approx a^2/R$, which allows us to write the force due to the adhesion energy F_{adh} as

$$F_{\text{adh}} \approx \pi R G_0. \tag{7.2.5}$$

This additional force means that, even in the absence of an external load, the hemisphere will be distorted and there will be a finite radius of contact a. The modified relation between the radius of contact and the applied load may be found by more detailed analysis (Johnson *et al.* 1971). Equation (7.2.3) is modified to

$$a^3 = \tfrac{3}{4}kR\{P + 3G_0\pi R + [6G_0\pi RP + (3G_0\pi R)^2]^{1/2}\}. \tag{7.2.6}$$

Two obvious consequences follow from this equation; firstly, in the absence of an external load, the contact radius takes a non-zero value a_0 given by

$$a_0^3 = \tfrac{9}{2}\pi^2 kG_0 R^2. \tag{7.2.7}$$

Moreover, a remains positive even if a small negative load is applied. One can find the minimum negative load – the pull-off force – by noting that, in order for equation (7.2.6) to have a real solution, $6G_0\pi RP + (3G_0\pi R)^2 \geqslant 0$. This gives for the pull-off force $F_{\text{pull-off}}$

$$F_{\text{pull-off}} = \tfrac{3}{2}G_0\pi R. \tag{7.2.8}$$

In principle, then, one could determine the adhesion energy G_0 simply from measuring the pull-off force of a hemispherical cap of known radius. However, what is more usually done in practice is to use an optical microscope to measure the contact radius a for a sample, which is loaded either directly or by imposing a displacement via a translation stage and measuring the resulting force with an analytical balance. In this way curves of the contact radius versus the load are obtained, which can be fitted to equation (7.2.6) to yield the adhesion energy G_0. This is illustrated in figure 7.14, in which data for a PDMS hemisphere in contact with a silicon wafer coated with grafted PDMS chains are shown (Deruelle *et al.* 1995). The deduced adhesion energy in this case is 45.1 ± 0.9 mJ m^{-2}. The surface energy of PDMS is 21.7 mJ m^{-2}, so it is

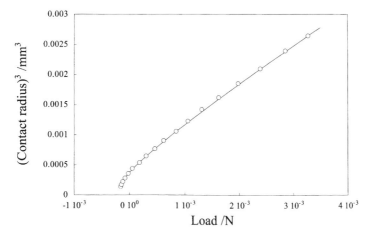

Figure 7.14. Data obtained from the JKR technique, showing the cube of the contact radius plotted against the load as increasing loads are applied. The sample is a hemisphere of PDMS elastomer and the substrate is a silicon wafer coated with a thin grafted PDMS layer. The solid line is a fit to equation (7.2.6), yielding a value for the adhesion energy of 45.1 ± 0.9 mJ m^{-2}. After Deruelle *et al.* (1995).

apparent that, within experimental error, we find that the adhesion energy measured by the JKR technique is indeed in this case equivalent to the ideal thermodynamic work of adhesion 2γ.

Although this approach seems to yield satisfactory agreement with the JKR theory for experiments in which the load on the sample is increased with time, very often a substantial hysteresis is found when the sample is unloaded. The origin of this hysteresis is not fully understood and may well be different for different situations. In one set of experiments hysteresis was attributed to interfacial chemical reactions (Silberzan *et al.* 1994), whereas in other cases the hysteresis has been associated with physical interpenetration of the network by dangling chains (Brown 1993, Creton *et al.* 1994b, Deruelle *et al.* 1995).

The JKR approach can be extended to analyse these kinds of time-dependent phenomena, using fracture mechanics (Maugis and Barquins 1978). This allows one to write an expression for the strain energy release rate or fracture energy G

$$G = \frac{k}{8\pi a^3} \left(\frac{4a^3}{3kR} - P \right)^2. \tag{7.2.9}$$

Experimentally this equation is used by first applying a load to the sample and recording the contact radius as a function of the load in this static case to determine the elastic constant k. The load is then released and the contact radius is measured as a function of time as the lens relaxes. Equation (7.2.9) is then used to convert the contact areas into values of the fracture energy as a function of time. Finally the data of contact radius versus time allow one to produce curves of the fracture energy as a function of the crack velocity. The advantage of this technique for measuring the fracture energy as a function of crack speeds over, for example, the peeling tests described above is that extremely low values of the crack velocity – down to 1 Å s^{-1} – are attainable.

This technique has been used to study the effect of tethered chains on adhesion between a substrate and an elastomer (Creton *et al.* 1994b). Figure 7.15 shows the results; a brush layer was formed by depositing a thin film of a styrene/isoprene copolymer on to a silicon wafer coated with a grafted poly-styrene brush. The elastomer was crosslinked polyisoprene. In the absence of any copolymer, there is a rather weak dependence of the fracture energy on the crack speed. The value of the adhesion energy G_0 can be deduced by extrapolating the fracture energy G to a zero rate of crack growth; a value of $0.17 \pm 0.03 \text{ J m}^{-2}$ is found. This is rather more than the thermodynamic work of adhesion, estimated as 0.065 J m^{-2}. The copolymer has two effects; increasing coverages of copolymer result in substantial increases in the adhesion

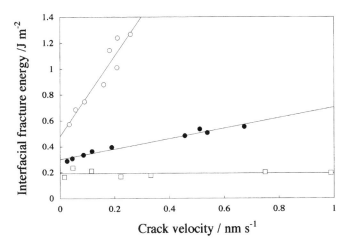

Figure 7.15. The fracture energy against the crack velocity for a polyisoprene elastomer in contact with a polystyrene substrate, in the presence and absence of a layer of styrene–isoprene copolymer. The relative molecular masses of the styrene and isoprene blocks were 60 000 and 66 000, respectively. $\bigcirc$, $\Sigma = 0.053$; $\bullet$, $\Sigma = 0.01$; and $\square$, $\Sigma = 0$. After Creton *et al.* (1994).

energy G_0 and in a much more rapid increase in fracture energy with increasing crack velocity. At higher rates still, a sharp change of slope was observed, a much lower gradient of the fracture energy against crack velocity plot being obtained.

A natural interpretation of this kind of experimental result is that the polymer brush formed by the isoprene block of the copolymer is able to penetrate the network during the initial loading stage. The fact that these chains need to be pulled out from the network before the crack can propagate will have two effects; firstly, even at zero rate, the adhesion energy will be greater than the ideal thermodynamic work of adhesion and secondly the increased energy dissipation associated with the pulling out of the chains from the network will lead to a greater increase in fracture energy with increasing crack velocity. A theoretical model of this situation has been developed by Raphaël and de Gennes (1992).

7.3 References

Andrews, E. H. and Kinloch, A. J. (1973). *Proceedings of the Royal Society (London)* A **332**, 385.
Ashby, M. F. and Jones, D. R. H. (1980). *Engineering Materials.* Oxford, Pergamon Press.
Brown, H. R. (1993). *Macromolecules* **26**, 1666.

Brown, H. R. (1989). *Macromolecules* **22**, 2859.
Brown, H. R. (1991a). *Annual Review of Materials Science* **21**, 463.
Brown, H. R. (1991b). *Macromolecules* **24**, 2752.
Brown, H. R., Char, K., *et al.* (1993). *Macromolecules* **26**, 4155.
Char, K., Brown, H. R. *et al.* (1993). *Macromolecules* **26**, 4164.
Creton, C., Brown, H. R. *et al.* (1994a). *Macromolecules* **27**, 1774.
Creton, C., Brown, H. R. *et al.* (1994b). *Macromolecules* **27**, 3174.
Creton, C., Kramer, E. J. *et al.* (1992). *Macromolecules* **25**, 3075.
Dai, C.-A., Dair, B. J. *et al.* (1994a). *Physical Review Letters* **73**, 2472.
Dai, K. H., Norton, L. J. *et al.* (1994b). *Macromolecules* **27**, 1949.
Deruelle, M., Léger, L. *et al.* (1995). *Macromolecules* **28**, 7419.
Ferry, W. D. (1980). *Viscoelastic Properties of Polymers*. New York, J. Wiley.
de Gennes, P. G. (1992). p55 in *Physics of Polymer Surfaces and Interfaces*. Ed. I. C. Sanchez. Boston, Butterworth-Heinemann.
Gent, A. N. and Lai, S.-M. (1994). *Journal of Polymer Science: Part B: Polymer Physics* **32**, 1543.
Gent, A. N. and Schultz, J. (1972). *Journal of Adhesion* **3**, 281.
Hobbs, S. Y., Bopp, R. C. *et al.* (1983). *Polymer Engineering and Science* **23**, 380.
Johnson, K. L., Kendall, K. *et al.* (1971). *Proceedings of the Royal Society (London)* A **324**, 301.
Kanninen, M. F. (1973). *International Journal of Fracture* **9**, 83.
Kinloch, A. J. (1987). *Adhesion and Adhesives*. London, Chapman and Hall.
Kramer, E.J., Norton, L. J. *et al.* (1994). *Faraday Discussions* **98**, 31.
Maugis, D. and Barquins, M. (1978). *Journal of Physics D: Applied Physics* **11**, 1989.
Norton, L. J., Smigolova, V. *et al.* (1995). *Macromolecules* **28**, 1999.
Raphaël, E. and de Gennes, P. G. (1992). *Journal of Physical Chemistry* **96**, 4002.
Schnell, R., Stamm, M. *et al.* (1998). *Macromolecules* **31**, 2284.
Sha, Y., Hui, C. Y. *et al.* (1995). *Macromolecules* **28**, 2450.
Silberzan, P., Perutz, S. *et al.* (1994). *Langmuir* **10**, 2466.
Washiyama, J., Kramer, E. J. *et al.* (1994). *Macromolecules* **27**, 2019.
Willett, J. L. and Wool, R. P. (1993). *Macromolecules* **26**, 5336.
Wool, R. P. (1995). *Polymer Interfaces*. Munich, Hanser.
Yeung, C., Balazs, A. C. *et al.* (1992). *Macromolecules* **25**, 1357.

8

Polymers spread at air/liquid interfaces

8.0 Introduction

Polymers spread at the interface between two fluid media are of importance in a number of areas, liquid–liquid extraction, stabilisation of emulsions and as model membrane systems being some examples. By far the most extensive body of work reported has been for cases in which one of the fluid media is air, the liquid phase generally being water. Polymers spread at the air/water interface can be viewed as pseudo-two-dimensional systems; thus there is fundamental interest in ascertaining whether theory can be adapted to explain the observations. One practical advantage of spread polymer films is that film compression leads to an increase in surface concentration, so it is relatively easy to change this quantity experimentally.

8.1 Surface pressure isotherms: classification, theory and scaling laws

The surface pressure (π) is the difference in surface tension between that of the pure sub-phase liquid surface and that when the surface is covered by a film. Surface pressure isotherms for materials of low relative molecular mass are values of π plotted as a function of the area of surface available per molecule. For spread films of materials of low relative molecular mass, the surface pressure isotherm may display a number of transitions as the area available is decreased. The classification of these isotherms into various types is discussed in various texts on surface and interfacial chemistry (Jaycock and Parfitt 1981, Adamson 1990, MacRitchie 1990). Spread polymer films do not show the same range of isotherm behaviour as that for compounds of low relative molecular mass. Generally only two types are observed, namely liquid expanded and condensed films. Schematic surface pressure isotherms of these two types are shown in figure 8.1; these are plotted in the manner most common for spread

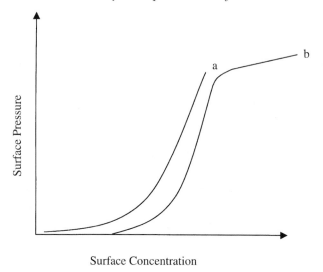

Figure 8.1. Schematic surface pressure isotherms of the two major types encountered in spread polymer films: (a) liquid expanded and (b) liquid condensed.

polymers, i.e. with the surface pressure as a function of the surface concentration since the relative molecular mass of polymers is generally not a unique value.

Since the surface pressure of a spread film is the two-dimensional equivalent of pressure, attempts have been made to set down an equation of state for the spread film. All derivations make use of the basic thermodynamic relationship between the surface pressure and the surface area (A):

$$\pi = (\partial \Delta G_s / \partial A)_{N,T}, \tag{8.1.1}$$

where ΔG_s is the change in surface Gibbs free energy between the clean liquid surface and the liquid covered by a spread film. Each of the derivations differs in the approach by which ΔG_s is calculated. The first of these equations of state was due to Singer (1948) and made the severe assumption that energetic interactions between solvent and polymer segments were negligible. Subsequent derivations incorporated such interactions and allowed for partial submersion of the polymer molecules rather than the loosely anchored molecules on the aqueous substrate surface envisioned by Singer. The various derivation methods notwithstanding, all the equations of state for surface spread polymer films can be written as a virial expansion

$$\pi = \frac{RT\Phi}{A_0}(1 + A_{2s}\Phi + A_{3s}\Phi^2 + \cdots), \tag{8.1.2}$$

where Φ is the fractional coverage of the surface by the polymer molecules, A_0

is the molar area of the polymer molecule in its condensed state and A_{2s} and A_{3s} are the second and third surface virial coefficients. Since the fractional coverage is related to the surface concentration of polymer (Γ_s mass per unit area) and the number average relative molecular mass by $\Phi = \Gamma_s A_0 / \overline{M}_n$, then

$$\pi = \frac{RT\Gamma_s}{\overline{M}_n} \left(1 + A_{2s}\frac{\Gamma_s A_0}{\overline{M}_n} + A_{3s}\left(\frac{\Gamma_s A_0}{\overline{M}_n}\right)^2 \cdots \right). \qquad (8.1.3)$$

Hence equation (8.1.3) could be used to analyse surface pressure data to obtain number average relative molecular masses from the intercept at $\Gamma_s = 0$ of a plot of π/Γ_s as a function of Γ_s. The interpretation of the virial coefficients in molecular terms depends on which theory is used; Huggins (1965) has collected together the correspondences between the virial coefficients for each of the theories. For the case of the original Singer derivation

$$A_{2s} = \frac{n}{2}[1 - (2/z')(1 - 1/n)], \qquad (8.1.4)$$

where n is the degree of polymerisation of the polymer molecule and z' is approximately equal to the coordination number of the two-dimensional lattice used in the derivation. An interesting variation on Singer's original equation was proposed by Motomura and Matuura (1963), who incorporated energetic effects of cohesion between polymer molecules in the spread film.

Application of equations of state to surface pressure isotherm data to obtain number average relative molecular masses requires that very small values of Γ_s be used: the spread polymer film must be dilute. Surface pressures in this region are very small and sensitive determinations of π need to be performed. Successful determinations of $\overline{M}_n$ have been confined to very few cases only. What all the equations of state show is that the polymer's relative molecular mass has little influence on the surface pressure when the surface concentration exceeds the range usable to determine $\overline{M}_n$. This is in contrast to the expression of Motomura and Matuura in which a strong influence of relative molecular mass on the energy of interaction (cohesion) between neighbouring segments of the polymers is predicted. Applications of this equation of state to surface pressure data for polymethyl acrylate and polyvinyl acetate are shown in figure 8.2; here the cohesive energies (in units of $k_B T$) are 0.62 and 0.35, respectively. However, both of these polymers have rather expanded isotherms and application to more condensed layers does not produce so good a fit.

Although numerous surface pressure isotherms have been reported for many different polymers (and polymer mixtures) their discussion has been mainly of a qualitative nature and confined to the reporting of limiting surface areas. One aspect that is not always made evident is the tacticity of the polymer, despite its strong influence on the surface pressure isotherm having been noted many

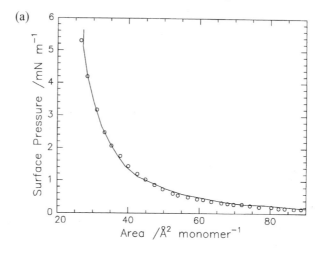

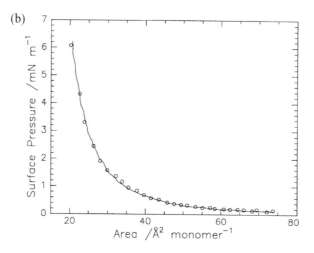

Figure 8.2. Surface pressure isotherms for (a) polyvinyl acetate and (b) polymethyl acrylate spread at the air/water interface. Lines are fits of the equation of state of Motomura and Matuura.

years ago (Beredijk 1963). It was remarked above that determinations of $\overline{M}_n$ from surface pressure data have been successful in only very few cases. One reason for this may be the existence of surface pressure gradients that are set up in the spread film when it is compressed. Barnes and Peng (Peng and Barnes 1990, 1991) have noted such pressure gradients in polymethyl methacrylate and polyvinyl stearate spread films. For the polymethyl methacrylate films, the gradient became steeper as both the compression rate and the relative molecular mass of polymer increased. They concluded that, where pressure gradients existed, equation of state theories (which presume equilibrium surface arrangement) were not applicable. It was also noted that, although the gradient may decay slightly after compression is halted, thereafter it remains constant. As a consequence of this finding the thermodynamic state functions for spreading calculated for polymethyl methacrylate need not pertain to the equilibrium surface configuration and consequently the values should be viewed cautiously.

Spread layers of polymers mixed either with molecules of low relative molecular mass or with other polymers have also been much investigated but apart from the work of Gabrielli *et al.* (1982) the discussion is mainly in qualitative terms. The work on polymer–polymer mixtures should be compared with results for copolymers since the systems are related, being composed of chemically distinct groups. Many of the copolymer systems have been synthesised for specific purposes, e.g. preparation of Langmuir–Blodgett films as exemplified by the work of Tredgold and Hodge (Vickers *et al.* 1985, Tredgold *et al.* 1987, Hodge *et al.* 1990, 1992, Young *et al.* 1990, Aliadib *et al.* 1991). For such binary mixtures of segments the Goodrich–Gaines (Gaines 1966) thermodynamic model can be applied and in favourable circumstances information obtained on the miscibility of the two components in the polymer mixture or copolymer which forms the spread film. If the two polymeric components in the surface spread film mix to form an ideal solution, then the average area per monomer unit at a given surface pressure, π, is given by

$$\langle A_\pi \rangle_{\text{ideal}} = X_1 A_{1\pi} + X_2 A_{2\pi}, \tag{8.1.5}$$

where X_i is the mole fraction of polymer i which has an area per monomer unit of A_i at a surface pressure π. This relation also applies when the two components are immiscible. The excess area on mixing is the difference between this average ideal value calculated from the surface pressure data of the two homopolymers and the value recorded for the polymer mixture or copolymer

$$\langle A_\pi \rangle_{\text{excess}} = A_{\pi \, \text{expt}} - \langle A_\pi \rangle_{\text{ideal}}. \tag{8.1.6}$$

The excess Gibbs free energy of mixing of the mixed polymer layer at the surface pressure π can be obtained from

$$\Delta G_\pi^{\text{excess}} = \int_0^\pi \langle A_\pi \rangle_{\text{excess}} \, \mathrm{d}\pi. \tag{8.1.7}$$

Thus when $\langle A_\pi \rangle_{\text{excess}}$ values are negative, the Gibbs free energy of the mixed systems is less than that predicted from the classical Gibbs free energy of mixing equation and the monolayers will be stable and mixed. When $\langle A_\pi \rangle_{\text{excess}}$ values are positive, the excess Gibbs free energy of mixing is positive and, depending on the magnitude of this excess Gibbs free energy relative to the

classical contribution, polymers in the spread films need not be in a molecularly mixed state. Gabrielli *et al.* (1982) have shown that spread films of
mixtures of polyvinyl acetate with various polymethacrylates exhibit immiscibility for the higher methacrylates, polymethyl methacrylate–polyvinyl acetate
mixtures forming miscible spread films. Polymethyl methacrylate formed a
miscible spread film with poly(n-butyl methacrylate) whereas polyethyl methacrylate–polybutyl methacrylate spread films showed positive excess areas.
Figure 8.3 shows surface pressure isotherms for polyethylene oxide and polymethyl methacrylate homopolymers and block copolymers with differing
amounts of ethylene oxide. The surface pressures calculated using equation
(8.1.5) are also shown. Clearly the excess areas (area $\propto 1/$surface concentration) are negative at low surface concentrations but become positive at higher
surface concentrations, suggesting that phase transitions may take place in the
film if the excess Gibbs free energy becomes sufficiently positive. The dependences of the excess Gibbs free energy calculated in this way on composition for
a series of linear diblock copolymers of polyethylene oxide and polymethyl
methacrylate are shown in figure 8.4.

A slightly different analysis of mixed polymer films has been proposed
(Runge and Yu 1993) to apply when there is no interaction between the two
components. It is a two-dimensional analogue of Dalton's law of partial
pressures, in that the surface pressure of the mixture at a particular area is the

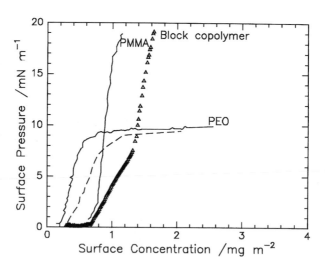

Figure 8.3. Surface pressure isotherms for polymethyl methacrylate (PMMA), polyethylene oxide (PEO) and a linear diblock copolymer (mole fraction of ethylene oxide
about 0.5) and the calculated surface pressure (dashed line) using equation (8.1.5).
After Rochford (1995).

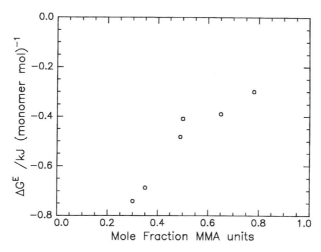

Figure 8.4. The excess Gibbs free energy of mixing as a function of the composition for a series of linear diblock copolymers of PEO and PMMA. After Gabrielli *et al.* (1982).

sum of the surface pressures of the two homopolymers at the same area. This appeared to describe spread films of mixtures of polydimethyl siloxane and polyvinyl acetate well. Evidently it will not apply to the block copolymer of polyethylene oxide and polymethyl methacrylate described earlier. Summarising, it appears that, although detailed equation-of-state theories have been developed for polymer films at the air/water interface, their application to experimental data has been limited. Surface pressure gradients of long duration may also prevent application of such equations since the spread film is in a non-equilibrium state. For polymer mixtures and copolymers surface pressure data can be used to obtain excess Gibbs free energies of mixing and thus allow conclusions about the stability of mixed films to be drawn. However, these may also be susceptible to the same criticism when surface pressure gradients exist. The overwhelming proportion of surface pressure isotherms for spread polymer layers have usually been discussed in a qualitative way and much of this work has been summarised (Kawaguchi 1993). Notable exceptions to this rather sweeping conclusion are the work of Shuler and Zisman (1970) on polyethylene oxide films and that of Granick (Granick 1985, Granick *et al.* 1985, 1989, Kuzmenka and Granick 1988a, b) on a range of polymers including polyethylene oxide.

The most significant recent development in the theory of polymer solutions has been the development of scaling theory and the application of renormalisation group methods to obtain relations between relative molecular mass, concentration etc. and measurable properties of the dissolved polymer such as the osmotic pressure, radius of gyration and correlation length (de Gennes 1979). In a bulk polymer solution regimes of different behaviour can be identified in the temperature concentration plane where the scaling relationships depend on the concentration and the distance of the solution from the theta temperature, as is described in section 5.2. Strictly scaling laws only apply to polymers of infinite relative molecular mass but experimental data on 'real' polymers are unable to detect any deviations from the theoretical predictions. The paper of Daoud and Jannink (1975) should be consulted for the range of detailed scaling laws available; only expressions for the surface pressure and the two-dimensional second virial coefficient are of concern to us here. The scaling relationship between the surface pressure and the surface concentration, Γ_s, is

$$\frac{\pi}{T} \sim \Gamma_s^{\nu d/(\nu d-1)} \tau^{(\nu-\nu_\theta)d(\nu d-1)\psi_\theta}, \tag{8.1.8}$$

where ν is the scaling exponent in the relation between the bulk radius of gyration of the polymer, R_g, and its degree of polymerisation, N, i.e. $R_g \sim N^\nu$; ν_θ is the same exponent at the theta temperature, θ. The dimensionality, d, is separated out in equation (8.1.8), and ψ_θ is an exponent related to the crossover between different behavioural regimes. The temperature dependence is expressed through the reduced temperature τ $(=1-\theta/T)$. For a fixed temperature and for the Γ_s which places the system in the semi-dilute region, equation (8.1.8) can be simplified to

$$\pi \sim \Gamma_s^{2\nu/(2\nu-1)} = \Gamma_s^y. \tag{8.1.9}$$

If the sub-phase is thermodynamically favourable for the polymer then in two dimensions $\nu = 0.75$. For ν_θ a range of values between 0.505 and 0.59 has been proposed, but a generally accepted value appears to be 0.571. When the spread polymer film collapses, ν has a value of 0.5. Consequently, in a double-logarithmic plot of the surface pressure as a function of the surface concentration one expects clear differences in the slope depending on the nature of the interaction between the polymer and the sub-phase. For thermodynamically favourable situations then $y = 3$, for the theta state y should take on a higher value with 101 and 6.6 being the limits of the range of values from the range predicted for ν_θ.

Relatively few polymers have been subjected to a scaling analysis of their isotherms, probably because relatively few homopolymers form a stable film at the air/water interface. Ober and Vilanove (1977) applied an early form of scaling analysis to polyvinyl acetate at the air/water interface; the value of y obtained was 2.7 for relative molecular masses in the range 4000–450 000.

Subsequently more detailed results on polyvinyl acetate and polymethyl methacrylate at a temperature of 16.5 °C were reported (Vilanove and Rondelez 1980). No influence of relative molecular mass on π was observed when the spread films were in the semi-dilute state and from the double-logarithmic plots the values of v obtained were 0.79 and 0.56 for polyvinyl acetate and polymethyl methacrylates, respectively. These are values close to those expected for an excluded-volume-influenced molecule (polyvinyl acetate) and for a polymer near the 2D theta state (polymethyl methacrylate). Kawaguchi (Takahashi *et al.* 1982, Kawaguchi *et al.* 1983) reported surface pressure studies both of polymethyl acrylate and of polymethyl methacrylate. For the acrylate polymer a theta temperature of 18.5 °C was suggested on the basis of the A_{22} values obtained from surface pressure data. Surface pressure data at this temperature in the semi-dilute region gave 0.51 as the value for v. This value of v has been questioned by work of Poupinet *et al.* (1989), who obtained $v = 0.7$ over a temperature range of 5–30 °C. Additionally, this same group made very careful measurements of π at very small surface concentrations for a series of polymethyl methacrylate samples of various relative molecular masses. For a temperature of 15.5 °C a value for v of 0.55 was obtained both from the dependence on the relative molecular mass of the two-dimensional virial coefficient A_{22} and from the concentration dependence of π above Γ_s^*. These findings are important for two reasons. Firstly they show that excluded volume exponents obtained from data in the semi-dilute region are as accurate as those obtained from surface pressure data when the surface concentration is less than Γ_s^*, for which much greater sensitivity in the measurement of π is required. Secondly, since the A_{22} values were negative, these data support the notion that v_θ must be greater than 0.505 and indicate that a value of 0.57 is more acceptable.

A factor that has been only rarely considered, but was pointed out by Vilanove *et al.* to be of importance, is the microstructure of the polymer stereochemistry. This is particularly relevant to methacrylates since their tacticity is determined by the synthetic method used. Isotactic polymethyl methacrylate displays expanded liquid isotherm behaviour whereas the syndiotactic polymer has an isotherm of the condensed liquid type. Given that this difference in behaviour was first remarked on some 30 years ago, the lack of appreciation of the importance of polymer stereochemistry in determining surface layer behaviour is surprising. Values of the exponent v at 25 °C obtained from the isotherms for the two stereoisomers of polymethyl methacrylate are 0.53 for the syndiotactic polymer and 0.78 for the isotactic polymer (Henderson *et al.* 1991). An additional factor here is that the relative molecular mass distribution for the isotactic polymer is generally *very* much broader than that of the syndiotactic polymer. Poly-ethylene oxide also has an isotherm of the expanded liquid type and at 25 °C, from a double-logarithmic plot of the surface pressure above Γ^*, the value of the exponent ν is 0.75 (Henderson *et al.* 1993a, b, c). Figure 8.5 shows surface pressure isotherms for both stereoisomers of polymethyl methacrylate spread on water at 25 °C. Note that the plateau value of the surface pressure for polyethylene oxide (figure 8.3) is much lower than that of the methacrylate polymers and no evidence of film collapse is observed

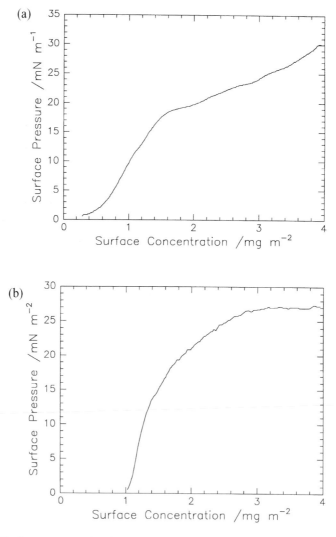

Figure 8.5. Surface pressure isotherms at 298 K for (a) isotactic polymethyl methacryl-ate and (b) syndiotactic polymethyl methacrylate. After Henderson *et al.* (1991).

as the surface concentration is increased. This aspect will be commented on later.

A sophisticated discussion of the surface pressure isotherms for spread films of amphiphilic block copolymers on liquid surfaces has been developed for the situation in which the solvated block forms a brush layer at the air/liquid interface that is anchored to the interface by the insoluble block, analogous to the 'anchor–buoy' systems discussed for solid/liquid interfaces in section 6.1. If the anchoring block is sufficiently small then it makes no contribution to the surface pressure. The surface pressure can be written in terms of the grafting density σ at the surface as

$$\pi = \sigma \left(\frac{\partial \Delta G_{\mathrm{L}}}{\partial \sigma} \right)_{T,A,\mu} - \Delta G_{\mathrm{L}}, \qquad (8.1.10)$$

where ΔG_{L} is the excess free energy in the layer. Carignano and Szleifer (1994) obtained an analytical expression for the surface pressure of such a grafted brush. Others (Wijmans *et al.* 1992, Bijsterbosch *et al.* 1995) have obtained values of π as a function of σ by numerical solution of the self-consistent field lattice theory. The predicted surface pressure variation is shown schematically in figure 8.6 for brush-like chains the segments of which are also able to absorb at the interface. As σ increases the surface pressure increases rapidly due to the adsorption energy of the segments. In this initial region the contribution to π from the osmotic pressure of the brush is small. The pseudo-plateau is due to the interface becoming packed with polymer segments. At higher grafting densities the surface pressure increases again at a much more rapid rate, which is attributed to the contributions from the osmotic pressure of the brush-like layer. Surface pressure isotherms of the type shown in figure 8.6 have been observed for spread films of amphiphilic linear diblock copolymers

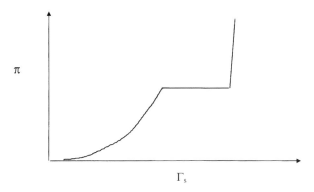

Figure 8.6. A schematic surface pressure isotherm for a polymer at the air/water interface that forms a brush-like layer in the aqueous phase.

on water. Identically shaped isotherms have also been predicted on the basis of classical thermodynamics by Israelachvili (1994), however, in this case the plateau formation is due to surface micelles forming. Since amphiphilic copolymers are also able to form micelles when they are dispersed in the bulk aqueous phase and because the degree of interpenetration of the polymer layer by the aqueous sub-phase is generally extensive (see below), the possibility of such surface micelles forming is bound to be finite. Eisenberg (Li *et al.* 1993) cites such surface micelles as being prevalent for the partially polyelectrolyte block copolymers he has investigated.

8.2 Organisation of polymers at the air/liquid interface

Investigation of the organisation of polymers spread at air/liquid interfaces has generally been undertaken for one or other of two reasons. These are enquiries into the intrinsic nature and arrangement of the polymer molecules at the interface and the use of model polymers (usually a block copolymer) to test theories for grafted polymer brushes. It is appropriate to summarise the main results for grafted brushes here; the details are set out in chapter 6. In its classical definition, a grafted brush consists of a number of inter-acting polymer chains, each attached by one end to a solid surface, with the rest of the molecule being surrounded by solvent. This type of model has been the focus of attention both theoretically and experimentally for a number of years and has the names of Scheutjens and Fleer associated with a self-consistent field model (Scheutjens and Fleer 1979, 1980). A highly detailed treatment of such systems and a full discussion of the Scheutjens–Fleer model is available in a monograph (Fleer *et al.* 1993). If the form of the volume fraction profile normal to the surface is known over a sufficient distance such that the equilibrium bulk polymer volume fraction is reached, then a number of parameters can be calculated. Figure 8.7 shows a schematic profile and indicates various parameters. The surface excess, z^*, is given by

$$z^* = \int_0^\infty \mathrm{d}z \, (\phi(z) - \phi_\infty). \tag{8.2.1}$$

σ, the grafting density in number of molecules per unit area, is

$$\sigma = \frac{z^* \rho N_\mathrm{A}}{Nm} \tag{8.2.2}$$

and L, the layer thickness, is the first moment of the profile,

$$L = \frac{1}{z^*} \int_0^\infty z(\phi(z) - \phi_\infty) \, \mathrm{d}z, \tag{8.2.3}$$

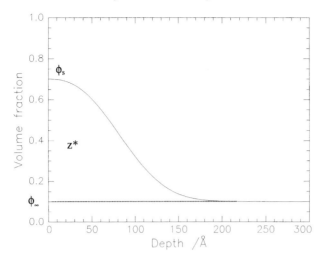

Figure 8.7. A schematic volume fraction profile for a brush-like layer at an interface. ϕ_s and ϕ_∞ are the surface volume fraction and equilibrium bulk volume fraction of the polymer, respectively.

where ϕ_∞ is the equilibrium bulk volume fraction, N the degree of polymerisation, ρ the polymer density and m the relative molecular mass of a monomer unit.

In section 6.2 we described some of the early attempts at determining the nature of the segment density profile normal to a surface, using small-angle neutron scattering (SANS) either from colloidal dispersions or from porous solids to which polymer had been either physically absorbed or chemically grafted. The success of these experiments depends very greatly on obtaining the exact contrast conditions required. If we take a particle that has a layer of polymer attached to its surface then the scattered intensity is given by

$$I(Q) = Kn_p[(\rho_p - \rho_s)^2 F_p(Q)^2 + (\rho_L - \rho_s)^2 F_L(Q)^2$$

$$+ 2(\rho_p - \rho_s)(\rho_L - \rho_s)F_P(Q)F_L(Q)], \qquad (8.2.4)$$

where $F_p(Q)^2$ is the structure factor for the particle, $F_L(Q)^2$ the structure factor for the layer and $F_P(Q)F_L(Q)$ the cross partial structure factor, ρ_i is the scattering length density of species i, where p, s and L represent particle, solvent and polymer layer, respectively. Equation (8.2.4) clearly has a direct relationship to the partial structure factor description of reflectivity outlined in chapter 3. Note that the use of equation (8.2.4) has to be preceded by the complete subtraction of any background scattering. To obtain $F_L(Q)^2$ then $\rho_p - \rho_s$ must be made equal to zero (contrast matching). This can be obtained by using mixtures of hydrogenous and deuterated solvents, but slight variations in exact contrast matching can lead to a large contribution to the observed scattered intensity from the cross partial structure factor. Additionally, there has been some discussion about the need to account for an additional contribution to the scattering from concentration fluctuations within the layer (Auvray and de Gennes 1986). Even when exact contrast matching is obtained, a form for the segment density distribution in the layer has to be assumed so that $F_L(Q)^2$ can be obtained from its Fourier transform. Consequently, the use of SANS is an *indirect* way of obtaining the adsorbed layer profile. Nonetheless, some results obtained are very informative (Cosgrove *et al.* 1987a, b, c, Auroy *et al.* 1991) and it is clear that considerable care and effort has been expended to obtain such data. Fairly complete discussions of the method and results obtained are given in Fleer *et al.* (1993) and in chapters 3 and 8. However, an additional disadvantage of the use of dispersed particles or solid walls to which polymers are attached as models for polymer brushes is that the grafting density σ is not easily controllable if at all! It is this disadvantage that can be overcome by using copolymers at the air/liquid interface as brush models.

If a block copolymer is spread at the air/liquid interface such that the liquid sub-phase is a solvent for one block, then the grafting density of the solvated block may be controllable over a wide range by simply compressing the spread polymer film. Evidently, to approach the conditions under which the solvated block can be viewed as an end-attached or grafted molecule, the surface tension of the polymer must be greater than that of the sub-phase otherwise an absorption or Gibbs layer would be formed (*vide infra*). This approach has been used by Lee and co-workers for a linear diblock copolymer of polydimethyl siloxane (PDMS) and polystyrene (PS) spread on ethyl benzoate, a good solvent for PS but a non-solvent for PDMS (Kent *et al.* 1992). The first experiments were not 'ideal' reflectivity experiments in that they were forced to use hydrogenous polymer and deuterated sub-phase, i.e. the segment density profile corresponds to 'absences' in the density profile of the deuterated sub-phase in the neighbourhood of the surface. Figure 8.8 shows the reflectivity obtained divided by the Fresnel reflectivity (equation (3.1.11)) calculated for the pure sub-phase. Depletion of sub-phase is clearly evident and the best fit to the data appears to be a parabolic volume fraction profile. Values of the maximum extension from the surface, h^*, were evaluated for a range of values of the normalised grafting density σ^*. This normalised grafting density is

$$\sigma^* = \frac{\sigma}{\sigma_{ol}}, \tag{8.2.5}$$

where σ_{ol} is the grafting density at which overlap becomes evident ($\sim (R_g^2)^{-1}$). There was remarkably little variation in h^* over a range of σ^* from 1 to 4

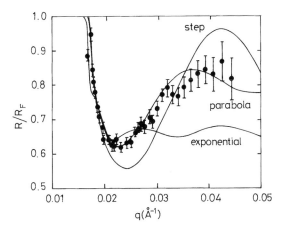

Figure 8.8. Calculated reflectivities compared with various profiles for a polydimethyl siloxane–polystyrene linear diblock copolymer spread at the surface of ethyl benzoate. After Kent *et al.* (1992).

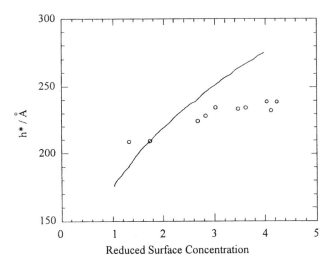

Figure 8.9. The thickness of the PS block of a linear PDMS–PS block copolymer in the ethyl benzoate sub-phase as a function of σ^*. The line is the scaling law prediction $h^* \sim \sigma^{*1/3}$. After Kent *et al.* (1992).

(figure 8.9). This raises the possibility that the transition from isolated grafted molecules to interacting, stretched molecules takes place over a *range* of σ^* values.

This aspect was mentioned by de Gennes in his original paper. Comparison with values of h^* obtained for PS by others showed remarkable agreement, the use of various solvents and solid substrates notwithstanding. However, the exponent in the dependence of h^* on σ^* for these data is about 0.12, far less than the 1/3 predicted both by scaling law and by self-consistent field theories. In a subsequent paper (Kent *et al.* 1995), a polydimethyl siloxane–deutero-polystyrene copolymer was spread on ethyl benzoate. The neutron reflectometry data were of much improved quality. They used a variety of volume fraction profiles to fit the data and they pointed out that an equally good fit is obtained whether the main profile was a parabola, an error function or a stretched exponential. However, in all cases they were forced to include a depletion region near $z = 0$ and an exponentially decaying tail region at large z. The variation of the normalised root mean square height of the tethered polystyrene layer as a function of the reduced areal density of tethered chains, σ^*, is shown in figure 8.10; evidently $\sigma^* = 2$ appears to be a critical value for the onset of chain stretching.

The dependence of the tethered layer's thickness both on the relative molecular mass and on the areal density is shown in figure 8.11. Also shown is the prediction of scaling theories, the solid line through the data being described by $h^* \sim \sigma^{0.22} M^{0.86}$.

The dependence of h^* on M and σ is rather weaker than that predicted by Alexander–de Gennes scaling laws and by the self-consistent field theory of Milner, Witten and Cates ($h^* \sim \sigma^{0.33} M$). A self-consistent field theory of Baranowski and Whitmore (1995) predicts weaker dependences of the layer thickness on the areal density and the relative molecular mass, the exponents

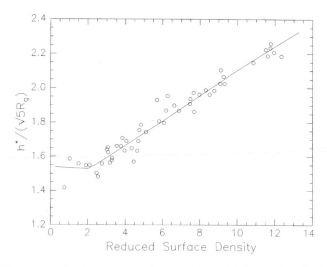

Figure 8.10. The normalised root mean square layer thickness for the PS block as a function of the grafting density of a spread film of a PDMS–PS linear diblock copolymer at the surface of ethyl benzoate. After Kent *et al.* (1995).

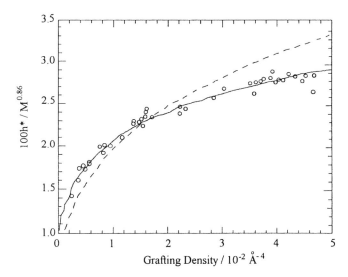

Figure 8.11. The dependence of the brush layer thickness normalised for the relative molecular mass of a spread film of a PDMS–PS linear diblock copolymer at the surface of ethyl benzoate. The dashed line is the scaling law prediction. After Kent *et al.* (1995).

being very close to those given above. Moreover, the volume fraction profiles obtained by Baranowski and Whitmore are very nearly identical to those used in the analysis of the reflectivity profiles, figure 8.12.

Note that the sharpness of the transition in the change on going from the depletion zone to the parabolic zone is due to limitations in the analytical function and does not reflect real transition behaviour. The more gentle transitions indicated in the theoretical SCF profile are more realistic. In the self-consistent field calculation a lattice model is not presumed; the volume fraction of the tethered chains is calculated from a diffusion equation that involves polymer propagators and a (z-dependent) potential function that includes enthalpic interactions between the two copolymer blocks and between each block and the solvent. Initially the potential function is set to zero, the polymer propagators are calculated and then the volume fraction variation of the tethered block. A new potential is calculated from this volume fraction profile and the process reiterated until the difference in volume fraction profiles calculated by sequential iterations is smaller than some defined tolerance. The approach bears similarities to the SCF approach of Shull (1991) but makes no allowance for the 'dry' brush case, i.e. that in which the relative molecular mass of the solvent approaches that of the tethered polymer molecule.

Scaling laws of bulk solutions of polymers are closely associated with a

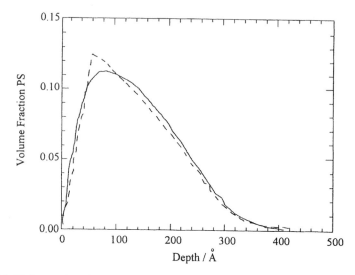

Figure 8.12. Volume fraction profiles for the PS brush layer of a PDMS–PS block copolymer at the air/ethyl benzoate interface. The dashed line is the best fit to neutron reflectometry data. The solid line is the volume fraction profile predicted by the self-consistent field theory of Baranowski and Whitmore (1995). After Kent *et al.* (1995).

'state diagram', showing crossovers between regions of different dependences of the correlation length and radius of gyration on the concentration and excluded volume parameter. In a similar manner, de Gennes (1980) proposed a 'state diagram' for grafted molecules immersed in a solution of free polymer molecules of the same chemical type. This diagram of different regimes is drawn in the plane of σ and Φ, where Φ is the volume fraction of free polymer in solution. Figure 8.13 reproduces the original diagram of de Gennes.

Subsequently, similar regimes of behaviour have been obtained using self-consistent field theory. The diagram is drawn for a free polymer degree of polymerisation, P, which is less than that of the grafted polymer. When the volume fraction of free polymer is less than that in the grafted layer, the concentration profile is essentially unchanged from that in the absence of the free polymer. This corresponds to the unstretched, unmixed and strongly stretched, unmixed regimes. When the volume fraction of the chains is greater than the threshold volume fraction of grafted chains at which stretching begins, a decrease in the layer thickness is predicted, but this can be achieved in two ways. In the first, the free chain volume fraction is much greater than that of the grafted chains and the mobile chains dominate; in this weak stretching mixed regime

$$L \sim \sigma^{1/3} N P^{-1/3} \Phi^{-5/12}. \tag{8.2.5}$$

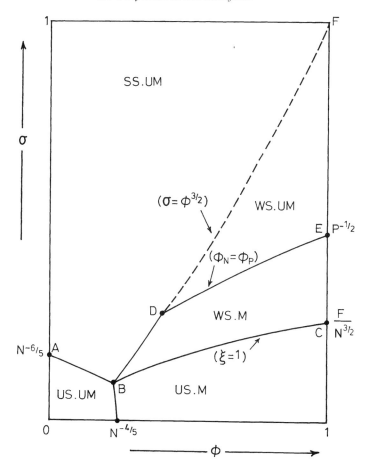

Figure 8.13. Regimes of behaviour of brush-like polymer molecules in the presence of a solution of polymer. S, stretched, SS, strongly stretched, WS, weakly stretched, US, unstretched, M, mixed with solution polymer, UM, not mixed with solution polymer. After de Gennes (1980).

When the grafted layer volume fraction exceeds that of the mobile polymer in the same spatial region, there is a regime of unmixed, weakly stretched behaviour, in which $L \sim \sigma N \Phi^{-1}$. The crossover from this region to the strongly stretched unmixed region is defined by a line describing the threshold volume fraction in the layer for strong stretching to become evident, i.e. $(\sigma/\sigma_m)^{2/3}$; this crossover is expected to be gradual. In contrast at lower values of (σ/σ_m) the transition line BD defines an abrupt decrease in the value of L.

Attempts to identify these regions experimentally have been made using a linear diblock copolymer of PDMS and deuterated PS spread on ethyl benzoate

(Lee *et al.* 1994). Hydrogenous polystyrene was dissolved in the sub-phase and three separate relative molecular masses were investigated, 8×10^3, 43×10^3 and 400×10^3 g mol^{-1}. Reflectivity data were fitted using a profile that had an initial parabolic depletion layer, a different parabola describing the main part of the profile and an exponential tail for the distal end of the profile (figure 8.14). The presence of free polymer clearly decreases the layer thickness (figure 8.15) and the boundary lines between some of the regimes were defined. Thus the crossover between strongly stretched unmixed and weakly stretched unmixed was taken to correspond to the volume fraction of monomer units in the grafted chain when it was surrounded by pure solvent and this volume fraction was determined from values of h^* and σ. The crossover line between the weakly stretched, unmixed (grafted chain dominant) and weakly stretched, mixed (free chain dominant) regions was defined by the volume fraction of grafted chains when free chains had just begun to penetrate the layer.

The majority of studies of spread polymer layers have been concerned with polymers at the air/water interface. Even where one component of the polymer film may be soluble in the aqueous sub-phase, the high surface tension of water generally results in a 'pancake' layer being formed, i.e. there is little stretching of the molecules into the sub-phase. Although in recent years the interpretation of surface pressure isotherms has become a little more quantitative, there is currently no general theory of organisation of polymers at the air/water

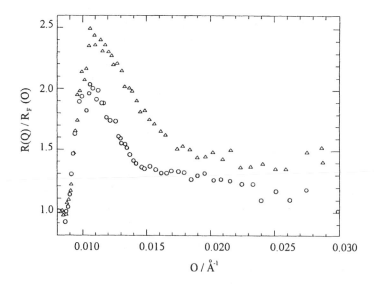

Figure 8.14. The reflectivity (normalised by the calculated Fresnel reflectivity) for a PDMS–PS block copolymer spread at the surface of ethyl benzoate: △, 0% dissolved PS; and ○, 20% dissolved PS. After Lee *et al.* (1994).

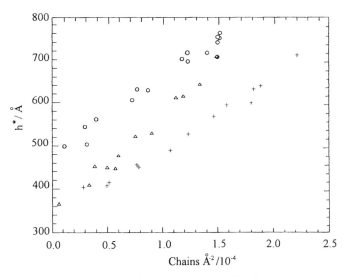

Figure 8.15. The PS layer thickness in the presence of various amounts of dissolved PS in the ethyl benzoate sub-phase (○, solvent; △, 6% polymer; and +, 10% polymer). After Lee *et al.* (1994).

interface which can be compared with experimental data. Nonetheless, certain general questions can be addressed; these include queries regarding the thickness and composition of the spread films and the influence of the surface concentration on these parameters. Where the spread polymer film has an amphiphilic nature questions concerning the distribution in space of the hydrophobic and hydrophilic regions arise. Apart from the work involving polymethyl methacrylate and polyethylene oxide, there has been no concerted, systematic examination of the organisation in spread polymer films. This may be due, in part, to the apparently distinctive behaviour of individual polymers manifested in their surface pressure isotherms and the very great influence that such factors as stereochemistry have on the isotherms. Ellipsometry has been used in some cases but it is often admitted that assumptions regarding the refractive index of the spread polymer film have to be made in order to extract the spread film's thickness. We set out here the conclusions drawn concerning the organisation of homopolymers of polymethyl methacrylate (PMMA) and polyethylene oxide (PEO) at the air/water interface. Extensions to a linear diblock copolymer and a graft copolymer are discussed and the added information obtainable using the kinematic approximation in neutron reflectometry set out. This last aspect is reinforced in the case of a simple homopolymer with a large substituent. Finally, we summarise studies on other polymers and attempt a general summary.

The dependence of surface pressure isotherms of polymethyl methacrylate on the polymer stereochemistry has been referred to earlier. However, this strong dependence does not appear to have been recognised in the scaling law analyses of the isotherms referred to earlier except by Rondelez. Initial analysis of spread monolayers of both isotactic and of syndiotactic polymethyl methacrylate using neutron reflectometry showed that, although the reflectivity of the syndiotactic polymer was greater, the thickness was the same and remained constant at about 18 Å over a surface concentration range of $0.2-2.0 \, \text{mg m}^{-2}$ (Henderson *et al.* 1991). Volume fraction compositions of the spread films were obtained from these data also and are shown in figure 8.16.

Surprisingly, in view of the difference in the scaling exponents, the water content for both stereotactic isomers is approximately equal. It is the air content of the spread films which is very different, being much higher for the isotactic polymer. Surface concentrations calculated from the layer thickness and scattering length density for the syndiotactic polymer agreed very well with the amount actually dispensed on the aqueous sub-phase. For the isotactic polymer, the calculated surface concentration was always less than the amount spread and above a spread concentration of about $0.5 \, \text{mg m}^{-2}$ the discrepancy was very significant. It was suggested that this was evidence of large loops and tails being formed in the sub-phase that were so diluted by the sub-phase that they made an insignificant contribution to the reflectivity.

Although insufficient combinations of polymer and sub-phase were used to permit a full application of the partial structure factor method, a limited use was made to obtain parameters for the spread polymer layer and the near surface water layer (Henderson *et al.* 1993a, b, c). The assumption made was that, for the hydrogenous polymer spread on heavy water, the scattering length density of the polymer could be approximated to zero. The partial structure factors both for the polymer and for the water could be fitted by uniform layer models and the thickness of each layer was approximately equal. In the absence of any information about the separation between the layers, the similarity of thicknesses was interpreted as evidence for a uniform distribution of water throughout the spread polymer film.

This reduced partial structure factor analysis was also applied to the isotactic polymer. For surface concentrations less than $0.8 \, \text{mg m}^{-2}$ a uniform layer model fitted the partial structure factor data for the spread polymer very well. Above this value, a Gaussian partial structure factor model fitted best; however, even though the standard deviations of the Gaussians were reasonably large, the surface concentrations calculated from the parameters of the distribution were still less than the amount dispensed. This was again taken as evidence that some of the polymer penetrates the sub-phase and becomes 'lost' insofar as the

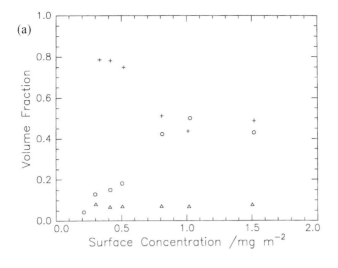

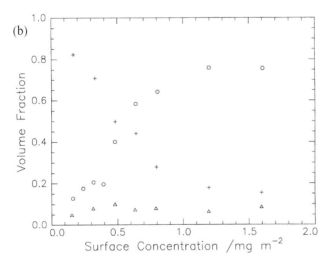

Figure 8.16. The volume fraction composition of polymethyl methacrylate films spread at the air/water interface for (a) isotactic PMMA and (b) syndiotactic PMMA (○, polymer; △, water; and +, air). After Henderson *et al.* (1991).

neutron reflectivity is concerned. It should be noted that, whereas the Gaussian distribution model did not fit well (if at all) to the lower surface concentrations, the uniform layer model could be reasonably well fitted to *all* the surface concentrations of isotactic polymethyl methacrylate used. This is because the reflectometry data did not extend beyond a Q value of about 0.2 Å^{-1} due to background contributions and this is an insufficient range to distinguish between the two models.

Polyethylene oxide is a water-soluble polymer that also forms a stable spread film at the air/water interface. Both solutions and spread layers of the polymer have been studied over many years, beginning with the surface tension studies of Glass (1968a, b). From surface quasi-elastic light scattering data and interpretation of isotherm data (Henderson *et al.* 1993a, b, c, Richards and Taylor 1996), it has been suggested that the spread polymer film undergoes some form of structural change at a surface concentration of 0.4 mg m^{-2}. Partial structure factor analysis of neutron reflectometry data confirms that an organisational change in spread polyethylene oxide films does indeed take place at 0.4 mg m^{-2}. Below this concentration the reflectivity data both for polymer film and for near-surface water are best fitted by single uniform layer models. For higher surface concentrations of polymer, a Gaussian segment distribution of polymer segments appears more viable, with a water distribution described by a tanh function. Figure 8.17 shows the distribution in terms of the number density of PEO molecules at three different surface concentrations (note that the polymer is highly diluted by the sub-phase). Again, because of the absence of all the necessary contrast combinations, the separation between the polyethylene oxide layer and the sub-phase could not be obtained. Addition of magnesium sulphate to the solution produces less favourable thermodynamic conditions (0.39 M aqueous MgSO$_4$ at 315 K is a theta solvent for polyethylene oxide). For a spread

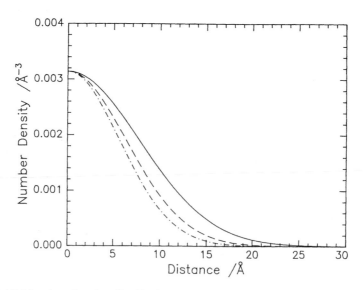

Figure 8.17. Number density distribution of polyethylene oxide spread at the air/water interface at three different surface concentrations: ——, $\Gamma = 0.8$ mg m^{-2}; - - -, $\Gamma = 0.6$ mg m^{-2}; –·–·, $\Gamma = 0.5$ mg m^{-2}.

polyethylene oxide layer with a surface concentration of $0.8 \, \text{mg m}^{-2}$, i.e. such that there is penetration of the sub-phase by the polymer, addition of the salt to the sub-phase results in compression and densification of the surface layer. It appears as though the polymer which penetrates the sub-phase is being adsorbed on to the air/water interface.

A full application of the partial structure factor analysis of neutron reflectometry data from polymers has thus far been applied only to two polymers. One of these is a block copolymer of polymethyl methacrylate and polyethylene oxide such that each block could be separately deuterated (Richards *et al.* 1994). Although complete labelling does allow a full description of the organisation of the block copolymer spread film to be obtained, unambiguous conclusions can only be drawn if the composition of each of the labelled species is identical. The use of anionic polymerisation methods notwithstanding, composition is difficult to control for methyl methacrylate–ethylene oxide block copolymers. Early results showed that the polymethyl methacrylate and polyethylene oxide blocks occupied different spatial regions at the surface, the polyethylene oxide being essentially submerged in the aqueous sub-phase. The block copolymer system is not quite as simple as is implied by the above statement. Subsequent results on a much lower surface concentration of block copolymer showed that demixing of the two blocks takes place above some critical surface concentration. Distributions of block segments were modelled using Gaussian distributions and the near-surface water layer was described by a tanh distribution. Figure 8.18 shows the distributions obtained from the partial structure factor analysis when the centre-to-centre separation has been obtained from the cross partial structure factors and the equations for these terms given by equations (3.1.33) and (3.1.34).

Spread films of polylauryl methacrylate (polydodecyl methacrylate) are also susceptible to investigation by neutron reflectometry and partial structure factor analysis because the main chain backbone and dodecyl substituent can be separately deuterated (Reynolds *et al.* 1995). From such analysis the distribution of dodecyl substituents of backbone segments and water molecules shown in figure 8.19 was obtained. A notable feature of these distributions is that the methacrylate backbone is completely immersed in the sub-phase, in contrast to the arrangement of polymethyl methacrylate. Secondly, the hydrocarbon substituent penetrates deeper into the sub-phase than does the immersed backbone. At first glance this seems most unlikely; however, all of this behaviour can be attributed to the dodecyl substituent. Its hydrophobic nature forces the backbone into the sub-phase. To relieve short-range steric interactions between substituents, rotations about main chain bonds force the dodecyl units slightly deeper into the sub-phase before sufficient bonds in the substituent can rotate

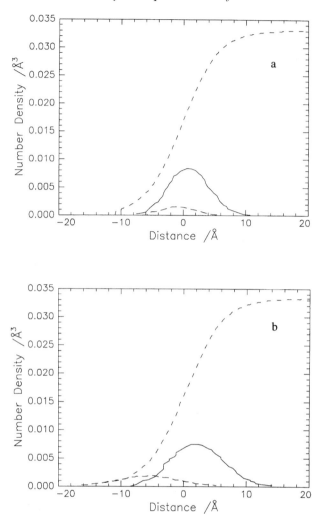

Figure 8.18. Number density distributions of the surface layer components for a linear diblock copolymer of polymethyl methacrylate and polyethylene oxide spread at the air/water interface: (a) a low surface concentration and (b) a higher surface concentration. Solid line, PEO; short dashes, water; and long dashes, PMMA. After Richards *et al.* (1994).

to direct the dodecyl unit out of the aqueous phase. A change in direction of almost 180° requires rotation of about five bonds. Immersion of hydrophobic species in spread copolymers of styrene and maleic anhydride has also been observed (Hodge *et al.* 1992). Here again it appears that the varying degrees of hydrophobicity determine which of the units on the polymer backbone are in the aqueous phase and which are in the air phase.

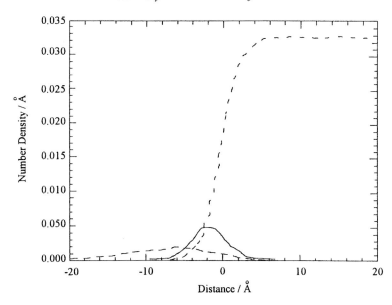

Figure 8.19. Distributions of water (short dashed line), methacrylate backbone (long dashed line) and dodecyl substituent (solid line) for poly(lauryl methacrylate) spread at the air/water interface. After Reynolds *et al.* (1995).

All of the information concerning the organisation of spread polymer films obtained from neutron reflectometry has come from specular reflection. Off-specular or diffuse reflection responds to the in-plane structure of the surface layer. In only one case has off-specular reflection been observed and that is for a polystyrene of very low relative molecular mass (about eight monomer units) terminated at one end by an alcohol unit (Saville *et al.* 1994). When the surface film was compressed to the collapse region, a significant off-specular signal was seen (figure 8.20) and was attributed to multilayer 'islands' being formed in this collapse region of the isotherm.

Polymer solutions are binary systems (we assume the polymer is monodisperse in relative molecular mass) and the variation of surface tension with composition is governed by the Gibbs equation in the same manner as it is for molecules of low relative molecular mass. In principle the hypothetical dividing surface is placed so that each phase either side is uniform up to the surface. In practice, because liquid surfaces are diffuse (due to evaporation processes and capillary waves), the dividing surface is usually placed so that the surface excess of solvent is zero. Figure 8.21 illustrates this and also defines the surface excess of solute.

The surface excess of solute, Γ_2, is given by $A_2 - A_1$, and assuming that

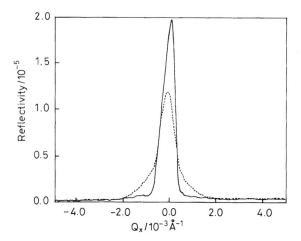

Figure 8.20. The reflectivity profile for hydroxyl-ended polystyrene spread on water. The broadening at the higher surface concentration (dashed line) is due to off-specular reflection. After Saville *et al.* (1994).

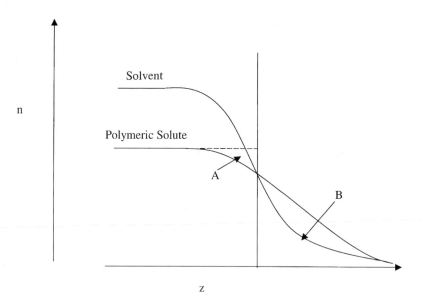

Figure 8.21. Number density variations of solvent and polymeric solute at the inter-face. The areas between the solvent curve each side of the vertical are equal. The surface excess of solute is the difference in area of the B region between the solute curve and the vertical and the region marked A on the bulk solution side of the dividing plane indicated as the vertical.

non-ideality of solution behaviour can be ignored, then solute activity can be replaced by concentration. Thus Gibbs' equation becomes

$$\Gamma_2 = -\frac{1}{RT}\frac{\mathrm{d}\gamma}{\mathrm{d}\ln c}. \tag{8.2.6}$$

Consequently, if the surface tension of a polymer solution decreases, there will be a positive surface excess, i.e. in the region of the air/solution interface the polymer will be at a greater concentration than it will in the bulk. Conversely, if the surface tension increases with concentration, a depletion layer will be formed at the air/solution interface. Glass has reported an extensive measurement of surface tension on a wide range of water-soluble polymers and has noted considerable equilibration times before a constant surface tension was observed (Glass 1968). Other workers do not observe such effects and indeed the diffusion coefficients calculated by Glass from his data suggest that the adsorbed layer has the mobility characteristic of a polymer melt. One is forced to conclude that the equilibration times are artefacts of the technique used.

Mean-field theories of the surface tension of polymer solutions have been developed using the Cahn square gradient approach for interfacial properties of solutions and mixtures both for attractive and for repulsive air/liquid interfaces (Cahn and Hilliard 1958), in a way analogous to the treatment of surface segregation in polymer blends given in section 5.1. For situations in which a surface excess was formed, the volume fraction profile was a hyperbolic cotangent, whereas repulsive profiles were described by hyperbolic tangent functions. Values of the surface tension of semi-dilute solutions of polystyrene in toluene (a depletion layer) and polydimethyl siloxane in toluene (an attractive interface, a surface excess formed) were well described by this theory.

The surface excess layer for a polymer solution is closely related to the situation of polymer adsorption at a polymer/solid interface and, as described in more detail in section 5.2, both mean-field and scaling descriptions of the organisation of such layers have been put forward. A mean-field description gives

$$\phi(z) = \phi_{\mathrm{b}}\coth^2\left[\left(\frac{\phi_{\mathrm{b}}}{\phi_{\mathrm{s}}}\right)^{1/2}\left(\frac{2+D}{D}\right)\right], \tag{8.2.7}$$

where $D = a/(3\nu\phi_{\mathrm{s}})$ with a the monomer size, ϕ_{s} the surface volume fraction in a solution in which the bulk volume fraction is ϕ_{b} and ν the excluded volume parameter of the polymer in solution ($=1/2 - \chi$). For a dilute solution or where there is strong absorption to the liquid/air interface, a region of length scale L in the centre of the surface excess region can be defined such that $L = a/(3\nu\phi_s)$ and in this case

$$\phi(z) = \phi_s[D/(z+D)]^2 \quad (0 < z < L). \tag{8.2.8}$$

Scaling theory divides the profile into three regions.

i. The proximal region, $0 \leqslant z \leqslant D$, where $D = a\delta^1$ and $-\delta k_B T$ is the absorption energy between a monomer unit and the surface. For $z \leqslant a$, $\phi(z) = \phi_s = \delta$; for $a \leqslant z \leqslant D$ then

$$\phi(z) = \frac{a}{D}\left(\frac{a}{z}\right)^{1/3}. \tag{8.2.9}$$

ii. The central region, $D \leqslant z \leqslant L$:

$$\phi(z) = (a/z)^{4/3}. \tag{8.2.10}$$

iii. The distal region, $z \geqslant L$:

$$\phi(z) \propto \phi_b \exp(-kz). \tag{8.2.11}$$

Guiselin *et al.* (1991a, b, Lee *et al.* 1991a, b) found that solutions of polydimethyl siloxane in toluene were best described by the scaling laws. In the central region the surface excess layer has a 'self-similar' structure determined by repulsions between monomer units of the loops and tails of the polymer molecules which constitute this region. The reflectometry data on very dilute solutions for a range of polymer relative molecular masses gave an exponent of $4/3$ in the central region and values of the cut-off length L were in very close agreement with the radii of gyration of the polymers. For these polymers the volume fraction profile could be represented by

$$\phi(z) = 0.96 \text{ for } 0 < z/\text{Å} \leqslant 5, \tag{8.2.12}$$

$$\phi(z) \simeq \left(\frac{5.2}{z+0.1}\right)^{1.3} \text{ for } 5 \leqslant z/\text{Å} \leqslant L, \tag{8.2.13}$$

$$L \sim 0.13 M_w^{0.57}. \tag{8.2.14}$$

Note that, although a surface excess layer is formed, the volume fraction rapidly falls to values close to that of the bulk solution. Thus, although L may be of the order of the radius of gyration, the majority of the polymer is organised more like a pancake than like a mushroom.

Toluene solutions of polystyrene have a depletion layer because the surface tension of toluene is smaller than that of polystyrene. For semi-dilute solutions of polystyrene in toluene, reflectometry data were fitted by the mean-field expression for the volume fraction profile

$$\phi(z) = \phi_b \tanh^2(z/d), \qquad (8.2.15)$$

where d is the thickness of the depletion layer. Scaling theory predicts that depletion layers should be the thickness of the radius of gyration in dilute solution, decreasing to the correlation length, ξ, as the solution becomes semi-dilute. Since d is related to ξ, then it should display the same scaling relation. Experimentally (Lee *et al.* 1991a, b), it was found that (figure 8.22)

$$\frac{d}{R_g} \sim \left(\frac{\phi_b}{\phi^*}\right)^{-0.75}, \qquad (8.2.16)$$

as predicted by the scaling relations with ϕ^* the overlap volume fraction of polystyrene in toluene.

8.3 Interfacial dynamics of polymers at fluid interfaces

Our attention here is focused on the high-frequency dynamics of polymer surface films that are generated by the thermal fluctuations of the liquid surface, i.e. the capillary waves; we do not consider the much lower frequency waves generated either by mechanical means or by electrocapillarity. As set out

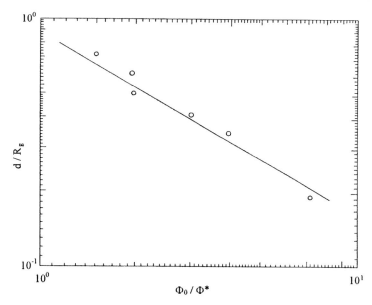

Figure 8.22. The normalised depletion layer thickness as a function of the normalised concentration for polystyrene in toluene solution. After Lee *et al.* (1991a, b).

in chapter 3, capillary wave parameters can be obtained from surface quasi-elastic light scattering (SQELS). Although Goodrich (1962, 1981) defines five surface moduli for a film-bearing fluid interface, only the transverse (capillary) mode produces a change in polarisation that thus generates scattered light. Furthermore, since only the capillary and dilation modes appear to be coupled, it is only for these two modes that values of parameters can be obtained in principle. The words in principle have been used in the last sentence since there has been some debate (Langevin 1992) on the viability of obtaining the four visco-elastic parameters from SQELS data. For SQELS from materials of low relative molecular mass it now appears to be accepted that, provided that the SQELS data are of sufficiently high signal-to-noise ratio, all four parameters can be extracted from the data. Having made this remark, it is appropriate to point out that, if the dilational modulus obtained by SQELS is a large value, then the accuracy of the dilational viscosity values will be compromised.

Relatively few spread polymer films have been investigated by SQELS and it is difficult to draw universal conclusions from these data due to the varying sets of assumptions made in obtaining the visco-elastic parameters. Some of the assumptions made are quite insupportable if objective, unbiased values of the visco-elastic parameters are to be obtained. Furthermore, the almost universal use of a Lorentzian form for the power spectrum of the surface scattered light probably means that the damping values quoted may be somewhat in error. Finally, many of the data reported are restricted to a narrow q (and hence capillary wave frequency) range. It is clear from the work of Earnshaw and McLaughlin (1991, 1993) that a wide frequency range of the data is required and it seems that, whatever values of the surface visco-elastic parameters are obtained, the dispersion behaviour should always be compared with that predicted from the dispersion equation, especially when unexpected behaviour is obtained. Unfortunately, many of the data have been obtained over a restricted q range and this precludes a dispersion-based discussion. Consequently, in the subsequent discussion we will focus on surface wave frequency and damping values in the main since this avoids biasing by the use or neglect of any assumptions in the determination of the surface visco-elastic parameters. Furthermore, due to the current absence of any molecularly based theories either of capillary wave dispersion or of surface visco-elastic properties of polymer films many of the data reported in the literature are discussed in a necessarily speculative way. Few corollaries have been drawn with more general treatments of vibrations and oscillations as expounded in classic texts (Ingard 1988, Pippard 1989, Main 1993). Particularly informative insight, albeit at a phenomenological level, can be drawn from the discussions by Pippard of mechanical models.

Although few polymers have been investigated by SQELS, detailed consideration of the data on each would be somewhat confusing because the results are by no means identical. Therefore we have selected a few polymers where a particularly complete investigation has been pursued or where results from different analytical procedures can be compared.

8.3.1 Polyvinyl acetate

Despite the fact that monomodal relative molecular mass distributions of polyvinyl acetate have not been used, polyvinyl acetate is without doubt the polymer to have been most extensively studied by SQELS. The surface pressure isotherm of the polymer when it is spread on water is of the liquid expanded type and the SQELS data were collected as power spectra over a rather restricted q range (about $250-450$ cm^{-1}) (Kawaguchi *et al.* 1986). Each of these power spectra was fitted to a Lorentzian function (see comments regarding this in chapter 3). The capillary wave frequency is initially constant at low surface concentrations (Γ_s) and then falls rapidly above Γ_s values of 0.5 mg m^{-2}. At the point where the frequency decreases, a sharp maximum is observed in the damping coefficient, then a second maximum is observed at $\Gamma_s \sim 1.5$ mg m^{-1}. No consideration was given to possible resonance situations, and from the data as presented it is impossible to obtain ε_0/γ_0 and test whether the classical resonance condition is traversed over the concentration range investigated. For this first paper all four surface visco-elastic parameters were obtained, albeit in a semi-empirical way. Power spectra were generated from the theoretical expression (equation 3.1.46)) and then this generated spectrum was fitted by a Lorentzian. The parameters in the power spectrum expression were altered until the frequency and damping were reproduced to within $1-5\%$ of the experimental values. The objectivity of this method is open to question. Nonetheless, the values obtained (figure 8.23) are not wildly unreasonable and it was acknowledged that, in the region where the dilational modulus was high, the dilational viscosity values were extremely unreliable. Notably, the SQELS-determined dilational modulus was larger than the static (zero frequency) value, clearly indicating some form of visco-elastic relaxation, but no attempt was made to analyse the data by such means. Attempts to explain the difference between static surface pressure and SQELS surface pressure were centred around the occurrence of a different effective density for water, for which little justification was given.

SQELS data on polyvinyl acetate were re-analysed in a subsequent paper presenting a compendium of results for polymers that were compared pairwise (Kawaguchi *et al.* 1989). In this later paper it was assumed that the

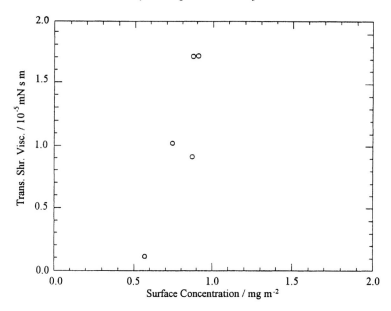

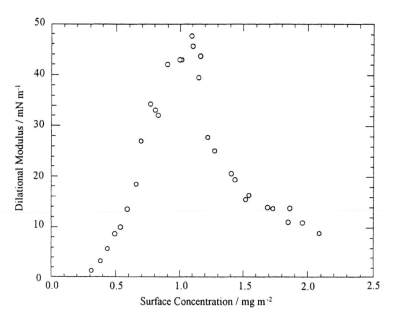

Figure 8.23. SQELS-determined values of the transverse shear viscosity, dilational modulus and dilational viscosity for polyvinyl acetate spread at the air/water interface. After Kawaguchi *et al.* (1986).

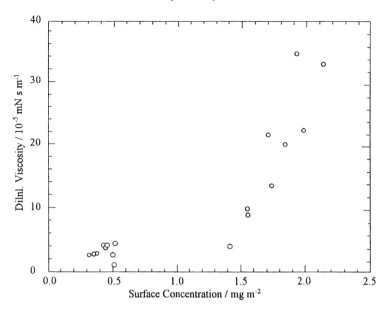

Figure 8.23. (*cont.*)

surface tension obtained from SQELS was the same as the static value and that the transverse shear viscosity was zero and the dispersion equation was solved to obtain ε_0 and ε'. Surprisingly in view of the reduced damping inherent in setting γ' to zero, the dilational moduli were about half the values recorded in the earlier paper and there was almost complete agreement with the static value of ε_0. Values of the dilational viscosity were large, about 10^{-4} mN s m^{-1}, but this time there were apparently no difficulties in obtaining the values even though ε_0 was still rather large. By combining the data in both papers it is possible to estimate values of ε_0/γ_0 obtained from surface pressure isotherms. For the two maxima observed in the damping values as a function of concentration, neither corresponds to the classical resonance condition. In a later paper in which the temperature dependence of the surface quasi-elastic light scattering from spread polyvinyl acetate layers was reported (Yoo and Yu 1989), the difficulties of attempting to extract four parameters from only two observables were explicitly stated and the authors emphasised that the values they reported were not to be taken as definitive. The values obtained were calculated by assuming that the SQELS-determined surface tension was equal to the static value and that ε_0 and ε' were independent of the frequency. It will be seen later that such assumptions cannot always be made; however, faced with only two observables, one is forced to make some assumptions. Earnshaw (1993) has commented on this in several places. To obtain the required frequency independence of the dilational parameters it was necessary to increase γ', the transverse viscosity, which reduced the value of the dilational moduli as it should if the experimental value of the damping coefficient is to be maintained due to the coupling between capillary and dilational modes. This coupling possibility is briefly discussed but there was no attempt to ascertain whether the resonance condition was obtained. As the temperature of the spread film and sub-phase increased, the capillary wave frequency decreased, as did the observed damping, the latter being accompanied by a shift of the maximum to a lower surface concentration of polymer. To first order one expects that the frequency should increase with increasing temperature and the damping probably remain little altered for pure fluids. Hence the experimental observations are attributable to the surface visco-elastic parameters of the spread film. Unfortunately, due to the restrictions inherent to calculating these parameters it is difficult to draw conclusions from their variation with temperature.

8.3.2 Polyethylene oxide

Polyethylene oxide also forms films of the expanded liquid type when it is spread at the air/water interface and it is pertinent to point out here that the spread film differs in its organisation from the surface excess layer that forms in solutions of polyethylene oxide. The evidence for this is now quite clear and the long-held belief that the surface excess and the spread film have the same structure is not tenable. From surface pressure isotherm studies on spread films of polyethylene oxide and some early SQELS work it had been suggested that there was some change in the nature of the spread film at surface concentrations in excess of about $0.4 \, \mathrm{mg \, m^{-2}}$. Kawaguchi (Kawaguchi *et al.* 1989) suggested a looping of molecules into the sub-phase, Kuzmenka and Granick (1988a, b) suggested that, at low surface concentrations, the molecules were confined to two dimensions. Both viewpoints indicate that an increase in the entropy occurred above the critical surface concentration.

Detailed SQELS analyses of spread films of polyethylene oxide have been reported by two groups (Sauer *et al.* 1987, Richards and Taylor 1996). Both (fortunately) observe the same broad features in the frequency and damping of the capillary waves, i.e. an initial plateau in frequency followed by a fall at a surface concentration of $0.2 \, \mathrm{mg \, m^{-2}}$ before settling at a roughly constant frequency at a surface concentration of about $0.4 \, \mathrm{mg \, m^{-2}}$. The damping coefficient shows a broad maximum for surface concentrations in the range 0.2–$0.6 \, \mathrm{mg \, m^{-2}}$ for a surface wave number of about $240 \, \mathrm{cm^{-1}}$, the damping being about $1000 \, \mathrm{s^{-1}}$, which is greater than that of water in this region.

Thereafter the two analyses part company. Sauer *et al.* were forced to set the surface tension to the static value and assumed that the transverse shear viscosity is zero to obtain ε_0 and ε'. The values of ε_0 coincided with the static values up to a surface concentration of about 0.4 mg m^{-2}, thereafter the SQELS values were slightly larger. The dilational viscosity values were about 1×10^{-5} mN s m^{-1} and had a broad maximum coinciding with the maximum in damping. No comment was made concerning the existence of visco-elastic relaxation in the region where the dilational moduli values differ. Additionally, it was not pointed out that, from the static values of ε_0 and γ_0, the resonance condition was traversed at a surface concentration of about 0.5 mg m^{-2}. Richards and Taylor used the spectral fitting method of Earnshaw (Earnshaw *et al.* 1990) which does *not* require any assumptions to be made about the surface visco-elastic parameters; additionally they made measurements over a wider q range for selected surface concentrations, one of these being at the resonance condition of which they were aware. The SQELS-determined surface tension was consistently greater than that of the static value, which was clearly indicative of a relaxation process. Transverse shear viscosities reached a maximum of about 5×10^{-4} mN s m^{-1} at a surface concentration of 0.6 mg m^{-2}. Dilational moduli from SQELS were roughly consistent with static values up to a surface concentration of about 0.6 mg m^{-2}, whereafter they fell to *very* small values (<1 mN m^{-1}). It was in the dilational viscosity variation with surface concentration that the most remarkable behaviour was observed (figure 8.24), *negative* values being obtained abruptly at concentrations of 0.6 mg m^{-2}, which became more positive as the concentration increased yet further.

Using the data obtained for various q values, the frequency dependence of the surface tension could be described by a Maxwell fluid model (Ferry 1980) with a single exponential relaxation process. For this model the storage ($G'(\omega)$) and loss ($G''(\omega)$) are given by

$$G'(\omega) = G_{\mathrm{E}} + \frac{G_{\mathrm{i}}\omega_0^2\tau^2}{1 + \omega_0^2\tau^2} = \gamma_0(\omega_0), \qquad (8.3.1)$$

$$G''(\omega) = \frac{G_{\mathrm{i}}\omega_0\tau}{1 + \omega_0^2\tau^2} = \gamma'(\omega_0)\omega_0, \qquad (8.3.2)$$

where τ is the relaxation time, G_{e} is the $\omega_0 = 0$ (static) value of the surface tension and G_{i} is the magnitude of the relaxation process. Thus, from the frequency dependence of the surface tension or transverse shear viscosity, a value for the relaxation time at a fixed concentration can be obtained. For a

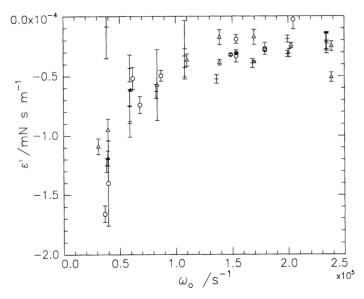

Figure 8.24. Dilational viscosity values as a function of the surface concentration for spread films of polyethylene oxide at the air/water interface. $q \approx 200 \text{ cm}^{-1}$. After Richards and Taylor (1996).

fixed value of q, the two equations can be manipulated to give an expression for the relaxation time:

$$\tau = \frac{\gamma_0(\omega_0) - \gamma_0(\omega_0 = 0)}{\omega_0^2 \gamma'(\omega_0)}. \qquad (8.3.3)$$

Figure 8.25 shows the variation in relaxation time for the spread polyethylene oxide film and a distinct change in value is observed at a surface concentration of 0.6 mg m^{-2}, precisely where the change in dilational viscosity from positive to negative values is seen.

Such an abrupt change suggests a phase transition; neutron reflectometry results indicate that this phase transition is due to the penetration of the subphase by the polymer becoming significant, i.e. the polymer is no longer being confined to two dimensions. The occurrence of negative dilational viscosities would appear to be unphysical at first sight and we return to this aspect of surface visco-elastic parameters later.

8.3.3 Polymethyl methacrylate

Polymethyl methacrylate forms a condensed film when it is spread at the air/water interface and this polymer has been analysed by SQELS by the same two

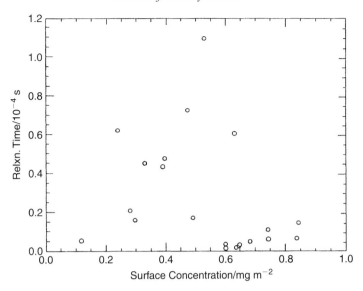

Figure 8.25. Relaxation times of spread polyethylene oxide films at the air/water interface. After Richards and Taylor (1996).

groups as for the polyethylene oxide work discussed above (Kawaguchi *et al.* 1989, Richards and Taylor 1996). Again the concentration dependences of the capillary wave frequency and damping were broadly consistent with each other. Values of ε_0 and ε' obtained by Sauer *et al.* were extremely large and with very large errors and, because of these large values and the low q values, these values should not be taken as physically realistic. The same problem was observed by Richards and Taylor, who disregarded any values of dilational parameters obtained since they were subject to huge errors. The SQELS-determined surface tension was identical to the static values over the surface concentration range of $0-2.5 \text{ mg m}^{-2}$ of polymethyl methacrylate. Transverse shear viscosities were small, being about $2 \times 10^{-5} \text{ mN s m}^{-1}$, of the same order as that observed for monolayers of materials of low relative molecular mass. The surface tension and transverse shear viscosity are sensibly independent of capillary wave frequency, which, together with the very irreproducible values of dilation modulus, suggests that the spread film is more akin to a Voigt solid than it is to a Maxwell fluid in behaviour. At lower surface concentrations of polymer the transverse shear viscosity has higher values (albeit with sizeable errors) and this is commensurate with the film containing water; however, it is also clear from the light scattering data at low concentrations that the polymer exists as discrete islands on the water surface. The transverse shear viscosity drops to a lower value at a surface concentration of 1.0 mg m^{-2} and neutron reflectometry shows that the air content of the spread 'film' has reached its minimum at this concentration. One concludes that it is at this concentration that the islands are all packed together since there is no change in layer thickness over this concentration range.

8.3.4 *Other polymer systems*

Several polymers spread at the air/water interface have been investigated using SQELS (Gau *et al.* 1993, May-Colaianni *et al.* 1993, Runge and Yu 1993, Runge *et al.* 1994a, b). In the majority of cases it has been assumed that the surface tension remains at the static value and that the transverse shear viscosity is zero. The first of these assumptions is not generally true; a sufficient number of systems has now been encountered that clearly indicates that making such an assumption is faulty. Furthermore, although the transverse shear viscosity *may* be negligibly small this is also not generally true and in particular need not be valid when there is considerable interaction between polymer and sub-phase. Consequently, the most unbiased data to be used from these reports are probably the frequencies and damping coefficient values for the capillary waves. These are generally defined to high resolution along the concentration axis. Unfortunately the range of capillary wave frequencies accessible at a fixed concentration of polymer is not large and it is abundantly clear from the work of Earnshaw that as wide a range of capillary wave frequency as possible should be investigated.

8.4 A generalised approach to capillary wave phenomena

We give here a more detailed consideration of capillary waves and their dispersion equation in the presence of an interfacial layer and attempt to generate a more generalised approach to the phenomenology rather than particular views that have been applied to each polymer system. For convenience we repeat the dispersion equation for the air/liquid interface in the presence of a surface layer:

$$D(\omega) = [\varepsilon q^2/\omega + i\eta(q + m)][\gamma q^2/\omega - \omega\rho/q + i\eta(q + m)] + [\eta(q - m)]^2$$

$$= 0. \tag{8.4.1}$$

The propagation frequency is complex ($=\omega_0 + i\Gamma$) and ε and γ are expanded as linear response functions to incorporate the surface viscosities (see equations (3.1.43) and (3.1.44)). The dispersion equation has two roots, corresponding to the capillary and dilational modes, respectively, and each of these roots can be factorised into a frequency and damping term. Thus the capillary wave frequency is

$$\omega_c = (\gamma q^3/\rho)^{1/2} + \mathrm{i}2q^2\eta/\rho \qquad (8.4.2)$$

and the dilational wave frequency is

$$\omega_d = \frac{(\sqrt{3}+\mathrm{i})}{2}\left(\frac{\varepsilon^2 q^4}{\eta\rho}\right)^{1/3}. \qquad (8.4.3)$$

Since dilational and capillary modes are coupled oscillators the phenomenon of resonance between them is possible. At resonance the frequencies of the two modes are equal and the momentum transfer at which this is observable is given by

$$q = \left(\frac{3}{4}\right)^3 \frac{\varepsilon\rho}{\gamma^3\eta^2}. \qquad (8.4.4)$$

Resonance is observed as a maximum in the damping, Γ, and frequency, ω_0, of the surface waves. Consequently, if the surface concentration of polymer at the surface is changed, the moduli (ε and γ) of the surface film are altered and resonance is noted at a particular values of (ε_0/γ_0). For low values of q, resonance is expected when this ratio is about 0.16, but this value increases as q increases, as shown in figure 8.26, the data for which were calculated with

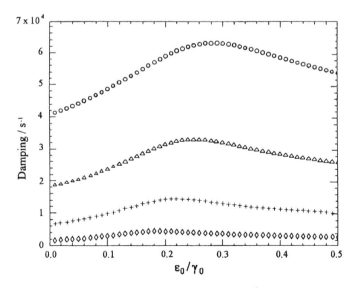

Figure 8.26. Capillary wave damping for $\gamma_0 = 50$ mN m^{-1} and q values of 1500 cm^{-1} ($\circ$), 1000 cm^{-1} ($\triangle$), 600 cm^{-1} ($+$) and 300 cm^{-1} ($\diamond$).

both surface viscosities equal to zero. Finite values of ε' and γ' will influence the exact value of (ε_0/γ_0) at which resonance occurs.

Generally, the surface tension determined by the classical Wilhelmy plate technique decreases as the surface concentration of polymer increases, an asymptotic value eventually being approached. The dilational modulus (ε_0) obtained from such data usually displays a maximum, therefore depending on the actual values of γ_0 and ε_0; two resonant frequencies may be observed by SQELS as the polymer surface concentration is increased. One resonance corresponds to a point where ε_0 is increasing, the second at a higher surface concentration is where ε_0 is decreasing having passed through the maximum. Such occurrences probably explain the observation of double resonances in polydimethyl siloxane at the air/water interface (Runge *et al.* 1994a, b).

Despite the knowledge that the two modes are coupled oscillators, little has been reported (for polymer systems) on attempting to explain the observed frequency and damping behaviour using classical theory. In classical theory the frequencies (ω) of the coupled modes are related to those of the uncoupled modes (the natural frequencies), Ω_1 and Ω_2, by the equation

$$(\omega - \Omega_1)(\omega - \Omega_2) = \kappa^2. \tag{8.4.5}$$

Note that all frequencies in the above equation are complex in the case of coupled *lossy* oscillators, which is the case for the coupled dilational and capillary modes; κ is the coupling constant. Equation (8.4.5) approximates to equation (8.4.1) with the coupling constant given by $\eta(q - m)$. Depending on the value of the coupling constant, the coupling is reactive or resistive (Pippard 1989). Reactive when κ is real, resistive coupling is associated with κ being imaginary. Since $m = q[1 + i\omega\rho/(\eta q^2)]^{1/2}$, the coupling between the two modes changes with q and frequency. Because the frequency is severely influenced by the surface concentration of polymer, it is evident that the coupling constant may change radically as the surface concentration alters.

An average of the two natural frequencies and their difference are defined as

$$\overline{\Omega} = (\Omega_1 + \Omega_2)/2,$$

$$\Delta\Omega = (\Omega_1 - \Omega_2)/2, \tag{8.4.6}$$

respectively. Since all frequencies are complex, we need to consider their behaviour in the complex plane. If we write $(\omega - \overline{\Omega})$ as $u + iv$ then

$$u^2 + v^2 = \kappa^2 + \mathrm{Re}\,[(\Delta\Omega)^2],$$

$$2uv = \mathrm{Im}\,[(\Delta\Omega)^2]. \tag{8.4.7}$$

When the coupling constant is 0, i.e. the two modes are decoupled, they have their natural frequencies (i.e. the capillary and dilational mode frequencies of equations (8.4.2) and (8.4.3)), but note that, in real situations, at the air/water interface the two modes are *never* decoupled; this concept is only adopted to aid clarity in the discussion), and thus $\Delta\Omega$ is a constant. Therefore $\omega - \overline{\Omega}$ forms a rectangular hyperbola in the complex plane of u and v, the axes being the asymptotic values of the curves formed by the solutions ω_1 and ω_2 to equation (8.4.5) as κ changes its value. Figure 8.27 shows this hyperbola.

As κ increases the imaginary parts of the two solutions ω_1 and ω_2 converge along the imaginary axis. At some point the imaginary parts of these two frequencies (i.e. the damping components) become equal and any further increase in κ causes the real parts of the natural frequencies (i.e. the capillary and dilational frequencies) to diverge along the real axis. This divergence manifests itself as an apparent change of the dependence of the capillary and dilational frequencies on various parameters of the system e.g. the dilational modulus or q values. What was initially a capillary wave acquires dilational wave behaviour above this critical κ value and the dilational wave becomes capillary-like. The two modes are *mixed* and it is strictly incorrect to identify them as capillary or dilational modes. Mode mixing is necessarily accompanied by resonance between the two modes, but resonance need not be accompanied by mode mixing. Mode mixing is certainly obtainable from reactive coupling

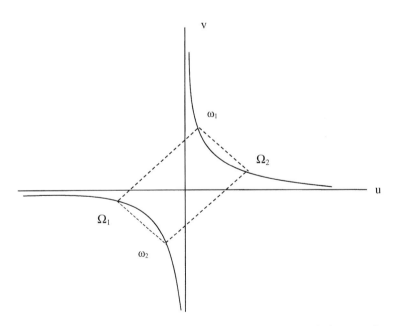

Figure 8.27. Relation between normal modes (Ω_1) and natural frequencies (ω) of coupled lossy oscillators.

of dilational and capillary modes, since coupling clearly causes the imaginary parts of the two modes to approach each other. Resistive coupling cause the real parts to approach each other and the imaginary components of the two mode frequencies to diverge; however, it does appear that, theoretically, a form of mode mixing is obtainable for resistive coupling (Earnshaw 1996) but the effect of this on observable properties is not known.

Given that the liquid phase density and viscosity remain constant, the coupling constant is influenced by q and ω. Changes in q automatically produce a change in ω; however, ω is also influenced by γ and ε. Convergence of the imaginary parts of the capillary and dilational frequencies requires that γ' or ε' change in particular ways. As γ' increases from zero (for ε' fixed), the capillary wave damping increases and the dilational mode damping decreases. Since dilational mode damping is generally greater than that for the capillary mode, this is the correct circumstance to bring about convergence of the two dampings. Figure 8.28 shows the change in frequencies for a particular set of surface visco-elastic parameters with $\gamma' = 0$ and 2.5×10^{-4} mN s m^{-1}, respectively. When the transverse shear viscosity is zero a resonance is observed at $\varepsilon_0 \approx 12$ mN m^{-1}, but no mode mixing. For the higher value of γ' a clear splitting of the two natural frequencies is evident, which is a symptom of mode mixing.

The second route by which the two damping factors can converge is if ε' is *negative*, and by judicious choice of ε' data identical to those in figure 8.28 can be obtained. Negative values of ε' are interpretable as a reduction in the dilational mode damping relative to that when ε' is zero. Negative values of ε' have been obtained both for polymeric and for low relative molecular mass surface films at the air/water interface. The generally accepted view is that such values are *effective* (Earnshaw and McCoo 1995) values and arise because the dispersion relation fails to account properly for all factors contributing to the surface modes. For solutions and surfactants plausible sources of additional energy contributions to the dilational mode can be identified (exchange between the surface layer and bulk solution (van den Tempel and Lucassen-Reynders 1983, Hennenberg *et al.* 1992), Marangoni effects (Scriven and Sternling 1960)), but for insoluble surface films no such sources can be cited. (Even for solutions the arguments and theory based on exchange processes have yet to be fully developed.)

Using classical coupled oscillator theory as outlined above (accepting that analogies with surface modes in liquid-borne layers are somewhat tenuous in view of all the influencing factors), a reasonably satisfying albeit phenomenological rationale can be obtained. Unfortunately in real life we cannot arrange that one surface visco-elastic parameter is changed leaving the others unaltered,

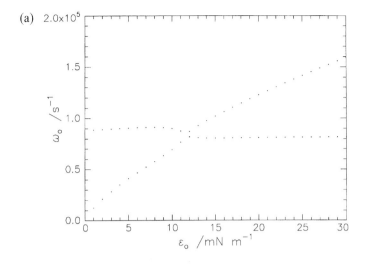

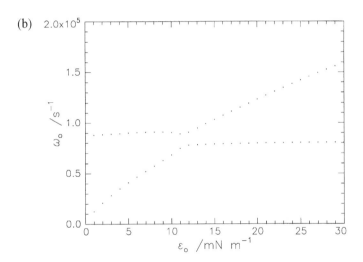

Figure 8.28. Capillary and dilational wave frequencies as a function of the dilational modulus. In each case $\gamma_0 = 72$ mN m^{-1} and $\gamma' = 0$ (a) and 2.5×10^{-4} mN m s^{-1} (b).

because they are all interconnected. What we are able to vary is q and hence the observed surface wave frequency, i.e. we can observe the dispersion of the visco-elastic parameters whilst the surface concentration of polymer is kept constant, by using the different reference beams in the SQELS apparatus. From the dependence of the damping on the surface wave frequency as mode mixing is approached, it is possible to ascertain whether convergence is due to large positive values of γ' or negative values of ε'.

The characteristics of each case are distinctive and can be best observed from dispersion curves in the normalised complex plane where the damping is divided by the damping calculated from the Kelvin equation ($\Gamma = 2\eta q^2/\rho$) and the frequency normalised by the frequency term of the same equation ($\omega_0 = (\gamma_0 q^3/\rho)$). Earnshaw and McLaughlin have discussed the phenomenology of mode mixing in detail (Earnshaw and McLaughlin 1991, 1993). We cannot better Earnshaw's description of the various behaviours and we reproduce his calculations in figure 8.29 (Earnshaw and McCoo 1995).

The capillary modes (indicated by C in figure 8.29) in the absence of strong coupling will be located near 1,1 in the plot due to the normalisation, whereas dilational modes will show much larger variation since the normalisation is not appropriate to them. In figure 8.29a for $\gamma' > 0$, the dilational mode crosses from right to left downwards whereas the capillary mode moves down more or less vertically. As the coupling between the modes becomes stronger, the two modes approach each other slightly. When mode mixing takes place above a critical value of γ' the capillary branch assimilates dilational character at large q values and the dilational branch moves towards 1,1. This would result in the observed capillary wave frequency undergoing a discontinuous increase at a critical value of q. For negative values of ε', figure 8.29b, the capillary mode initially goes vertically downwards as q increases before looping in a semi-circular way to the left. The dilational mode sweeps down and to the right. As the resonance condition is approached both capillary and dilational branches exhibit more curvature. At resonance between the two modes the capillary mode exhibits dilational characteristics. Experimentally this would be observed as a sudden jump in the capillary wave frequency at some critical q value following a vertical decrease as q increases. Distinct frequency jumps have been observed in surfactant solutions (Earnshaw and Sharpe 1996) but as yet no equivalent has been reported for polymer layers.

At this point it is appropriate to return to the negative dilational viscosities observed by Richards and Taylor for spread films of polyethylene oxide on water. The dispersion equation of the capillary waves used in all studies on spread films is also appropriate for surface excess layers. As remarked earlier the surface organisation of the surface excess in polyethylene oxide solutions has been well described. Analysis of the SQELS data from a solution of polyethylene oxide by the spectral fitting method results in negative dilational viscosities. The SQELS data for spread polyethylene oxide films at a surface concentration of 0.6 mg m^{-2} (resonance condition) when plotted in the complex plane certainly display the curved dependence for strong coupling and negative values of ε' predicted by Earnshaw and McLaughlin. Large values of

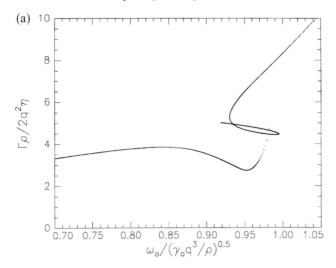

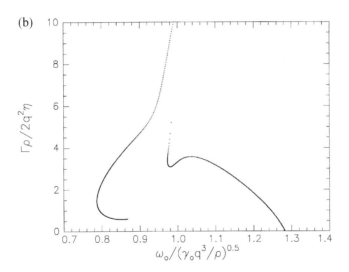

Figure 8.29. The normalised damping as a function of the normalised frequency plotted in the complex plane for two situations in which mode mixing occurs. (a) $\gamma_0 = 65$ mN m^{-1}, $\gamma' = 3 \times 10^{-5}$ mN s m^{-1}, $\varepsilon_0 = 15$ mN m^{-1}, $\varepsilon' = 0$; (b) $\gamma_0 = 70$ mN m^{-1}, $\gamma' = 0$ mN s m^{-1}, $\varepsilon_0 = 12.5$ mN m^{-1}, $\varepsilon' = -2 \times 10^{-5}$ mN s m^{-1}. After Earnshaw and McCoo (1995).

the transverse shear viscosity necessarily complicate such dispersion plots compared with those calculated by Earnshaw and the dispersion behaviour resulting from a combination of surface visco-elastic parameters can only be predicted by numerical solution of the dispersion equation.

8.5 Other effects and future applications

The absence of a wholly satisfactory explanation for negative dilational viscosities in spread polymer films notwithstanding, the overall behaviour thus far discussed is qualitatively similar to that of spread films and surface excess layers of materials of low relative molecular mass. It is also becoming evident that a much richer behaviour of surface visco-elastic parameters is displayed by polymer systems, which expose omissions in currently available theory of surface waves on liquids. A case in point is the behaviour of the dilational parameters for a series of graft copolymers of a polymethyl methacrylate backbone bearing methoxy-terminated polyethylene oxide grafts which has been reported by Peace (Peace *et al.* 1998). The grafts were randomly distributed along the backbone and each graft contained 54 ethylene oxide units. Typical dependences of ε_0 and ε' on the surface concentration are shown in figure 8.30 for one copolymer with an average of 24 grafts per molecule.

A clear absorption resonance is seen in ε_0, the real part of ε, whereas at the same point ε', the imaginary part, shows a discontinuity. The shapes of the curves in figure 8.30 are tantalisingly similar to Kramers–Kronig (Pippard 1989) curves for response functions of resonant transducers. Familiar examples in polymer visco-elasticity of this relation are the real and imaginary parts of the compliance of a polymer subjected to an oscillatory perturbation. In Kramers–Kronig curves the independent variable is generally the oscillation frequency; in figure 8.30 it is the surface concentration. Recall that, as the surface concentration increases, ε_0 and ε' of the film-covered surface will change and thus ω will also alter; thus changing the surface concentration effectively constitutes a change in surface wave frequency (Pippard 1985). The major difficulty in making a direct comparison with such properties as compliance is that the real component in figure 8.30 has the characteristics of the imaginary part of the compliance ($J''(\omega)$) and the imaginary part (ε' in figure 8.30) has the characteristics of the real compliance. Furthermore, both ε_0 and ε' are mirror images of the real and imaginary compliance behaviour, which is somewhat disturbing.

The compliance relates the time dependence of the mechanical displacement of a polymer to the applied force and is a particular example of a transfer function $\chi(\omega)$. A transfer function converts an input function (force, electric field, polarisation etc.) to the observed signal or response function. In many case the response function is a displacement or like property. Consider a simple harmonic oscillator of mass m and natural frequency, ω_0, which is subjected to an oscillatory force, $F \exp(-i\omega t)$, where ω is the frequency following applica-

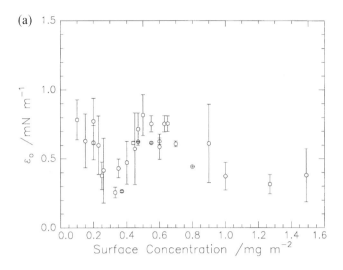

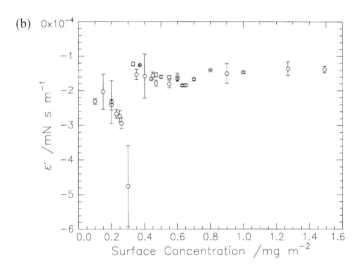

Figure 8.30. The dilational modulus (a) and dilational viscosity (b) as functions of the surface concentration for a graft copolymer of polymethyl methacrylate and polyethylene oxide spread at the air/water interface. After Peace *et al.* (1998).

tion of the force. The time dependence of the displacement of the mass, $\xi(t)$, is given by

$$\xi(t) = \frac{-F \exp(-i\omega t)}{m(\omega^2 - \omega_0^2 + 2i\omega''\omega)}, \tag{8.4.8}$$

where ω'' is the complex component of the frequency, i.e. its damping. In equation (8.4.8) the transfer function is identifiable as

$$\chi(\omega)\frac{-1}{m(\omega^2 - \omega_0^2 + 2\omega''\omega)}.$$

Multiplication of the top and bottom of this expression by the complex conjugate of the denominator term in parentheses allows factorisation of the transfer function into real and imaginary parts:

$$\mathrm{Re}\,(\chi(\omega)) = \frac{\omega_0^2 - \omega^2}{m[(\omega^2 - \omega_0^2)^2 + 4\omega''\omega^2]}, \qquad (8.4.9)$$

$$\mathrm{Im}\,(\chi(\omega)) = \frac{2\omega''\omega}{m[(\omega^2 - \omega_0^2)^2 + 4\omega''\omega^2]}. \qquad (8.4.10)$$

The variations of the real and imaginary parts of $\chi(\omega)$ are shown in figure 8.31; each goes through sharp changes when $\omega = \omega_0$, i.e. at resonance. As remarked earlier, such curves do not display the features of ε_0 and ε' observed in the graft copolymer. If, instead of the displacement, we consider the velocity response function of the mass m, the relevant transfer function is $-i\omega\chi(\omega)$, and the appropriate real and imaginary parts are obtained by multiplying equations (8.4.9) and (8.4.10) by ω. Figure 8.32 shows that in this case the real and imaginary components of the transfer function are qualitatively identical to the behaviours of both ε_0 and ε' in figure 8.30. Hence what is being observed here is a new resonance phenomenon, a resonance such that the response to the oscillatory perturbation of the system is observed as a velocity function of the dilational parameters. Although it appears reasonable to assume that the oscillatory perturbation derives from the capillary waves, at present there is no molecular theory for the property being acted on or how it couples to the dilational properties of the surface layer. Some attempts at providing a more soundly based dispersion equation using the perturbed brush layers as a model have been made and suggest that additional coupling mechanisms may be operating. However, these additional terms become significant only when the surface tension approaches zero and thus do not appear to be the source of the behaviour observed.

One area where SQELS can make a significant contribution is that of polymers at liquid/liquid interfaces. Since the viscosity and density of the two fluids are much closer the coupling between dilation and capillary waves is severely attenuated and the prospects for obtaining dilational modulus and viscosity are much reduced or nonexistent. Oil/water interfaces are particularly interesting when block copolymers are placed at the interface. Only one such system has so far been discussed; a block copolymer of styrene and ethylene oxide was placed at the interface between heptane and water (Sauer *et al.* 1987). There appears to be some evidence of a maximum in the damping of the interfacial capillary waves as the concentration of copolymer was increased.

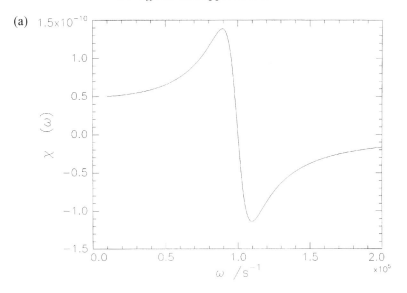

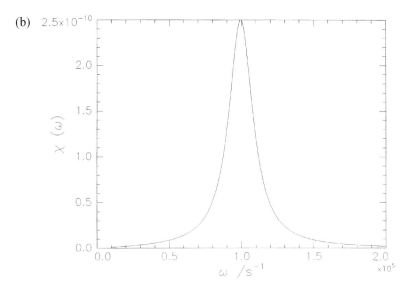

Figure 8.31. (a) The real part and (b) the imaginary part of the displacement transfer function.

Since no error bars were reproduced in the data it is impossible to say how significant this maximum is. If it is real then a coupling between modes is implied. This possibility was ignored in analysing the data and dilational contributions to the dispersion equation were ignored. The only parameter extracted was the interfacial tension and its dependence on the copolymer

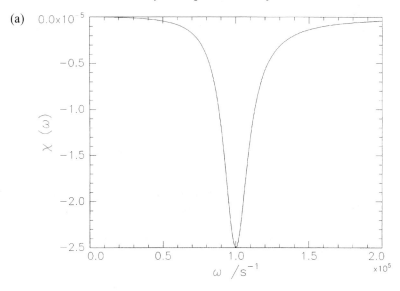

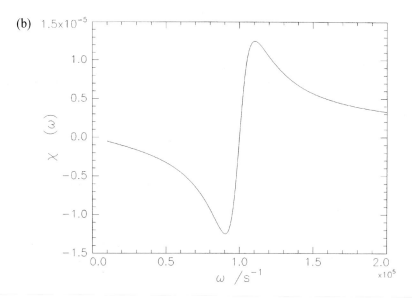

Figure 8.32. (a) The real part and (b) the imaginary part of the velocity transfer function.

concentration since the transverse viscosity was reported to be zero to within experimental error. Evidently this is an area that requires much more experimental attention before a proper consideration can be given to the results and their potential. The results already reported for polymers at the air/water interface notwithstanding, there are aspects that have yet to be investigated, e.g.

polyelectrolytes, mixed polymer systems, polymers of controlled architecture and flowing systems.

8.6 References

Adamson, A. W. (1990). *Physical Chemistry of Surfaces*. New York, Wiley-Interscience.

Aliadib, Z., Tredgold, R. H. *et al.* (1991). *Langmuir* **7**, 363.

Auroy, P., Auvray, L. *et al.* (1991). *Macromolecules* **24**, 2523.

Auvray, L. and de Gennes, P. G. (1986). *Europhysics Letters* **2**, 647.

Baranowski, R. and Whitmore, M. D. (1995). *Journal of Chemical Physics* **103**, 2343.

Beredijk, N. (1963). Monomolecular film studies of polymers. *Newer Methods of Polymer Characterisation*. B. Ke. New York, Wiley-Interscience.

Bijsterbosch, H. D., de Haan, V. O. *et al.* (1995). *Langmuir* **11**, 4467.

Cahn, J. W. and Hilliard, J. E. (1958). *Journal of Chemical Physics* **28**, 258.

Carignano, M. A. and Szleifer, I. (1994). *J. Chem. Phys.* **100**, 3210.

Cosgrove, T., Heath, T. G. *et al.* (1987a). *Macromolecules* **20**, 1692.

Cosgrove, T., Heath, T. G. *et al.* (1987b). *Macromolecules* **20**, 2879.

Cosgrove, T., Heath, T. G. *et al.* (1987c). *Polymer Communications* **28**, 64.

Daoud, M. and Jannink, G. (1975). *Journal de Physique* **37**, 973.

de Gennes, P. G. (1979). *Scaling Concepts in Polymer Physics*. Ithaca, NY, Cornell University Press.

de Gennes, P. G. (1980). *Macromolecules* **13**, 1069.

Earnshaw, J. C. (1993). In *Polymer Surfaces and Interfaces II*. W. J. Feast, H. Munro and R. W. Richards. Chichester, Wiley.

Earnshaw, J. C. (1996). Private communication.

Earnshaw, J. C. and McCoo, E. (1995). *Langmuir* **11**, 1087.

Earnshaw, J. C., McGivern, R. C. *et al.* (1990). *Langmuir* **6**, 649.

Earnshaw, J. C. and McLaughlin, A. C. (1991). *Proceedings of the Royal Society* A **433**, 663.

Earnshaw, J. C. and McLaughlin, A. C. (1993). *Proceedings of the Royal Society* A **440**, 519.

Earnshaw, J. C. and Sharpe, D. J. (1996). *Journal of the Chemical Society Faraday Transactions* II **92**, 611.

Ferry, J. D. (1980). *Viscoelastic Properties of Polymers*. New York, Wiley.

Fleer, G. J., Cohen-Stuart, M. A. *et al.* (1993). *Polymers at Interfaces*. London, Chapman and Hall.

Gabrielli, G., Pugelli, M. *et al.* (1982). *Journal of Colloid and Interfacial Science* **86**, 485.

Gaines, G. L. (1966). *Insoluble Monolayers at Liquid–Gas Interfaces*. New York, Wiley-Interscience.

Gau, C.-S., Yu, H. *et al.* (1993). *Macromolecules* **26**, 2525.

Glass, J. E. (1968a). *Journal of Physical Chemistry* **72**, 4450.

Glass, J. E. (1968b). *Journal of Physical Chemistry* **72**, 4459.

Goodrich, F. C. (1962). *Journal of Physical Chemistry* **66**, 1858.

Goodrich, F. C. (1981). *Proceedings of the Royal Society* A **374**, 341.

Granick, S. (1985). *Macromolecules* **18**, 1597.

Granick, S., Clarson, S. J. *et al.* (1985). *Polymer* **26**, 925

Granick, S., Kuzmenka, D. J. *et al.* (1989). *Macromolecules* **22**, 1878.

Guiselin, O., Lee, L.-T. *et al.* (1991a). *Journal of Chemical Physics* **95**, 4632.
Guiselin, O., Lee, L.-T. *et al.* (1991b). *Physica* B **173**, 113.
Henderson, J. A., Richards, R. W. *et al.* (1991). *Polymer* **32**, 3284.
Henderson, J. A., Richards, R. W. *et al.* (1993a). *Macromolecules* **26**, 65.
Henderson, J. A., Richards, R. W. *et al.* (1993b). *Acta Polymerica* **44**, 184.
Henderson, J. A., Richards, R. W. *et al.* (1993c). *Macromolecules* **26**, 4591.
Hennenberg, M., Chu, X.-L. *et al.* (1992). *Journal of Colloid and Interface Science* **50**, 7.
Hodge, P., Davis, F. *et al.* (1990). *Philosophical Transactions of the Royal Society of London Series A – Physical Sciences and Engineering* **330**, 153.
Hodge, P., Towns, C. R. *et al.* (1992). *Langmuir* **8**, 585.
Huggins, M. L. (1965). *Die Makromolekulare Chemie* **87**, 119.
Ingard, K. U. (1988). *Fundamentals of Waves and Oscillations*. Cambridge, Cambridge University Press.
Israelachvili, J. (1994). *Langmuir* **10**, 3774.
Jaycock, M. J. and Parfitt, G. D. (1981). *Chemistry of Interfaces*. Chichester, Ellis Horwood.
Kawaguchi, M. (1993). *Progress in Polymer Science* **18**, 341.
Kawaguchi, M., Sano, M. *et al.* (1986). *Macromolecules* **19**, 2606.
Kawaguchi, M., Sauer, B. B. *et al.* (1989). *Macromolecules* **22**, 1735.
Kawaguchi, M., Yoshida, A. *et al.* (1983). *Macromolecules* **16**, 956.
Kent, M. S., Lee, L.-T. *et al.* (1992). *Macromolecules* **25**, 6240.
Kent, M. S., Lee, L.-T. *et al.* (1995). *Journal of Chemical Physics* **103**, 2320.
Kuzmenka, D. J. and Granick, S. (1988a). *Polymer Communications* **29**, 64.
Kuzmenka, D. J. and Granick, S. (1988b). *Macromolecules* **21**, 779.
Langevin, D. (1992). *Light Scattering by Liquid Surfaces and Complementary Techniques*. New York, Marcel Decker.
Lee, L.-T., Factor, B. J. *et al.* (1994). *Faraday Discussions* **98**, 139.
Lee, L.-T., Guiselin, O. *et al.* (1991a). *Macromolecules* **24**, 2518.
Lee, L.-T., Guiselin, O. *et al.* (1991b). *Physical Review Letters* **67**, 2838.
Li, S., Hanley, S. *et al.* (1993). *Langmuir* **9**, 2243.
MacRitchie, F. (1990). *Chemistry at Interfaces*. San Diego, Academic Press.
Main, I. G. (1993). *Vibrations and Waves in Physics*. Cambridge, Cambridge University Press.
May-Colaianni, S., Gandhi, J. V. *et al.* (1993). *Macromolecules* **26**, 6595.
Motomura, K. and Matuura, R. (1963). *Journal of Colloid Science* **18**, 52.
Ober, R. and Vilanove, R. (1977). *Colloid and Polymer Science* **255**, 1067.
Peace, S. K., Richards, R. W. *et al.* (1998). *Langmuir* **14**, 667.
Peng, J. B. and Barnes, G. T. (1990). *Langmuir* **6**, 578.
Peng, J. B. and Barnes, G. T. (1991). *Langmuir* **7**, 1749.
Pippard, A. B. (1985). *Response and Stability*. Cambridge, Cambridge University Press.
Pippard, A. B. (1989). *The Physics of Vibration*. Cambridge, Cambridge University Press.
Poupinet, D., Vilanove, R. *et al.* (1989). *Macromolecules* **22**, 2491.
Reynolds, I., Richards, R. W. *et al.* (1995). *Macromolecules* **28**, 7845.
Richards, R. W., Rochford, B. R. *et al.* (1994). *Faraday Discussions* **98**, 263.
Richards, R. W. and Taylor, M. R. (1996). *Journal of the Chemical Society – Faraday Transactions* II **92**, 601.
Rochford, B. R. (1995). Ph.D. Thesis, University of Durham.

Runge, F. E., Kent, M. S. *et al.* (1994a). *Langmuir* **10**, 1962.

Runge, F. E. and Yu, H. (1993). *Langmuir* **9**, 3191.

Runge, F. E., Yu, H. *et al.* (1994b). *Berichte der Bunsengesellschaft. Physikalische Chemie* **98**, 1046.

Sauer, B. B., Kawaguchi, M. *et al.* (1987). *Macromolecules* **20**, 2732.

Sauer, B. B., Yu, H. *et al.* (1987). *Macromolecules* **20**, 393.

Saville, P. M., Gentle, I. R. *et al.* (1994). *Journal of Physical Chemistry* **98**, 5935.

Scheutjens, J. M. H. M. and Fleer, G. J. (1979). *Journal of Physical Chemistry* **83**, 1619.

Scheutjens, J. M. H. M. and Fleer, G. J. (1980). *Journal of Physical Chemistry* **84**, 178.

Scriven, L. E. and Sternling, C. V. (1960). *Nature* **187**, 186.

Shuler, R. L. and Zisman, W. A. (1970). *Journal of Physical Chemistry* **74**, 1523.

Shull, K. R. (1991). *Journal of Chemical Physics* **94**, 5723.

Singer, S. J. (1948). *Journal of Chemical Physics* **16**, 872.

Takahashi, A., Yoshida, A. *et al.* (1982). *Macromolecules* **15**, 1196.

Tredgold, R. H., Young, M. C. J. *et al.* (1987). *Thin Solid Films* **151**, 441.

van den Tempel, M. and Lucassen-Reynders, E. H. (1983). *Advances in Colloid and Interface Science* **18**, 281.

Vickers, A. J., Tredgold, R. H. *et al.* (1985). *Thin Solid Films* **134**, 43.

Vilanove, R. and Rondelez, F. (1980). *Physical Review Letters* **45**, 1502.

Wijmans, C. M., Scheutjens, J. M. H. M. *et al.* (1992). *Macromolecules* **25**, 2657.

Yoo, K.-H. and Yu, H. (1989). *Macromolecules* **22**, 4019.

Young, M. C. J., Lu, W. X. *et al.* (1990). *Electronics Letters* **26**, 993.

Index

ABS (acrylonitrile–butadiene–styrene) 127
acronyms, unpronounceable 56
adhesion 53, 57, 105, 137, 268, 280, 293–315
  between glassy polymers 295–309
  between rubbery polymers 309–15
adsorption, polymer (*see also* segregation,
    surface) 59, 76, 78, 187–8, 199–200, 209–27
  at air/liquid interfaces 328–30, 345
  critical 227–8, 235
  experimental results for solid/liquid
    interfaces 111–13, 115–16, 220–7
  from poor solvents 226–7, 236–8
  from solution of end-functionalised polymers and
    block copolymers 249–51
  from melts of end-functionalised polymers and
    block copolymers 266–8
  of copolymers at polymer/polymer
    interfaces 269–71, 272–4
  scaling theory of 215–19, 345–7
  self-consistent field theory of 212–15, 217–20
attenuated total reflection IR (ATR-IR) 57, 102,
    223–4, 236–7
atomic force microscopy (AFM), *see* scanning force
    microscopy (SFM)

brushes, polymer 111, 113–14, 116, 122, 200, 222,
    245–68, 272–5
  and adhesion 314–15
  at air/liquid interfaces 327, 329–36
  in melts 261–8
  in poor solvents 257–61
  in solution, scaling theory of 245–9
  in solution, self-consistent field theory 253–6
  lateral phase separation 258–61

capillary rise 15
capillary waves
  classical phenomenology of 356–68
  thermally excited 45–6, 78–86, 150–2, 224, 297,
    343, 347–68
coexistence curve 129–31, 133, 228–9, 231–2
cohesive forces 8–12, 25, 183
conformation, polymer, near interfaces 46–9

contact angle 19, 21–3, 50–1, 57, 122, 230, 235,
    279
copolymer, block 44, 66, 68, 105, 116, 122, 244–5,
    250–2
  at polymer/polymer interfaces 268–77, 298–306
  at air/liquid interface 327–8, 329–36, 341–2,
    366
  phase behaviour of 282–90
  surface-induced ordering of 287–90
corresponding states, principle of 41–4
crazing 294–5, 301–2, 303–5, 309
critical adsorption 227–8, 235
critical exponents 32
critical micelle concentration 274–6
critical point 25, 31–2, 130, 133–5, 227–8, 233,
    235, 289
crystallinity 49, 57
cyclic polymer 220–1

depletion layers
  equilibrium 200, 213, 332–4, 345–6
  kinetic 234, 237
Derjaguin approximation 108
dewetting 52, 122, 235
dextran
  grafted at solid/liquid interface 259–60
diffusion
  at early times 165–71, 309
  case II 165
  mutual 157–65, 174, 180, 234, 238
  of polymers 152–65
  self-diffusion 154–7
  tracer 160–5
dimethyl siloxane/styrene block copolymer
  at polystyrene/poly(dimethyl siloxane)
    interface 272
  spread at air/liquid interface 330–36
dynamics, surface 50–4

Edwards equation (*see also* self-consistent field
    theory) 145, 212
elastic recoil detection (ERDA) *see* forward recoil
    spectrometry (FRES)

electron energy loss spectroscopy 102, 106–7
ellipsometry 52, 76–8, 92, 151, 337
equation of state
    Flory, Orwoll and Vrij 40–1, 43, 135
    Poser and Sanchez 38
    surface 318–19
ethyl/ethyl ethylene random copolymer
    blends of, surface segregation 236
ethylene oxide/methyl methacrylate block
        copolymer
    spread at air/liquid interface 322–3, 341–2
ethylene oxide/styrene block copolymer
    at liquid/liquid interface 366
evanescent wave 49, 66, 70, 102, 225

Flory–Huggins theory (*see also* interaction
        parameter, Flory–Huggins) 131–6, 158, 183,
        191, 201, 262
forward recoil spectrometry (FRES) 87, 97–100,
        155, 164, 193, 269–71
fracture energy, interfacial
    definition 294
    for failure by crazing 303–5
    measurement 295–7, 310–14
    rate dependence of 303, 310–11
fracture toughness, *see* fracture energy
free energy of mixing 128–9, 131–2, 191, 201,
        262, 321, 323
Fresnel law of reflection 61–4, 69

glass transition
    in bulk polymer 50, 156, 161–2, 165, 180
    near interfaces and in thin films 51–4, 77
grazing incidence x-ray diffraction 49–50

Hamaker constant 11–12, 151

interaction parameter, Flory–Huggins 132, 134–6,
        142, 146, 158, 164, 176–7, 183–4, 191, 201,
        204–5, 232, 262, 267, 272, 275, 277–8, 280,
        284, 286–7, 298
interfacial rheology (*see also* surface
        viscoelasticity) 51–2, 117–20, 123, 366–9
interfacial tension between polymers (*see also*
        polymer interfaces)
    experimental values of 150
    measurement of 17, 22
    modification of by block copolymers 268,
        270–1, 275–7
    modification of by small molecules 277–8
    modification of by polydispersity 278–9
    modification of by random and graft
        copolymers 280–2
    and phase separation 177–8, 180, 282, 284
    scaling theory for 137
    self-consistent field theory for 144–9
    square gradient theory of 138–9
    and wetting 229–33
interfacial width between polymers (*see* polymer
        interfaces)
Ising model 28

isoprene/2-vinyl pyridine block copolymer
    adsorbed at liquid/solid interface 252
isoprene/styrene block copolymer
    at melt/solid interface 267

JKR (Johnson–Kendall–Roberts)
    experiment 310–14

latent heat 10
lattice fluid model
    of polymers 33–4, 131, 135, 190, 201
    of small-molecule fluids 28–31, 138
    Poser and Sanchez 38
line tension 23
liquid surface, density profile at 25–7, 32
Lifshitz–Slyozov law 180, 239
low-energy-ion scattering spectrometry 87, 96–7

methyl methacrylate/styrene block copolymer
    at polystyrene/poly(methyl methacrylate)
        interface 271, 273
    at poly(methyl methacrylate)/poly(xylenyl ether)
        interfaces 300
    ordering in thin films 68, 287–90
microemulsion, polymer 275–7
microphase separation 282–90

neutron reflectivity 52, 57, 58–76, 150–2,
        168–9, 194–5, 197, 220–2, 225–6, 256,
        258, 264–5, 271, 287, 290, 330–2, 334,
        336–7
neutron reflectivity, Born approximation for 71–3
nuclear reaction analysis 87, 100–1, 140–1,
        235–6, 267
nucleation and growth 173, 176–8, 239

optical reflectivity 76

pendant drop 16–18
phase diagram 131, 133, 172, 228, 286
phase portrait 191–2, 230–1
phase separation (*see also* microphase
        separation) 129–31, 133–4, 159, 171–81, 188,
        212, 229, 238–41, 257, 283
polybutadiene
    blend with deuterated polybutadiene; spinodal
        decomposition 177
    crosslinked, fracture energy 311
    interface with poly(dimethyl siloxane) 282
    neutron-scattering length density 64
polycapralactone
    blend with polycaprolactone; mutual
        diffusion 166–7
poly(dimethyl siloxane)
    adsorption at liquid/air interface 224–5, 346
    adsorption at solid/liquid interface 220
    crosslinked, adhesion to grafted layer 313
    grafted at solid/liquid interface 249
    interface with polybutadiene; interfacial
        tension 282
    surface tension 36

polyethylene (*see also* hydrogenated polybutadiene)
   surface tension 36–7, 39–40
   Zisman plot 22
poly(ethylene oxide) (*see also*
      poly(oxyethylene)-diol)
   adsorption at air/liquid interface 86, 225–7
   adsorption at solid/liquid interface 111–13,
      220, 222
   neutron-scattering length density 64
   spread at air/liquid interface 322, 340–1,
      352–4, 362
poly(ethylene propylene)
   blend with deuterated poly(ethylene propylene),
      surface segregation from 71, 194, 197
   blend with deuterated poly(ethylene propylene),
      surface-directed spinodal decomposition
      in 238–9
poly(ethylene terephthalate)
   static secondary ion mass spectrometry
      spectrum 89
   surface tension 36
polyimide
   surface ordering 49, 54
polyisobutylene
   surface tension 39–40
polyisoprene
   interface with polystyrene; fracture
      energy 314–15
poly(lauryl methacrylate)
   spread at air/liquid interface 343
polymer/polymer interfaces
   block copolymers at 268–77, 298–306
   effect of polydispersity on 278
   graft copolymers at 280–1, 305–8
   random copolymers at 280, 305, 308
   scaling argument for interfacial tension and
      width 137
   self-consistent field theory of interfacial tension
      and width 144–52
   square gradient theory of 136–44
polymer melt surface
   density profile at 41, 44
   roughness of (*see also* capillary waves, thermally
      excited) 65
polymer solutions, scaling theory of 205–9
poly(methyl acrylate)
   spread at air/liquid interface 320, 325
poly(methyl methacrylate)
   adsorption at solid/liquid interfaces 223–4
   glass transition of thin films 52
   interface with polystyrene, diblock copolymers
      at 271, 273
   interface with polystyrene, fracture energy 297
   interface with polystyrene, interfacial
      tension 149–50
   interface with polystyrene, random copolymers
      at 308
   interface with polystyrene, width 150–2, 297
   interface with poly(xylenyl ether), diblock
      copolymers at 300
   neutron-scattering length density 64

spread at air/liquid interfaces 75, 320, 322,
   325–6, 338–9, 354–6
surface tension 36
x-ray photoelectron spectrum of 104
poly(methyl methacrylate)/ethylene oxide graft
      copolymer
   spread at air/liquid interface 364–5
poly(*p*-methyl styrene)
   interface with polystyrene, fracture energy and
      width 299
poly(oxyethylene)-diol
   surface tension 36
poly(phenylene vinylene)
   dopant profile in 95
polypropylene
   surface tension 20
polypropylene (atactic)
   surface tension 36
poly(pyromellitic dianhydride oxydianiline)
      (PMDA-ODA) (*see* polyimide)
polystyrene
   adsorption at solid/liquid interface 116, 120,
      221–3, 236–7
   blend with deuterated polystyrene, mutual
      diffusion 159–60
   blend with deuterated polystyrene, surface
      segregation 69, 98–9, 107, 193–6, 199
   blend with deuterated polystyrene,
      thermodynamics 136
   blend with poly(xylenyl ether), diffusion 164
   blend with poly(methyl methacrylate), thin film
      morphology 241
   blend with poly(vinyl methyl ether), surface
      segregation 188
   blend with poly(vinyl methyl ether),
      thermodynamics 136
   depletion at air/liquid interface 346–7
   diffusion at early times 168–71
   glass transition of thin films 52–3
   interface with deuterated polystyrene
      (immiscible), width 141
   interface with polybutadiene, interfacial
      tension 279
   interface with polyisoprene, fracture
      energy 314–15
   interface with poly(*p*-methyl styrene), fracture
      energy and width 299
   interface with poly(methyl methacrylate), diblock
      copolymer at 271, 273
   interface with poly(methyl methacrylate), fracture
      energy 297
   interface with poly(methyl methacrylate),
      interfacial tension 149–50
   interface with poly(methyl methacrylate), random
      copolymer at 308
   interface with poly(methyl methacrylate),
      width 150–2
   interface with poly(2-vinylpyridine), diblock
      copolymers at 269–71, 299–301
   interface with poly(2-vinylpyridine), fracture
      energy 298–301

polystyrene (*cont.*)
  interface with poly(2-vinylpyridine), width 297
  interface with styrene/ *p*-hydroxy styrene
    copolymer 277
  neutron-scattering length density 64
  self-diffusion coefficient 155
  static secondary ion mass spectrometry
    spectrum 91
  surface dynamics 51
  surface tension 36, 39–40
polystyrene sulphonate
  adsorption at air/liquid interface 238
polystyrene, end-functionalised
  contact angle with water 51
  grafted at liquid/solid interface 114, 118, 120,
    250–1, 256, 258
  grafted at melt/solid interface 264–6
  grafted at polymer/polymer interface 307–8
  segregated at melt/solid interface 93
  spread at air/liquid interface 343–4
poly(tetrafluoroethylene)
  surface tension 20, 36
poly(tetramethylene oxide)
  surface tension 38
poly(vinyl acetate)
  spread at air/liquid interface 320, 324–5, 349–52
  surface tension 39–40
poly(vinyl chloride)
  blend with polycapralactone, mutual
    diffusion 166–7
poly(vinyl methyl ether)
  blend with polystyrene, surface segregation 188
  blend with polystyrene, thermodynamics 136
poly(2-vinyl pyridine)
  contact angle on polystyrene 24
  interface with polystyrene, diblock copolymers
    at 269–71
poly(xylenyl ether)
  blend with polystyrene, mutual and tracer
    diffusion 164
  interface with poly(methyl methacrylate), diblock
    copolymers at 300

random phase approximation (RPA) 138, 143, 285
refractive index 59–61, 77
reptation, theory of 99, 156–7, 161, 168–71
Roe functional 143–4
Rutherford backscattering (RBS) 87, 94–6

scanning force microscopy 24, 51–2, 57, 108, 110–
  20, 259–60, 290
scattering, neutron, incoherent 71, 76
Scheutjens–Fleer theory (*see also* self-consistent
  field theory) 209–15, 217–21, 253, 256, 328
secondary ion mass spectrometry (SIMS) 57,
  87–93
  dynamic 92–3, 151, 170, 194, 196, 281, 289
  static 88–91
  time of flight 88, 90
segregation
  chain end 48–9, 170

surface 49, 66, 69, 71, 93, 98–9, 105, 187–99,
  200, 212, 227, 230, 266–8, 275–6, 345
self-consistent field theory
  of polymer adsorption 212–15, 217–20
  of polymer brushes 253–6, 263–8
  of polymer interfaces 144–9
  of surface segregation 196
small-angle neutron scattering 183, 220–1, 248–9,
  329–30
spinodal decomposition 173–80, 238
spinodal decomposition, surface-directed 237–40
spinodal line 130–1, 133, 174, 176, 238
spread polymer films, at air/liquid interface 317–69
square gradient theory
  of a lattice fluid 27–32
  of block copolymers 285, 287
  of polymer adsorption 212, 345
  of polymer interfaces 136–44, 149, 278
  of a polymer surface 33, 38–41
  of surface segregation 191, 196, 227
  of the wetting transition 227–32
steric stabilisation 108, 282
structure factors, partial 72–3, 329–30, 338, 340–1
styrene/2-vinyl pyridine block copolymer
  adsorbed at liquid/solid interface 252
  at polystyrene/poly(2-vinyl pyridine)
    interface 269–71, 299–301
  at polystyrene/styrene/ *p*-hydroxy styrene
    copolymer interface 277
styrene/4-vinyl pyridine block copolymer
  adsorption at liquid/solid interface 122
surface-enhanced Raman spectroscopy 102, 106
surface energy (*see* surface tension)
surface forces apparatus 57, 108, 110–20, 249,
  251–3
surface pressure 317–28, 336–7
  scaling theory of 323–6
surface quasi-elastic light scattering 78–86,
  348–56, 358, 361, 366–9
surface tension
  and capillary waves 45–6, 80
  and surface pressure 317
  and wetting 229–33
  correlation with latent heat 10
  critical surface tension 21–3
  definition 9, 12–14
  measurement of 14–20
  of miscible polymer blends 188–90
  of polymer melts, dependence on relative
    molecular mass 37
  of polymer melts, dependence on
    temperature 38–40
  of polymer melts, theory of 33–6, 38–44
  of polymer melts, values for 36–40
  of polymer solutions 345, 353, 358
  relation to surface energy 8–12, 25
  van der Waals theory of 25–31
surface viscoelasticity (*see also* interfacial
  rheology) 80–2, 85, 238, 347–69

thermodynamics of interfaces 12–14

van der Waals 8, 11, 27–8, 134
Vogel–Fulcher equation (*see also* Williams–
    Landel–Ferry (WLF) equation) 156, 310

wetting 19–21, 227–41, 259
wetting transition 230–3, 240–1, 279
Wilhelmy plate 17–20, 358
Williams–Landel–Ferry (WLF) equation 156, 161,
    168, 310

x-ray photoelectron spectroscopy (XPS) 57, 101–6,
    138
x-ray reflectivity 58–9

Young–Dupré equation 21, 229–30
Young–Laplace equation 14

Zisman plot 21–3